Advances in Industrial Control

Other titles published in this Series:

Data-driven Techniques for Fault Detection and Diagnosis in Chemical Processes
Evan L. Russell, Leo H. Chiang and Richard D. Braatz

Nonlinear Identification and Control
Guoping Liu

Digital Controller Implementation and Fragility
Robert S.H. Istepanian and James F. Whidborne (Eds.)

Optimisation of Industrial Processes at Supervisory Level
Doris Sáez, Aldo Cipriano and Andrzej W. Ordys

Applied Predictive Control
Huang Sunan, Tan Kok Kiong and Lee Tong Heng

Hard Disk Drive Servo Systems
Ben M. Chen, Tong H. Lee and Venkatakrishnan Venkataramanan

Robust Control of Diesel Ship Propulsion
Nikolaos Xiros

Hydraulic Servo-systems
Mohieddine Jelali and Andreas Kroll

Model-based Fault Diagnosis in Dynamic Systems Using Identification Techniques
Silvio Simani, Cesare Fantuzzi and Ron J. Patton

Strategies for Feedback Linearisation
Freddy Garces, Victor M. Becerra, Chandrasekhar Kambhampati and Kevin Warwick

Robust Autonomous Guidance
Alberto Isidori, Lorenzo Marconi and Andrea Serrani

Dynamic Modelling of Gas Turbines
Gennady G. Kulikov and Haydn A. Thompson (Eds.)

Control of Fuel Cell Power Systems
Jay T. Pukrushpan, Anna G. Stefanopoulou and Huei Peng

Fuzzy Logic, Identification and Predictive Control
Jairo Espinosa, Joos Vandewalle and Vincent Wertz

Optimal Real-time Control of Sewer Networks
Magdalene Marinaki and Markos Papageorgiou

Process Modelling for Control
Benoît Codrons

Rudder and Fin Ship Roll Stabilization
Tristan Perez
Publication due May 2005

Adaptive Voltage Control in Power Systems
Giuseppe Fusco and Mario Russo
Publication due August 2005

Control of Passenger Traffic Systems in Buildings
Sandor Markon
Publication due November 2005

Ajoy K. Palit and Dobrivoje Popovic

Computational Intelligence in Time Series Forecasting

Theory and Engineering Applications

With 66 Figures

Springer

Dr.-Ing. Ajoy K. Palit
Institut für Theoretische Elektrotechnik und Microelektronik (ITEM),
Universität Bremen, Otto-Hahn-Allee-NW1, D-28359, Bremen, Germany

Prof. Dr.-Ing. Dobrivoje Popovic
Institut für Automatisierungstechnik (IAT), Universität Bremen,
Otto-Hahn-Allee-NW1, D-28359, Bremen, Germany

British Library Cataloguing in Publication Data
Palit, Ajoy K.
Computational intelligence in time series forecasting: theory and engineering applications. – (Advances in industrial control)
1. Time-series analysis – Data processing 2. Computational intelligence
I. Title II. Popovic, Dobrivoje
519.5′5′0285

Advances in Industrial Control series ISSN 1430-9491
ISBN-13: 978-1-84996-970-3
e-ISBN-13: 978-1-84628-184-6
Springer Science+Business Media
springeronline.com

Printed in the United States of America
69/3830-543210 Printed on acid-free paper

Advances in Industrial Control

Professor Emeritus O.P. Malik
Department of Electrical and Computer Engineering
University of Calgary
2500, University Drive, NW
Calgary
Alberta
T2N 1N4
Canada

Professor K.-F. Man
Electronic Engineering Department
City University of Hong Kong
Tat Chee Avenue
Kowloon
Hong Kong

Professor G. Olsson
Department of Industrial Electrical Engineering and Automation
Lund Institute of Technology
Box 118
S-221 00 Lund
Sweden

Professor A. Ray
Pennsylvania State University
Department of Mechanical Engineering
0329 Reber Building
University Park
PA 16802
USA

Professor D.E. Seborg
Chemical Engineering
3335 Engineering II
University of California Santa Barbara
Santa Barbara
CA 93106
USA

Doctor I. Yamamoto
Technical Headquarters
Nagasaki Research & Development Center
Mitsubishi Heavy Industries Ltd
5-717-1, Fukahori-Machi
Nagasaki 851-0392
Japan

Writing a book of this volume involves great strength, devotion and the commitment of time, which are lost for our families. We are, therefore, most grateful to our wives, Mrs. Soma Palit and Mrs. Irene Popovic, for their understanding, patience and continuous encouragement, and also to small Ananya Palit who missed her father on several weekends and holidays.

A. K. Palit and D. Popovic

Series Editors' Foreword

The series *Advances in Industrial Control* aims to report and encourage technology transfer in control engineering. The rapid development of control technology has an impact on all areas of the control discipline. New theory, new controllers, actuators, sensors, new industrial processes, computer methods, new applications, new philosophies…, new challenges. Much of this development work resides in industrial reports, feasibility study papers and the reports of advanced collaborative projects. The series offers an opportunity for researchers to present an extended exposition of such new work in all aspects of industrial control for wider and rapid dissemination.

Computational Intelligence is a newly emerging discipline that, according to the authors Ajoy Palit and Dobrivoje Popovic, is about a decade old. Obviously, this is a very young topic the definition and content of which are still undergoing development and change. Nonetheless, the authors have endeavoured to give the topic a framework and demonstrate its procedures on challenging engineering and commercial applications problems in this new *Advances in Industrial Control* monograph, *Computational Intelligence in Time Series Forecasting*.

The monograph is sensibly structured in four parts. It opens with an historical review of the development of "Soft Computing" and "Computational Intelligence". Thus, Chapter 1 gives a fascinating insight into the way a new technology evolves and is consolidated as a self-evident discipline; in this case, proposals were made for constituent methods and then revised in the light of applications experience and the development of new methodologies which were added in to the core methods. No doubt the debate will continue for a few more years before widely accepted subject definitions appear, but it is very useful to have a first version of a "Computational Intelligence" technology framework to consider.

In Part II, the core methods within Computational Intelligence are presented: neural networks, fuzzy logic and evolutionary computation – three neat self-contained presentations of the building blocks for advanced development. It is in Part III that new methods are developed and presented based on hybridisation of the three basic routines. These new hybrid algorithms are demonstrated on various application examples. For the practicing engineer, chapters in Part II and III should almost provide a self-contained course on Computational Intelligence methods.

The current and future development of Computational Intelligence methods are the subject of Chapter 10 which forms Part IV of the monograph. This chapter balances the historical perspective of Chapter 1 by attempting to identify new development areas that might be of significant interest to the engineer. This is not an easy task since even a quick look at Chapter 10 reveals an extensive literature for a rapidly expanding field.

This volume on Computational Intelligence by Dr. Palit and Dr. Popovic is a welcome addition to the *Advances in Industrial Control* monograph series. It can be used as a reference text or a course text for the subject. It has a good opening historical review and a nice closing chapter looking to the future. Most usefully, the text attempts to present these new algorithms in a systematic framework, which usually eases comprehension and will, we hope, lead the way to a new technology paradigm in industrial control methods.

M.J. Grimble and M.A. Johnson
Industrial Control Centre
Glasgow, Scotland, U.K.

Preface

In the broad sense, computational intelligence includes a large number of intelligent computing methodologies and technologies, primarily the evolutionary, neuro and fuzzy logic computation approaches and their combinations. All of them are derived through the studies of behaviour of natural systems, particularly of the connectionist and reasoning behaviours of the human brain/human being.

The computational technology was evolved, in fact, from what was known as soft computing, as defined by Zadeh in 1994. Also, soft computing is a multidisciplinary collection of computational technologies still representing the core part of computational intelligence. The introductory chapter of this book is dedicated to the evolutionary process from soft computing to computational technology. However, we would like to underline that computational intelligence is more than the routine-like combination of various techniques in order to calculate "something"; rather, it is a goal-oriented strategy in describing and modelling of complex inference and decision-making systems. These soft computing approaches to problem formulation and problem solution admit the use of uncertainties and imprecisions. This, to a certain extent, bears a resemblance to artificial intelligence strategies, although these emphasize knowledge representation and the related reasoning rather than the use of computational components.

Computational intelligence, although being not more than one decade old, has found its way into important industrial and financial engineering applications, such as modelling, identification, optimization and forecasting required for plant automation and making business decisions. This is due to research efforts in extending the theoretical foundations of computationally intelligent technologies, exploiting their application possibilities, and the enormous expansion of their capabilities for dealing with real-life problems.

Although in the near past books on computational intelligence and soft computing have been published, today there is no other book dealing with the systematic and comprehensive expositions of methods and techniques for solving the forecasting and prediction problems of various types of time series, *e.g.* nonlinear, multivariable, seasonal, and chaotic. In writing this book our intention was to offer researchers, practising engineers and applications-oriented professionals a reference volume and a guide in design, building, and execution of forecasting and prediction experiments, and this includes from the collection and structuring of time series data up to the evaluation of experimental results.

The fundamental knowledge and the methodologies of computationally intelligent technologies were drawn from various courses for advanced students and from the experimental studies of Ph.D. candidates at the Institute of Automation Technology of University of Bremen, the Control Engineering Laboratory of Delft University of Technology, and from our experience in co-operation with industry. The material presented in the book is therefore suitable to be used as a source in structuring the one-semester course on intelligent computational technologies and their applications.

The book is designed to be largely self-contained. The reader is supposed to be familiar with the elementary knowledge of neural networks, fuzzy logic, optimum search technique, and probability theory and statistics. The related chapters of the book are written so that the reader is systematically led to the deeper technology and methodology of the constituents involved in computational intelligence and to their applications. In addition, each chapter of the book is provided with a list of references that are intended to enable the reader to pursue individual topics in greater depth than that has been possible within the space limitations of this book. To facilitate the use of the book, an index of key terms is appended.

The entire book material consists of 10 chapters, grouped into four parts, as described in the following.

Part I of the book, containing the first two chapters, has the objectives of introducing the reader to the evolution of computational intelligence and to the traditional formulation of the time series forecasting problem and the approaches of its solution.

The evolution of computational intelligence is presented in the introductory Chapter 1, starting with the soft computing as developed by Zadeh in 1994 up to the present day. During this time, the number of constituents of computational intelligence has grown from the fuzzy logic, neurocomputing, and probabilistic reasoning as postulated by Zadeh, with the addition of genetic algorithms (GAs), genetic programming, evolutionary strategies, and evolutionary programming. Particular attention is paid to the achievements of hybrid computational intelligence, which deals with the parameter tuning of fuzzy systems using neural networks, performance optimization of neural networks through monitoring, and parameter adaptation by fuzzy logic systems, *etc.* The chapter ends with the application fields of computational intelligence today.

The ensuing Chapter 2 is devoted to the traditional definition and solving of the time series forecasting problem. In the chapter, after the presentation of the main characteristic features of time series and their classification, the objective of time series analysis in the time and frequency domains is defined. Thereafter, the problem of time series modelling is discussed, and the linear regression-based time series models that are mostly used in time series forecasting are presented, like the ARMA, ARIMA, CARIMA models, *etc.*, as well as some frequently considered models, such as the multivariate, nonlinear, and chaotic time series models. This is followed by the discussion of model estimation, validation, and diagnostic checks on which the acceptability of the developed model depends. The core part of the chapter, however, deals with the forecasting approaches of time series based on

Box-Jenkins methods and the approaches using exponential smoothing, adaptive smoothing, and the nonlinear combination of forecasts. The chapter ends with an example in control engineering from the industry.

In Part II of the book, which is made up of Chapters 3, 4, and 5, the basic intelligent computational technologies, *i.e.* the neural networks, fuzzy logic systems, and evolutionary computation, are presented.

In Chapter 3 the reader is introduced to neuro-technology by describing the architecture, operating principle, and the application suitability of the most frequently used types of neural network. Particular attention is given to various network training approaches, including the training acceleration algorithms. However, the kernel part of the chapter deals with the forecasting methodology that includes the data preparation, determination of network architecture, training strategy, training stopping and validation, *etc.* This is followed by the more advanced use of neural networks in combination with the traditional approaches and in performing the nonlinear combination of forecasts.

Chapter 4 provides the reader with the foundations of fuzzy logic methodology and its application to fuzzy modelling on examples of building the Mamdani, relational, singleton, and Takagi-Sugeno models, suitable for time series modelling and forecasting. Special attention is paid to the related issues of optimal shaping of membership functions, to automatic rules generation using the iterative clustering from time series data, and to building of a non-redundant and conflict-free rule base. The examples included deal with chaotic time series forecasting, and modelling and prediction of second-order nonlinear plant output using fuzzy logic systems. Also here, the advantage of nonlinear combination of forecasts is demonstrated on temperature prediction in a chemical reactor.

In Chapter 5 the main approaches of evolutionary computations or intelligent optimal solution search algorithms are presented: GAs, genetic programming, evolutionary strategies, evolutionary programming, and differential evolution. Particular attention is paid to the pivotal issues of GAs, such as the real-coded GAs and the optimal selection of initial population and genetic operators.

Part III of the book, made up of Chapters 6 through to 9, presents the various combinations of basic computational technologies that work in a cooperative way in implementing the hybrid computational structures that essentially extend the application capabilities of computational intelligence through augmentation of strong features of individual components and through joint contribution to the improved performance of the overall system.

The combination of neuro and fuzzy logic technology, described in Chapter 6, is the earliest experiment to generate hybrid neuro-fuzzy and fuzzy-neuro hybrid computational technology. The motivation for this technology merging, which in the mean time is used as a standard approach for building intelligent control systems, is discussed and the examples of implemented systems presented. Two major issues are pointed out: the training of typical neuro-fuzzy networks and their application to modelling nonlinear dynamic systems. In order to demonstrate the improved capability and performance of neuro-fuzzy systems, their comparisons with backpropagation and radial basis function networks are presented. Finally, forecasting examples are given from industrial practice, such as short-term forecasting of electrical load, prediction of materials properties, correction of pyrometer readings, tool wear monitoring, as well as the examples on modelling and prediction of Wang data and on prediction of chaotic time series.

The subjects of the succeeding Chapter 7 are two most important, but very often neglected, and recently increasingly considered issues of model transparency and the interpretability of data-driven automated fuzzy models. Here, strong emphasis is placed on making the reader familiar with the compact and transparent modelling schemes that include the model structure selection, data clustering, similarity-based simplification, and model validation. In addition, the similarity-based rule base simplification through removing irrelevant fuzzy sets, removing redundant inputs, and the merging of rules are presented. In this chapter some formal techniques are proposed for regaining the interpretability and transparency of the generated fuzzy model, which helps in generating the "white-box-like" model, unlike the black-box model generated by a neural network.

Chapter 8 covers the application of GAs and evolutionary programming in evolution design of neural networks and fuzzy systems. This is a relatively new application field of evolutionary computation that has, in the past decade, been the subject of intensive research. The text of the chapter focuses on evolving the optimal application-oriented network architecture and the optimal values of their connection weights. Correspondingly, optimal selection of fuzzy rules and the optimal shaping of membership function parameters are on the agenda when evolving fuzzy logic systems.

Chapter 9, again, deals in a sense with the inverse problem, *i.e.* with the problem of adaptation of GAs using fuzzy logic systems for optimal selection and tuning of genetic operators, parameters, and fitness functions. In the chapter, the probabilistic control of GA parameters and - in order to avoid the prematurity of convergence - the adaptation of population size while executing of search process is discussed. The chapter closes with the example of dynamically controlled GA using a rule-based expert system with a fuzzy government module for tuning the GA parameters.

Part IV of the book, consisting of Chapter 10, introduces the reader to some more recently developed computationally intelligent technologies, like support vector machines, wavelet and fractal networks, and gives a brief outline about the development trends. In addition, the entropy and Kohonen networks-based fuzzy clustering approaches are presented and their relevance to the time series forecasting problem pointed out, for instance through the design of Takagi-Sugeno fuzzy model. In the introductory part of the chapter the reasons for selecting the above items of temporary computational intelligence are given. It is also indicated that the well advanced bioinformatics, swarm engineering, multi-agent systems, and fuzzy-logic-based data understanding are the constituents of future emerging intelligent technologies.

Finally, we would like to thank Springer-Verlag, London, particularly the AIC series editors, Professor M.A. Johnson and Professor M.J. Grimble, and Mr. Oliver Jackson, Assistant Editor, Springer-Verlag, London, for their kind invitation to write this book. Our special thanks also go to Mr. Oliver Jackson, for his cordial cooperation in preparing and finalizing the shape of the book.

Bremen, March 2005 Ajoy K. Palit and Dobrivoje Popovic

Contents

Part I Introduction

Part II Basic Intelligent Computational Technologies

Part I

Introduction

1

Computational Intelligence: An Introduction

1.1 Introduction

Within the artificial intelligence society the term computational intelligence is largely understood as a collection of intelligent computational methodologies, such as fuzzy-logic-based computing, neurocomputing, and evolutionary computing, that help in solving complex computational problems in science and technology, not solvable or at least not easily solvable by using the conventional mathematical methods.

1.2 Soft Computing

The research activity in the area of combined application of intelligent computing technologies was initiated by Zadeh (1994), who has coined the term soft computing, which he defined as a "collection of methodologies that aim to exploit the tolerance for imprecision and uncertainty to achieve tractability, robustness, and low solution cost". According to Zadeh, the principal constituents of soft computing are fuzzy logic, neurocomputing, and probabilistic reasoning.

The reason for the need of soft computing was, in Zadeh's opinion, that we live in a pervasively imprecise and uncertain world and that precision and certainty carry a cost. Therefore, soft computing should be seen as a partnership of distinct methods, rather than as a homogeneous body of concepts and techniques.

Initially, as the main partnership members of soft computing, also called its principal constituents, the following technologies have been seen:

- ***fuzzy logic***, which has to deal with the imprecisions in computing and to perform the approximate reasoning
- ***neurocomputing,*** which is required for learning and recognition purposes
- ***probabilistic reasoning***, which is needed for dealing with the uncertainty and belief propagation phenomena

Later, the initial partnership group was extended by adding

- ***evolutionary computation***
- ***belief theory***
- ***learning theory.***

Fuzzy logic, which is the most important part of soft computing, bridges the gap between the quantitative information (*i.e.* the numerical data) and the qualitative information (or the linguistic statements), which can be jointly processed using fuzzy computing. In addition, fuzzy logic operates with the concept of IF-THEN rules in which the antecedents and the consequents are expressed using linguistic variables. Neural networks, for their part, have the capability of extracting knowledge from available data, *i.e.* the capability of learning from examples, which fuzzy logic systems do not have. This capability is known as the connectionist learning paradigm.

The process of learning can take place in ***supervisory mode*** (when the ***backpropagation networks*** are used) or in ***unsupervised mode*** (when the ***recurrent networks/Kohonen networks*** are used). This is due to the ***computing neuron*** or the ***perceptron*** (Rosenblatt, 1962), the theoretical background of which was worked out by Minsky and Papert (1969). It is the multi-layer perceptron configuration that is capable of emulating human brain behaviour in learning and ***cognition***. The learning capability of multi-layer perceptrons, as proposed by Werbos (1974), should be obtained through a process of adaptive training on examples.

Dubois and Prade (1998) remarked that soft computing, because it was a collection of various technologies and methodologies with distinct foundations and distinct scopes, "lumped together" although each of the components has little in common with the other, could not form a scientific discipline in the traditional sense of the term. Therefore, they understand the term soft computing more as a "fashionable name with little actual contents". This is in fact a hard judgement, in view of the fact that in the meantime various combinations of the constituent technologies have been used to build hybrid computational systems, such as ***neuro-fuzzy systems***, ***fuzzy-neuro systems***, ***evolutionary neural networks***, ***adaptive evolutionary systems***, and others, that were extensively documented by Bonissone (1997 and 1999). This issue is the main subject of Part 3 of this book, where it will be shown that the individual components of soft computing are not ***mutually competitive,*** but rather are ***complementary*** and ***co-operative***. Jang *et al.* (1997) considered soft computing from the neuro-fuzzy point of view, rather than from the fuzzy set theory only, and pointed out that the neuro-fuzzy approach is to be seen as a technological revolution in modelling and control of dynamic systems, taking the ***adaptive network-based fuzzy inference system (ANFIS)*** as an example.

1.3 Probabilistic Reasoning

As the third principal constituent of soft computing, probabilistic reasoning is a tool for evaluating the outcome of computations affected by randomness and probabilistic uncertainties. To name a few, ***Bayesian belief networks*** and ***Dempster–Shafer theory*** belong to this kind of reasoning approach.

At this point a few words of clarification concerning the similarity between the terms ***probability*** and ***fuzziness*** could be of use, because it is still controversial. The reason is that probability theory as a formal framework for reasoning about uncertainty was "there earlier" than fuzzy reasoning, so that some doubts have been raised about the fuzzy reasoning: Is it really something new or only a clever disguise for probability? Bezdek (1992b) denied this. Zadeh (1995) has even seen probability and fuzzy logic as being complementary, rather than as competitive approaches. In the meantime, this is actually accepted consensusly within the soft computing community.

Probabilistic reasoning deals with the evaluation of the outcomes of systems that are subjects of probabilistic uncertainty. The reasoning helps in evaluating the relative certainty of occurrence of true or false values in random processes. It relies on sets described by means of some probability distributions. Therefore, probabilistic reasoning represents the ***possible worlds*** that are the solutions of an approximate reasoning problem and thus being consistent with the existing information and knowledge (Ruspini, 1996). Probabilistic reasoning methods are primarily interested in the ***likelihood,*** in the sense of whether a given hypothesis will be true under given circumstances.

Zadeh (1979) extended the reasoning component of soft computing by introducing the concepts of

- ***fuzzy reasoning***
- ***possibilistic reasoning***

which belong to the ***approximated reasoning***. According to Zadeh, approximate reasoning is the reasoning about ***imprecise propositions***, such as the ***chains of inferences*** in fuzzy logic. Similarly, the ***predicate logic*** deals with ***precise propositions***. Therefore, approximate reasoning can be seen as an extension of the traditional ***propositional calculus*** operating with the incomplete truth.

Fuzzy reasoning, with roots in fuzzy set theory, deals with the ***fuzzy knowledge*** as imprecise knowledge. Unlike the probabilistic reasoning, fuzzy reasoning deals with ***vagueness*** rather than with ***randomness***. Fuzzy reasoning is thus an ***approximate reasoning*** (Zadeh, 1979), in the sense that it is neither exact nor absolutely inexact, but only to a certain degree exact or inexact. Fuzzy reasoning schemes operate on chains of inferences in fuzzy logic, in a similar way to predicate logic reasons with precise propositions. That is why approximate reasoning is understood as an extension of traditional prepositional calculus dealing with uncertain or imprecise information, primarily with the elements of fuzzy sets, where an element belongs to a specific set only to some extent of certainty. The inference by reasoning with such uncertain facts produces new facts, with the degree of certainty corresponding to the original facts.

Possibilistic reasoning, which also roots in fuzzy set theory (Zadeh, 1965), as an alternative theory to ***bivalent logic*** and the traditional theory of probability, tends to describe possible worlds in terms of their similarity to other sets of possible worlds and produces estimates that should be valid in each given case and under all circumstances. Possibilistic reasoning produces solutions to the problems that bear the indication that the determination of validity is an impossible task.

Possibility theory is closely related to ***evidence theory*** and the ***theory of belief***. It deals with events relying on ***uncertain information***, such as fuzzy sets are, and it is a complementary alternative to the traditional probability theory. Therefore, the membership functions of a fuzzy set, which represent imprecise information, are to be considered as ***possibility distributions*** (Zadeh, 1978).

The issue of the relationship between ***fuzziness*** and ***probability*** was for many years on the agenda. Kosko (1990) considers that probability arose from the question of whether or not an event occurs, in the sense that the probability that an event at a certain time occurs ***or*** does not occur is the certainty. Similarly, the probability that a possible event at a certain time occurs ***and*** does not occur is impossible. ***Fuzziness*** measures the degree to which an event occurs, but not whether it occurs. Therefore, ***fuzzy probability*** extends the classical concept of probability, admitting the outcomes to belong at the same time to several event classes to different degrees (Dubois and Prade, 1993).

1.4 Evolutionary Computation

Evolutionary computation, which was later adjoined to the methodologies of soft computing as their new constituent, is a computational technology made up of a collection of ***randomized global search paradigms*** for finding the optimal solutions to a given problem. The term ***evolutionary*** is borrowed from the terminology introduced by Charles Darwin (1859), describing the process of adaptation of survival capabilities through natural selection, fitness improvement of individual species, *etc*. To achieve this, evolutionary computation tries to model the natural evolution process for a successful survival battle, where reproduction and fitness play predominant roles. Being an evolutionary process, it is essentially based on the genetic material of offspring inherited from the parents. Therefore, if this material is of bad quality then the offspring can not win the battle of survival.

The evolutionary process considers the ***population*** of individuals represented by ***chromosomes***, each chromosome bearing its characteristics called ***genes***. The genes are assigned their individual values. Through the process of ***crossover*** the offspring are generated by combining the gene values of their parents. During the combination, the genes can undergo a (low probability) ***mutation*** process consisting of random changes of gene value in a chromosome, in order to insert fresh genetic material into the chromosomes. Finally, the winner will be the offspring with the highest value of ***fitness,*** *i.e.* with the best characteristics inherited from the parents.

However, the evolutionary computation algorithms used in practice are not strictly confined to the natural evolutionary process described above. In the meantime, various evolutionary algorithms and their modifications are found. But still, the following variants are only considered as *basic* evolutionary algorithms:

- ***genetic algorithms***, which model genetic evolutionary processes in a generation of individuals

- ***genetic programming***, which is an extension of genetic algorithms to the population in which the individuals are themselves computer programs
- ***evolutionary strategies***, which deal with "***evolution of evolution***" by modelling the strategic parameters that control variations in evolutionary process
- ***evolutionary programming***, which models adaptive evolutionary phenomena

It is interesting to note that the algorithms of evolutionary computation listed above, although being structurally similar, have still been quite independently developed by different researchers without any contact between them.

Genetic algorithms, the first evolutionary algorithms, have been widely studied across the world and predominantly used for optimum random search. The basic version of genetic algorithm, originally proposed by Holland (1975), models the ***genetic evolution*** of a population of individuals represented by strings of binary digits. Based on this model, genetic evolution is simulated using the operations of ***selection***, ***crossover***, and ***mutation*** and monitoring and controlling the simulation performance using the ***fitness function***.

Genetic programming, developed by Koza (1992), extends the original version of genetic algorithms to the space of programs by representing the evolving individuals through individual programs to be evolved. While evolving the programs, genetic programming for each generation qualifies their fitnesses by measuring the performances. The qualifying one is used to find out the programs that at least approximately solve the problem at hand.

Evolutionary strategies have been formulated by Rechenberg (1973) for the direct solving of the engineering optimization problems. This is performed by emulation of the evolutionary process of self-optimization of biological systems in the given environments. It is similar to the case in biological evolutionary processes. Schwefel (1975) extended the concept of initially formulated evolutionary strategies and developed the ***evolution of evolution strategy***. In the latter, the individuals are represented by genetic building blocks and by a set of parameters related to the strategy and these are used to determine the behaviour of individuals in the given environment. The strategic parameters are simultaneously evolved while evolving the genetic characteristics of individuals. During the evolutionary process, the mutation operator is strictly permitted only if it directly improves the fitness value.

Evolutionary programming was introduced by Fogel *et al.* (1975) using the concept of ***finite-state automata***. In contrast to genetic algorithms, the algorithm deals with the development of adequate ***behavioural models***, rather than of ***genetic models***. Evolutionary programming was developed to simulate the adaptive behaviour of some real-life phenomena and by selecting the set of optimal behaviours using the fitness function as a measure of success. The substantial operative difference to genetic algorithms is that evolutionary programming does not use the crossover operator.

1.5 Computational Intelligence

According to the published sources, the term ***computational intelligence*** was coined and defined by Bezdek (1992a), in his attempt to study the relationship between neural networks, pattern recognition, and intelligence. He stated that computational intelligence deals with the numerical data provided by the sensors and does not deal with knowledge. This is different from ***artificial intelligence,*** which mainly deals with the non-numerical system knowledge.
Bezdek later attempted to classify the two kinds of intelligence, considering artificial intelligence as a "mid-level computation in the style of the mind", whereas computational intelligence was the "the low-level computation in the style of the mind". However, this classification and the definitions of two types of intelligence, viewed more or less from the aspect of pattern recognition and neural networks, remained as more of a personal view of the author than a general opinion.

A still different view on computational intelligence was presented by Poole *et al.* (1998), who considered computational intelligence as the study of ***intelligent agent*** design, *i.e.* capable of learning from experience and flexible to the changing environments and to the changing goals.

However, a most decisive step in defining the nature of computational intelligence was made during the 1994 IEEE World Congress of Computational Intelligence (WCCI), which brought together the International Conferences on Neural Networks, Fuzzy Systems, and Evolutionary Programming. On the eve of the WCCI, Marks (1993), in his Editorial to IEEE Transactions of Neural Networks entitled "Intelligence: Computational Versus Artificial," pointed out that "although seeking similar goals, computational intelligence has emerged as a sovereign field whose research community is virtually distinct from artificial intelligence". This indicated that there are two alternative intelligent technologies, the artificial and computational.

In the middle of the 1990s, some researchers advocated defining computational intelligence using the adaptivity concept. Eberhard *et al.* (1995) pleaded for a definition of computational intelligence as a methodology that exhibits the capability of learning and that comprises practical adaptation concepts, paradigms, algorithms, and implementations for facilitation of appropriate actions in complex and changing environments. Similarly, Fogel (1995) suggested that the intelligent technologies, *i.e.* neural, fuzzy, and evolutionary computation, brought together under the generic term computational intelligence should be viewed as a new research field holding the computational methodologies capable of adapting solutions to new problems without relying on human knowledge. Bezdeck went a step further and even viewed computational intelligence and adaptation as synonyms.

To sum up, in the last decade or so, we have witnessed a parallel evolution of two computational streams, soft computing and computational intelligence, both based on methods and tools of artificial intelligence (Popovic and Bhatkar, 1994), predominantly on neural networks, fuzzy logic, and evolutionary computation. Nowadays, because both soft computing and computational intelligence have integrated a large number of computational methodologies, it is difficult to draw a clear distinction between them. Tettamanzi and Tomassini (2001) rather view the scope of computational intelligence as the broader of the two methodologies, because computational intelligence encompasses most various techniques for describing and modelling of complex systems, which is not the case with the scope of soft computing. This is in accordance with the view of Zadeh (1993, 1996, 1999), which defines computational intelligence as the combination of soft computing and numerical processing. But still, Engelbrecht (2002) suggests conceiving soft computing as an extension of computational intelligence in the sense that the probabilistic methods are added to the paradigms of computational intelligence.

In fact, the boundary of the disciplines associated with computational intelligence are still not finally defined. They are still growing up to include new emerging disciplines. For example, the agenda of the 2002 IEEE World Congress on Computational Intelligence includes ***neuroinformatics*** and ***neurobiology*** as new constituents. In the meantime, computational intelligence is viewed as a *new-generation artificial intelligence* for human-like data and knowledge processing, professionally known as ***High Machine Intelligence Quotient*** (**HMIQ**) technology. Most recently, the convergence of the core computational technologies - neural networks, fuzzy systems, and evolutionary computation - to a common frontier has drawn strong attention from the computational intelligence society. A related term was coined: ***autonomous mental development*** (Wenig, 2003).

1.6 Hybrid Computational Technology

In the 1990s we witnessed a new trend in computational intelligence. A growing number of publications on its applications have been published reporting on successful combination of intelligent computational technologies – neural, fuzzy, and evolutionary computation – in solving advanced artificial intelligence problems. The hybrid computational technology created in this way is rooted mainly in integrating various computational algorithms in order to implement more advanced algorithms required for solving more complex problems. For instance, neural networks have been combined with fuzzy logic to result in neuro-fuzzy or fuzzy-neuro systems in which:

- Neural networks tune the parameters of the fuzzy logic systems, which are used in building of adaptive fuzzy controllers, as implemented in the Adaptive Network-Based Fuzzy Inference System (ANFIS) proposed by Jang (1993).
- Fuzzy logic systems monitor the performance of the neural network and adapt its parameters optimally, for instance in order to achieve the nonlinear mapping and/or the function approximation to any desired accuracy (Wang, 1992).
- Fuzzy logic is used to control the learning rate of neural networks to avoid the creeping phenomenon in the network when approaching the solution minimum (Arabshahi *et al.*, 1992).

Evolutionary algorithms have also been successfully used in combination with fuzzy logic in improving heuristic rules and in manipulating optimally the genetic parameters, particularly the crossover operator (Herrera and Lozano, 1994).

Neural networks, in combination with evolutionary algorithms, have profited in optimal evolution of network topology and in finding the optimal values of network weights directly, without network training (Maniezo, 1994). Finally, evolutionary algorithms have also profited through combinations with the traditional computing methods. For instance, in order to improve the efficiency and the accuracy of evolutionary computing algorithms in locating the global extremum, Renders and Bersini (1994) combined these algorithms with the conventional search methods, such as the hill climbing method. Renders and Flasse (1976) even simply integrated such a method in the crossover operator.

1.7 Application Areas

Computational intelligence and soft computing have proven to be very efficient and valuable tools for solving numerous problems in science and engineering that could not be solved using their individual constituents, *i.e.* neuro, fuzzy, or evolutionary computing alone. Although their constituents are themselves capable of solving problems that are difficult or even impossible to solve by traditional computation methods, the synergetic effect of aggregation of two or more constituents enlarges the number and the complexity of solvable problems. This holds not only for the so-called academic problems, but also for real-life problems, including the problems of industrial engineering. Moreover, application of soft computing and computational intelligence has provided the appropriate means for merging the vagueness (*e.g.* perceptions of human beings) and real-life uncertainty with a relatively simplified computational program. This has made them capable of participating in a variety of real-life applications in engineering and industry. For instance, the application of soft computing in engineering covers most areas of data handling, like:

- intelligent signal processing, which includes time series analysis and forecasting
- data mining
- multisensor data fusion, including intelligent pattern recognition and interpretation, performance monitoring and fault diagnosis
- systems engineering, to which belong system identification, system modelling, advanced systems control
- planning and design processes, like optimal path planning and engineering design

Intelligent signal processing solves the problems of adaptive signal sampling, analysis of sampled data, signal features extraction, *etc.* Of outstanding interest for engineering, commerce, and management here is the forecasting of time series data (Kim and Kim, 1997).

Data mining is a strategy for rapid collection, storage and processing of huge amounts of data (Mitra and Mitra, 2002) in some particular application areas, such as in production and financial engineering (Heider, 1996; Major and Riedinger, 1992), surgery (Blum, 1982), telecommunication networks (Pedrycz, Vasilakos, and Karnouskos, 2003/2004), Internet (Etzioni, 1996), *etc*.

Multisensor data fusion, again, is an advanced area of signal processing that deals with the simultaneous collection of multiple sensor values related to a physical system or to any observable phenomenon. It is the most useful technique for solving the problems of pattern recognition and pattern interpretation (Bloch, 1996). For instance, in analysis of remotely sensed satellite images the multisensor image interpretation plays a crucial role. Here, the reflected radiation values from different sensors build a feature vector, which subsequently undergoes the feature extraction and classification process (Bloch, 1996). In engineering, multisensor data fusion has been applied to solve the problems of systems performance monitoring and the problems of fault diagnosis of rotating machinery based on vibration measurements (Emmanouilidis *et al*., 1998). In addition, the multisensor data fusion approach has been particularly applied in monitoring of operability of individual sensors (Taniguchi and Dote, 2001). In recent years, on-line fault detection and diagnosis of dynamic systems based on a reliable model of the overall system behaviour under normal operating conditions have been the subjects of research by the soft computing experts. Remarkable results have been reported in this field of research by Akhimetiv and Dote, (1999).

In ***systems engineering***, the application of soft computing encompasses the activities that are essential for system study, optimal system design, and design of adaptive system control concepts: identification and model building of dynamic systems (Tzafestas, 1999; Zurada *et al*., 1994). Here, model building and parameter estimation of dynamic systems are the initial steps in the generation of a mathematical description of dynamic systems behaviour, based on experimental data. The methodology of computational intelligence helps generally in implementation of advanced neuro and fuzzy controllers and supports the evolving of adaptive controllers.

Optimal path planning is a soft computing application area widely needed in manufacturing, primarily in job-shop scheduling and rescheduling, in optimal routing in very large-scale integration layouts, and in robotics for optimal path planning of robots and manipulators.

As a systems designer's tool, computational intelligence helps in styling the circuit layout in microelectronics (Bosacci, 1997), optimal product shaping, *etc*.

1.8 Applications in Industry

In the industrial reality, there is a growing need for employing completed machine and process automation, which includes not only the motion or process control, but also their performance monitoring, diagnosis, and similar tasks. Owing to the increasing complexity of the tasks, advanced intelligent computational tools, such as soft computing and computational intelligence, are called upon to help in handling the execution of the tasks efficiently. The application capabilities of both intelligent computational tools presented above guarantee their successful use in solving the majority of high-complexity problems in the industrial world. This was demonstrated on a number of examples published in the last decade.

The earliest use of fuzzy logic in the process industry was recorded in Japan, where, in the late 1980s, fuzzy logic facilities capable of solving complex nonlinear and uncertainty problems of a chemical reactor were used to replace the skilled plant operator. Around the same time, neural networks were applied in statistical analysis of huge sets of acquired sensor data by time series analysis and forecasting. This application was later extended to include ***data mining*** for managing very large amounts of more complex data using the methodologies of soft computing based on pattern recognition and multisensor data fusion. This was helpful in better understanding the process behaviour through analysis and identification of essential process features hidden in data piles. In addition, it was also possible to solve some accompanying problems related to plant monitoring and diagnosis, product quality control, production monitoring and forecasting, plant logistics and various services, *etc*.

In the iron and steel industry, enormous progress was made after introducing intelligent computational approaches in process modelling, advanced process control, production planning and scheduling, *etc*. For more than three decades the steel producers have profited from advanced methods, starting with direct digital control and finishing with the glorious distributed computer control systems developed by systems and control engineers (Popovic and Bhatkar, 1990). With the advent of intelligent computational technologies, fuzzy logic control, neural networks-based modelling, intelligent sensing, evolutionary computing-based optimization at various process and plant levels, *etc*. have been on the agenda mainly because of high international competition in this industrial branch in producing high quality product at the lowest production cost.

However, it was the electronic industry that has to the most remarkable extent profited from the introduction of intelligent computational technology in chip design and production processes.

Computational intelligence has also found wide application in manufacturing, particularly in product design, production planning and scheduling, monitoring of tool wear, manufacturing control and monitoring of automated assembly lines, and product quality inspection (Dagli, 1994). The use of intelligent technologies in this area was particularly accelerated after the discovery and massive applications of the mechatronics approach in product development. This has also contributed to extending the application field of intelligent technology to include rapid prototyping, integration of smart sensors and actuators, design of internal communication links oriented systems, *etc*. (Popovic and Vlacic, 1999).

References

[1] Akhimetiv DF and Dote Y (1999) Fuzzy system identification with general parameter radial basis function neural network. In: Farinwata SS, Filev D, and Langari R (Eds) Fuzzy control synthesis and analysis, Wiley, Chichester, UK, Ch. 4.

[2] Arabshahi P, Choi JJ, Marks RJ, and Caudell TP (1992) Fuzzy control of backpropagation. In: IEEE Internat. Conf. on Fuzzy Systems, San Diego: 967-972.
[3] Bezdek JC (1992a) On the relationship between neural networks, pattern recognition and intelligence. Int. J. Approximated Reasoning, 6: 85-102.
[4] Bezdek JC (1992b) Computing with uncertainty. IEEE Commu. Magaz., Sept: 24-36.
[5] Bloch I (1996) Information combination operators for data fusion: a comparative review with classification. IEEE Trans. Syst. Man and Cybern. A26(1): 52-67.
[6] Blum RI. (1982) Discovery and representation of causal relationship from a large time-oriented clinical database: The RX project, Lecture Notes in Medical Informatics, vol. 19:23-36, Springer-Verlag, New York.
[7] Bonissone PP (1997) Soft computing: the convergence of merging reasoning technologies. Soft Computing 1: 6-18.
[8] Bonissone PP, Chen YT, Goebel K, and Khedkar PS (1999) Hybrid soft computing systems: industrial and commercial applications. Proc. of the IEEE 87(9): 1641-1667.
[9] Bosacci B (1997) On the role of soft computing in microelectronic industry. Soft Computing 1: 57-60.
[10] Dagli CH (ed.) (1994) Artificial neural networks in intelligent manufacturing. Chapman and Hall, London.
[11] Darwin C (1859) The origin of species. John Murray, London, UK.
[12] Dubois D and Prade H (1993) Fuzzy sets and probability: misunderstandings, bridges and gaps. Proc. of the Second IEEE Inter. Conf. On Fuzzy Systems, 2: 1059-1068.
[13] Dubois D and Prade H (1998) Soft computing, fuzzy logic and artificial intelligence. Soft Computing 2(1): 7-11.
[14] Eberhard R, Simpson P, and Dobbins R (1995) Computational intelligence PC tools. Academic Press, Boston, USA.
[15] Emmanoulidis C, MacIntyre J, and Coxs C (1998) Neurofuzzy computing aided machine fault diagnosis. Proc. of Joint Conf. on Information Sciences, 1:207-210.
[16] Engelbrecht AP (2002) Computational intelligence: an introduction, Wiley, NJ.
[17] Etzioni O (1996) The world-wide-web: Quagmire or goldmine? Communication, ACM, 39:65-68.
[18] Fogel DB (1995) Review of computational intelligence: imitating life (Zurada JM, Marks RJ, and Robinson CJ, Eds.) IEEE Trans. on Neural Networks, 6(6): 1562-1565.
[19] Fogel LLJ, Owens AJ, and Walsh (1966) Artificial intelligence through simulated evolution. Wiley, New York.
[20] Heider R (1996) Troubleshooting CFM 56-3 engines for the Boeing 737 using CBR and data-mining. LNCS, vol. 1168:512-523, Springer-Verlag, New York.
[21] Herrera F and Lozano M (1994) Adaptive genetic algorithm based on fuzzy techniques. In: Proc. of IPMU '96, Granada, Spain: 775-780.
[22] Holland JH (1975) Adaptation in natural and artificial Systems. The University of Michigan Press, Ann Arbor, Michigan.
[23] Jang JSR (1993) ANFIS: Adaptive-network-based-fuzzy-inference system. IEEE Trans. Syst. Man Cybern. 23(3):665-685.
[24] Jang J-SR, Sun C-T, and Mizutani E (1997) Neuro-fuzzy and soft computing. Prentice Hall, Upper Saddle River, NJ.
[25] Kim D and Kim Ch (1997) Forecasting time series with genetic fuzzy predictor ensemble. IEEE Trans. on Fuzzy Systems, 5(4): 523-535.
[26] Kosko B (1990) Fuzziness versus probability. Int. J. General Syst. 17(2/3): 211-240.
[27] Koza JR (1992) Genetic programming. The MIT Press, Cambridge, MA.
[28] Major JA and Riedinger DR (1992) EFD- A hybrid knowledge statistical –based system for the detection of fraud. Internat. J. Intelligent System, 7:687-703.
[29] Maniezzo V (1994) Genetic evolution of the topology and weight distribution of neural networks. IEEE Trans. on Neural Networks 5(1): 39-53.

[30] Marks RJ (1993) Intelligence: computational versus artificial. (Editorial) Trans. on Neural Networks, 4(5): 737-739.

[31] Minsky ML and Papert S (1969) Perceptrons. MIT Press, Cambridge, MA.

[32] Mitra S and Mitra P (2002) Data mining in soft computing framework: a survey. IEEE Trans. on Neural Networks, 13(1): 3-14.

[33] Pedrycz, Vasilakos, and Karnouskos (2003/2004) IEEE Trans. on Syst. Man and Cybern., special issue on computational intelligence in telecommunication networks and internet service. Pt.-I, 33 (3): 294-426; Pt.–II, 33(4): 429-501; Pt.-III, 34(1):1-96.

[34] Poole D, Mackworth, and Goebel R (1998) Computational intelligence: a logical approach. Oxford University Approach, New York.

[35] Popovic D and Bhatkar VP (1990) Distributed computer control for industrial automation. Marcel Dekker Inc., New York .

[36] Popovic D and Bhatkar VP (1994) Methods and tools for applied artificial intelligence. Marcel Dekker Inc., New York.

[37] Popovic D. and Vlacic Lj (1999) Mechatronics in engineering design and product development. Marcel Dekker Inc., New York.

[38] Rechenberg I (1973) Evolutionsstrategie: Optimierung technischer Systeme nach Prinzipien der biologischen Evolution. Fromman-Holzborg Verlag, Stuttgart.

[39] Renders YM and Bersini H (1994) Hybridizing genetic algorithms with hill climbing methods for global optimization: two possible ways. 1st IEEE-CEC: 312-317.

[40] Renders YM and Flasse SP (1970) Hybrid methods using genetics algorithms for global optimisation. IEEE Trans. Syst. Man Cyber., 26(2): 243-258).

[41] Rosenblatt F (1962) Principles of aerodynamics: perceptrons and the theory of brain mechanics. Spartan Books, Washington D.C.

[42] Ruspini EH (1996) The semantics of Approximated reasoning. In: Fuzzy logic and neural network handbook, Chen CH (Editor), McGraw-Hill, New York:5.1-5.28.

[43] Schwefel H-P (1975) Evolutionsstrategie und numerische optimierung. PhD Thesis, Technical University Berlin.

[44] Taniguchi S and Dote Y (2001) Sensor fault detection for uninterruptible power supply control systems using fast fuzzy network and immune network. Proc. of the SMC 2001: 7-10.

[45] Tettamanzi A and Tomassini M. (2001) Soft computing: integrating evolutionary, neural, and fuzzy systems. Springer-Verlag, Berlin.

[46] Tzafestas SG (1999) Soft computing in systems and control technology. World Scientific Series in Robotics and Intelligent Systems, Vol. 18.

[47] Wang LX (1992) Fuzzy systems are universal approximators. Proc. Intl. Conf. on Fuzzy Systems, San Francisco, CA: 1163-1172.

[48] Wenig J (2003) Autonomous mental development: A new frontier for computational intelligence, IEEE Connections. Nov 2003: 8-13.

[49] Werbos P (1974) Beyond regression: new tools for prediction and analysis in the behavioural science. PhD Thesis, Harvard University, Cambridge, MA.

[50] Zadeh LA (1965)Fuzzy sets. Information and Control, 8: 338-353.

[51] Zadeh LA (1979) A theory of approximate reasoning. In: Hayes P, Michie D, and Mikulich I, eds. : Machine Intelligence, Halstead Press, New York: 149-194.

[52] Zadeh LA (1993) Fuzzy logic, neural networks, and soft computing. Proc. IEEE Int. Workshop Neuro Fuzzy Control, Muroran, Japan: 1-3.

[53] Zadeh LA (1994) Soft computing and fuzzy logic. IEEE Software, Nov.: 48-58.

[54] Zadeh LA (1995) Probability theory and fuzzy logic are complementary rather than competitive. Technometrics 37: 271-276.

[55] Zadeh LA (1996) The role of soft computing: An introduction to fuzzy logic in the conception, design, and development of intelligent systems. Proc. IEEE Int. Workshop Soft Computing in Industry, Muroran, Japan: 136-137.

[56] Zadeh LA (1998) Fuzzy sets as a basis for a theory of possibility. Fuzzy Sets and Systems 1: 3-28.

[57] Zadeh LA (1999) From computing with numbers to computing with perceptions. Proc. IEEE Int. Workshop Soft Computing in Industry. Muroran, Japan: 221-222.

[58] Zurada JM, Marks RJ, Robinson CJ (1994) Review of computational intelligence: imitating life. IEEE Press, New York.

2

Traditional Problem Definition

2.1 Introduction to Time Series Analysis

The importance of time series analysis and forecasting in science, engineering, and business has, in the past, increased steadily and it is still of actual interest for engineers and scientists. In process and production industry, of particular interest is time series forecasting where, based on some collected data, the future data values are predicted. This is important in process and production monitoring, in optimal processes control, *etc*.

A time series is a time-ordered sequence of observation values of a physical or financial variable made at equally spaced time intervals Δt, represented as a set of discrete values $x_1, x_2, x_3, \ldots,$ *etc*. In engineering practice, the sequence of values is obtained from sensors by sampling the related continuous signals. Being based on measured values and usually corrupted by noise, time series values generally contain a deterministic signal component and a stochastic component representing the noise interference that causes statistical fluctuations around the deterministic values.

The analysis of a given time series is primarily aimed at studying it's internal structure (autocorrelation, trend, seasonality, *etc*.), to gain a better understanding of the dynamic process by which the time series data are generated. In process control, the predicted time series data values help in deciding about the subsequent control actions to be taken.

The broad term of time series analysis encompasses activities like

- definition, classification, and description of time series
- model building using collected time series data
- forecasting or prediction of future values.

For forecasting the future values of a time series a wide spectrum of methods is available. From the system-theoretical point of view they can be

- ***model-free***, as used in exponential smoothing and regression analysis

- ***model-based***, particularly used in modelling of time series data to capture the feature of long-time behaviour of the underlying dynamic system.

In the following, various traditional approaches to time series classification, modelling, and forecasting are considered and their application in engineering demonstrated on practical examples taken from process and production industry sectors. This should help in better understanding the modern approaches to time series analysis and forecasting using the methods and tools of artificial intelligence exposed in the chapters to follow. The items presented here should also serve as a source of definitions and explanations of terms used in this field of data processing. It will, however, be supposed that the time series, the model of which should be built, are homogeneous, made up of uniformly sampled discrete data values.

2.2 Traditional Problem Definition

Traditionally, time series analysis is defined as a branch of statistics that generally deals with the structural dependencies between the observation data of random phenomena and the related parameters. The observed phenomena are indexed by ***time*** as the only parameter; therefore, the name time series is used.

Basically, there are two approaches to time series analysis:

- ***time domain approach***, mainly based on the use of the covariance function of the time series
- ***frequency domain approach***, based on spectral density function analysis and Fourier analysis.

Both approaches are appropriate for application to a wide range of disciplines, but the time domain approach is mostly used in engineering practice. This is particularly due to the availability of the Box-Jenkins approach to time series analysis, which is primarily concerned with the linear modelling of stationary phenomena. However, Box and Jenkins have pointed out that their approach is also applicable to the analysis of nonstationary time series, after their differencings (trend removal).

2.2.1 Characteristic Features

The major characteristic features of time series are the ***stationarity***, ***linearity***, ***trend***, and ***seasonality***. Although a time series can exhibit one or more of these features, for presentation, analysis, and prediction of time series values each feature is rather treated separately.

2.2.1.1 Stationarity

This property of a random process is related to the mean value and variance of observation data, both of which should be constant over time, and the covariance between the observations x_t and x_{t-d} should only depend on the distance between the two observations and does not change over time, *i.e.* the following relationships should hold:

$$E\{x_t\} = \mu, \quad t = 1, 2, \ldots$$

$$\text{Var}(x_t) = E\{(x_t - \mu)^2\} = \kappa_0, \quad t = 1, 2, \ldots$$

$$\text{Cov}(x_t, x_{t-d}) = E\{(x_t - \mu)(x_{t-d} - \mu)\} = \kappa_d,$$

with $t = 1, 2, \ldots, \; d = \ldots, -2, -1, 0, 1, 2, \ldots$, and where μ, κ_0, and κ_d are some finite-value constants.

In statistical terms, a time series is stationary when the underlying stochastic process is in a particular state of statistical equilibrium, *i.e.* when the joint distributions of $X(t)$ and $X(t-\tau)$ depend only on τ but not on t. Consequently, the ***stationary model*** of a time series can be easily built if the process (or the dynamics generating the time series) remains in the equilibrium state for all times around a constant mean level.

It is difficult to verify whether a given time series meets the three stationarity conditions formulated above simultaneously. In earlier practice, the stationarity of a time series was roughly checked by inspection of the time series pattern. A given time series was recognized as stationary when it is represented by a flat-looking pattern, with no trend or seasonality, and with time-invariant variance and autocorrelation structure. When the time series model is available, the stationarity of the process generating the time series observation values can be easily checked. For instance, for the first-order autoregressive process

$$x_t = \theta x_{t-1} + \varepsilon_t$$

the stationarity condition requires that the condition

$$\text{Var}(x_t) = \text{Var}(x_{t+1})$$

or the equality

$$E\{[\theta x_{t-1} + \varepsilon t]^2\} = E\{[\theta x_{t-2} + \varepsilon(t-1)]^2\}$$

holds. Therefore, because of mutual independence of ε_t and x_t, the equality

$$\text{Var}(x_t) = \theta^2 \text{Var}(x_{t-1}) + \text{Var}(\varepsilon_t)$$

follows, and finally the equality

$$\kappa_0 = \theta^2 \kappa_0 + \sigma^2, \qquad |\theta| << 1$$

where κ_0 does not depend on time t.

Although for the majority of time series used in practice the stationarity is a common assumption, forecasting of ***nonstationary time series*** is still of considerable importance. For instance, in engineering, business, and economics the collected observation data are better represented through nonstationary time series. Also, nonstationary time series can be transformed into the equivalent stationary time series by taking the differences between the successive data values along the time series pattern, *i.e.* by simple or multiple differencing the given time series data. This approach is generally recommended, because some stationary looking time series can still be nonstationary. To resolve the stationarity problem experimentally, the time series should first be partitioned into two or more "long enough" segments that are apparently stationary, then the autocorrelation and spectrum properties of each segment are checked and the results compared.

2.2.1.2 Linearity

Linearity of a time series indicates that the shape of the time series depends on it's state, so that the current state determines the local time series pattern. If a time series is linear, then it can be represented by a linear function of the present value and the past values. Example of linear representations are the AR, MA, ARMA, and ARIMA models (see Section 2.5), based on autoregression and/or on a moving average technique. Nonlinear time series can be represented by the corresponding nonlinear or bilinear models.

Time series represented by the linear model

$$X_t = \sum_{i=-\infty}^{\infty} \psi_i Z_{t-i} ,$$

generally describe a linear process, where ψ_i is a set of constants that satisfies the condition

$$\sum_{i=-\infty}^{\infty} |\psi_i| < \infty ,$$

and $|Z_t|$ is white noise with a zero mean value and variance σ^2.

The multivariable form of a linear process is statistically defined by the relation

$$X_t = \sum_{i=-\infty}^{\infty} C_i Z_{t-i} ,$$

where $|C_i|$ represents a series of $n \times n$ matrices with the absolutely summable elements, and $|Z_t|$ is the white noise with zero mean value and covariance matrix Σ.

2.2.1.3 Trend

The ***trend component*** of a time series is its long-term feature that is manifested through the local or global increase or decrease of data values as a consequence of superposition of true time series values and a disturbance with upward or downward trend. The presence of a disturbing component is detectable by pursuing the changes in the mean values in certain successive time intervals across the time series pattern.

Trend analysis is important in time series forecasting. In practice, it is accomplished using linear and nonlinear regression technique that satisfactorily helps in identifying non-monotonous trend component in the time series. For instance, for identifying the character of the trend present in a time series, the linear, exponential, or polynomial relation

$$x_t = \alpha t + \beta + \varepsilon_t$$
$$x_t = \exp(\alpha t + \beta + \varepsilon_t)$$
$$x_t = \alpha t + \beta t^2 + \gamma + \varepsilon_t$$

is used for fitting the collected data.

2.2.1.4 Seasonality

The seasonality component of a time series is demonstrated through its periodically fluctuating pattern. This feature is more common in economic time series and in time series in which the observations are taken from real life, where the pattern may repeat hourly, daily, weekly, monthly, yearly, *etc*. Thus, the main objective of seasonal time series analysis is focused on the detection of the character of its periodical fluctuations and on their interpretation. In engineering, seasonal time series are found in the problems of power, gas, water, and other distribution systems, where the prediction of consumer demands represents the basic problem.

2.2.1.5 Estimation and Elimination of Trend and Seasonality

When two or more time series with different features are superimposed, or when a time series is superimposed by trend and/or seasonality component, ***decomposition analysis*** is needed to discriminate and separate individual components involved. More frequently, decomposition analysis is used for *de-**trending*** and ***de-seasonalizing*** the time series data. A classical decomposition example is complex decomposition, where a time series could be made up of various components, such as trend, random, seasonal, and cycling components. In this context, the seasonal component $S(t)$ is viewed as a periodic component with a fixed cycling period corresponding to the individual seasons. In practice, it is convenient to combine the trend and the cyclical components into a ***trend-cycle*** component $TC(t)$, so that the observed resulting value of the time series X at time t can be written as

$$X(t) = S(t) + R(t) + TC(t),$$

where $R(t)$ is the random component. This is the additive representation model of a multi-component time series. The corresponding multiplicative representation model is

$$X(t) = S(t) \times R(t) \times TC(t) .$$

Both models are useful because, in some real-life cases, time series made up of values collected in trade or in commerce, the seasonal and trend-cycle components can add their values to the main component or to multiply them as interrelated factors.

Anyhow, to make a proper forecast when a multi-component time series is given, it must first be identified to what extent the individual components are present in the time series data. This needs the decomposition of time series data to identify and extract the partial data superimposed to the main time series data. The time series decomposition process can be presented as shown in Figure 2.1.

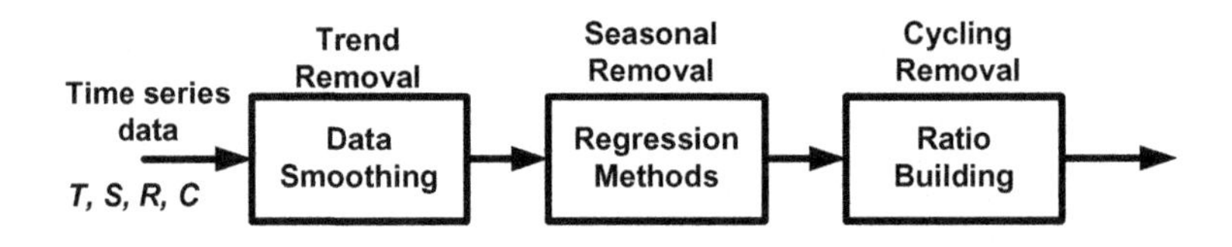

Figure 2.1. Time series decomposition process

For solving the decomposition problem, two methods have been mostly used.

- ***Census I method***, to eliminate the variability within the individual seasons. This uses the moving average windows for calculating the average time series values within the windows. The windows have a width equal to the length of the season. This enables the removal of both the seasonal and random components. Depending on the representation model used, moving-average values are subtracted from the time series values (when an additive model is used) or the time series values are divided by the moving average values (when the multiplicative model is used). In the first case the seasonal component is calculated as the average value.
- ***Census II method***, an extended and improved Census I method. This is predominantly used in financial engineering, trading, and econometrics. It also relies on additive and multiplicative representation models, but it is very data-table oriented.

2.3 Classification of Time Series

Depending on the character of data that they carry, the time series could be

- stationary and nonstationary
- seasonal and non-seasonal
- linear and nonlinear
- univariate and multivariate

- chaotic.

Time series encountered in practice can have two or more of the properties listed above. For instance, linear time series can be stationary, seasonal, and can have the trend component incorporated. In the following we will mainly focus on linear, nonlinear, univariate, multivariate, and chaotic time series.

2.3.1 Linear Time Series

Linear time series are generated through observation of ***linear processes***, mathematically defined by ***linear models*** of the form

$$y(t) = \sum_{j=-\infty}^{\infty} \alpha_j x(t-j),$$

where the coefficients α are subjected to the restriction

$$\sum_{i=-\infty}^{\infty} |\alpha_i| < \infty$$

Linear time series could be generated by second-order stationary processes that are generally linear processes or they can be transformed to linear processes using ***World's decomposition*** (Brockwell and Davis, 2002) technique for elimination of its deterministic component.

2.3.2 Nonlinear Time Series

Many time series in engineering and macroeconometrics require nonlinear modelling (see Section 2.5.8). Some of them are represented as *bilinear time series*, modeled as

$$x_t = z_t + \sum_{i=1}^{p} a_i x_{t-i} + \sum_{j=1}^{q} b_j z_{t-j} + \sum_{i=1}^{r} \sum_{j=1}^{s} c_{ij} x_{t-i} z_{t-j}.$$

2.3.3 Univariate Time Series

The term ***univariate time series*** refers to time series obtained by sampling a single observation pattern, for instance the values of a single physical variable or of a single time-dependent signal at equal time intervals. Thus, in univariate time series the time is an implicit variable that is usually replaced by an ***index variable***. If the data sampling is equispaced then the index variable can be omitted.

Time series presented here in the majority of cases are univariate time series. In the case where a univariate time series can be exactly represented by a mathematical model, the time series is said to be ***deterministic***. Otherwise, if the time series can only be represented in terms of a ***probability distribution function***, then the time series is said to be ***non-deterministic*** or ***stochastic***.

2.3.4 Multivariate Time Series

Multivariate time series are generated by simultaneous observation of two or more processes. The observation values collected are represented here as vector values. These kinds of observation are very common in engineering, where two or more physical variables (temperature, pressure, flow, *etc.*) have to be simultaneously sampled for building the model of a dynamic system.

Multivariate time series are best understood as being a set of simultaneously built time series, the value of each series – apart from their internal dependency within the series itself – also have an interdependency with the values of other component series. ***Multivariate analysis***, a branch of mathematical statistics qualified for processing of multidimensional sampled data, is used for their processing (Dillon and Goldstein, 1985; Johnson and Wichern, 1988).

2.3.5 Chaotic Time Series

Random components of a time series mainly fall into one of two categories:

- They are ***truly random***, *i.e.* the observations are drawn from the underlying probability distribution characterized by a statistical distribution function or by statistical moments of data, such as mean, variance, skew, *etc.*
- They are ***chaotic***, characterized by values that appear to be randomly distributed and non-periodic, but are actually resulting from a completely deterministic process.

The main feature of ***chaotic time series*** is that they have no definite periodicity, *i.e.* they can be represented by the values that may be randomly repeated several times without maintaining any definite periodicity. A typical example of a chaotic signal generator is the nonlinear dynamic oscillating system

$$d^2x/dt^2 = -\left\{x + 0.5\left(dx/dt\right)^2\right\},$$

which is sensitive to its initial conditions. This can be presented geometrically by the trajectory of the system in the phase plane, in which the trajectory of ***non-dissipative systems*** make up a set of nested closed curves, whereas those of ***dissipative nonlinear systems*** for all initial conditions lead to trajectories that either lie on a single surface or converge to individual points in phase space. The set of surfaces and points in the phase space to which all trajectories of the system converge is called the ***attractor*** of the system. The attractors of a chaotic system can have a non-integral, *i.e.* fractal, dimensions and are called ***strange attractors***. Such attractors are very important for forecasting of chaotic time series.

2.4 Time Series Analysis

Time series analysis deals with the problems of identification of basic characteristic features of time series, as well as with discovering - from the observation data on which the time series is built - the internal time series structure.

2.4.1 Objectives of Analysis

The main objectives of time series analysis are

- ***building of input-output models*** that represent the equivalent transfer functions of processes behind the time series
- ***forecasting the future time series values*** from the past values using the models developed
- ***control systems design***, based on the result of analysis.

Depending on the origin of the observation data, forecasting of future values of time series can also provide support in efficient process and production monitoring and failure diagnosis, in product quality inspection, *etc.*, using the time-domain or frequency-domain approach.

Once the time series model has been developed and tested it can be used for forecasting the future time series values at various time distances d. Of course, the forecasting does not deliver the exact future values of data that the given time series will really have, but rather their estimates. For example, using the auto-regressive model

$$x_t = \phi_1 x_{t-1} + \phi_2 x_{t-2} + \varepsilon_t$$

based on a one-step movement along the time series

$$x_{t+1} = \phi_1 x_t + \phi_2 x_{t-1} + \varepsilon_{t+1},$$

we can formally write the predicted value to be

$$\hat{x}_{t+1} = \phi_1 x_t + \phi_2 x_{t-1}.$$

For the ***two-steps ahead prediction***, based on a two-steps movement along the time series, we can also formally write

$$x_{t+2} = \phi_1 x_{t+1} + \phi_2 x_t + \varepsilon_{t+2},$$

or

$$x_{t+2} = \phi_1 (\phi_1 x_t + \phi_2 x_{t-1} + \varepsilon_{t+1}) + \phi_2 x_t + \varepsilon_{t+2},$$

and the predicted value to be

$$\hat{x}_{t+2} = \phi_1 \hat{x}_{t+1} + \phi_2 x_t$$

or

$$\hat{x}_{t+2} = \phi_1 (\phi_1 x_t + \phi_2 x_{t-1}) + \phi_2 x_t .$$

2.4.2 Time Series Modelling

In engineering, modelling of dynamic phenomena has long been seen as a valuable support tool for winning a deep insight into the structure and behaviour of dynamic systems. Much research and development efforts have been made in development and application of system models. In control engineering, system models have been widely used for design and implementation of advanced control strategies, such as adaptive, predictive, and self-tuning control. In business and financial engineering, as well as in water, gas, fuel, and electrical power distribution systems, the mathematical models have for a long time been used for quantity demand forecasting. This is, in fact, the most significant aspect of time series analysis, which also helps to reduce, or even to eliminate, the inherent disturbances or fluctuating components present in observed or in measured values.

2.4.3 Time Series Models

In statistics, two basic mathematical system models are used:

- ***deterministic models***, mathematically viewed as analytical models represented by deterministic relations like

  $$x_t = f(t),$$

  or by recurrence equations like

  $$x_t = f(x_{t-1}, x_{t-2}, \ldots)$$

- ***stochastic models***, statistically viewed as functions of ***random variables***.

Mathematical models used for time series analysis are generally

- *regression models*
- *time-domain models*
- *frequency-domain models*,

whereas, again, the time-domain models could be

- *transfer function models*

- *state-space models.*

In the following, various approaches for building stationary models of time series are presented.

2.5 Regressive Models

Regressive models are built using ***regression analysis***, which is a collection of methods for the study of relationships between the variables and for estimation and prediction of values of one variable using the values of other variables incorporated in a joint time series (Drapper and Smith, 1981). For instance, to implement an efficient predictor for a variable of interest, the measurable variables representing the strong indicators for the same variable should first be identified.

The most popular regression models in engineering are the

- *autoregression model* (AR)
- *moving-average model* (MA)
- *ARMA model*
- *ARIMA model*
- *CARIMA models.*

2.5.1 Autoregression Model

Autoregression models express the current value of a time series by a finite linear aggregate of previous values and by a shock μ_t

$$x_t = \alpha_1 x_{t-1} + \alpha_2 x_{t-2} + \ldots + \alpha_v x_{t-v} + \mu_t ,$$

where α_1 to α_v are the ***autoregression parameters***, μ_t is the white noise and v is the model order. The validity of an autoregressive model assumes that the time series to be modeled is stationary. Also, because of some possible internal cumulative effects, the autoregressive process will only be stable if the values of parameters α are within a certain range.

It is common to write the autoregressive equation in terms of deviations $\tilde{x}_t = x_t - \mu_t$, generally using the variable Z and its deviation $\tilde{Z} = Z - \mu$. The individual terms of the time series now become $\tilde{Z}_t, \tilde{Z}_{t-1}, \tilde{Z}_{t-2}, \tilde{Z}_{t-3}, \cdots$, resulting in the autoregressive model

$$\tilde{Z}_t = \phi_1 \tilde{Z}_{t-1} + \phi_2 \tilde{Z}_{t-2} + \phi_3 \tilde{Z}_{t-3} + \cdots + \phi_p \tilde{Z}_{t-p} + a_t,$$

where μ, ϕ_1, ϕ_2, ϕ_3,..., ϕ_q, σ_a^2 are unknown parameters to be estimated from the observation data. Introducing the autoregressive operator

$$\phi(B) = 1 - \phi_1 B - \phi_2 B^2 - \phi_3 B^3 - \cdots - \phi_p B^p$$

the autoregressive model can be written in the compact form

$$\phi(B)\tilde{z}_t = a_t.$$

The model contains (p+2) unknown parameters, *i.e.* p internal parameters and two additional parameters: the variance σ_a^2 and the white noise a_t.

A crucial problem in modelling of autoregressive time series is the selection of the order of the model to be built. A useful approach in this case is the analysis of the related ***partial autocorrelation function*** and the ***inverse autocorrelation function***, because using the autocorrelation function itself is computationally complicated in the case of building of higher order models. Alternatively, fitting the time series shape by models of progressively higher order can be used, along with the analysis of the ***residual sum of squares*** for each order.

2.5.2 Moving-average Model

Another approach frequently used in modelling of univariate time series is based on the ***moving-average model***

$$\tilde{z}_t = a_t - \theta_1 a_{t-1} - \theta_2 a_{t-2} - \theta_3 a_{t-3} - \ldots - \theta_q a_{t-q}$$

which expresses $\tilde{z}_t$ in terms of an infinite weighted linear sum of $a_t, a_{t-1}, a_{t-2}, \ldots, a_{t-q}$. Introducing the moving-average operator of order q

$$\theta(B) = 1 - \theta_1 B - \theta_2 B^2 - \theta_3 B^3 - \ldots - \theta_q B^q$$

the moving-average model can be written in the compact form as

$$\tilde{z}_t = \theta(B) a_t$$

The model contains (q+2) unknown parameters $\mu, \theta_1, \theta_2, \theta_3, \ldots, \theta_q, \sigma_a^2$ to be estimated from the observation data.

2.5.3 ARMA Model

The combination of the AR and MA models makes up the ARMA model

$$\tilde{z}_t = \phi_1 \tilde{z}_{t-1} + \phi_2 \tilde{z}_{t-2} + \ldots + \phi_p \tilde{z}_{t-p} + a_t - \theta_1 a_{t-1} - \theta_2 a_{t-2} - \ldots - \theta_q a_{t-q}.$$

Rewriting the model as

$$\tilde{Z}_t - \phi_1 \tilde{Z}_{t-1} - \phi_2 \tilde{Z}_{t-2} - ... - \phi_p \tilde{Z}_{t-p} = a_t - \theta_1 a_{t-1} - \theta_2 a_{t-2} - ... - \theta_q a_{t-q}$$

and rearranging it as

$$(1 - \phi_1 B - \phi_2 B^2 - ... - \phi_p B^p)\tilde{Z}_t = (1 - \theta_1 B - \theta\ B^2 - ... - \theta_q B^q)a_t$$

the model can finally be written in compact form as

$$\phi(B)\tilde{Z}_t = \theta(B)a_t,$$

where B is a delay operator. The derived compact model contains (p+q+2) unknown parameters $\mu, \phi_1, \phi_2, ..., \phi_p$ and $\theta_1, \theta_2, ..., \theta_q, \sigma_a^2$ that are to be estimated from the given time series data. In practice, for the representation of actually occurring stationary time series, it is frequently adequate enough to take p and q not greater than 2. The presence of both autoregressive and moving-average terms in the ARMA model enables the representation of complex time series with fewer parameters than would be needed using a corresponding AR model.

2.5.4 ARIMA Model

This Box-Jenkins variant of the ARMA model is predestinated for applications to nonstationary time series that become stationary after their differencing. ***Differencing*** is an operation by which a new time series is built by taking the successive differences of successive values, such as $X(t) - X(t\text{-}1)$ along the nonstationary time series pattern. In the acronym ARIMA, the letter I stands for *integrated.*

The widely accepted convention for defining the structure of ARIMA models is ARIMA(p, q, d), where p stands for the number of autoregressive parameters, q is the number of moving-average parameters, and d is the number of differencing passes. For instance, the ARIMA(2, 3, 1) model has two autoregressive parameters, three moving-average parameters, computed after the series have been differenced once.

A variety of time series encountered in industry and business exhibit nonstationary behaviour. In particular, they do not vary about a ***fixed mean*** because of the possible presence of a drift component. Such time series may, nevertheless, exhibit homogeneous behaviour of a kind. In particular, although the general level about which fluctuations are occurring may be different at different times, the broad behaviour of the series, when differences in level are allowed for, may be similar. It can be shown that such behaviour may be represented by a ***generalized autoregressive operator***

$$\varphi(B) = \phi(B)(1-B)^d,$$

where ϕ is the **stationary autoregressive operator** and φ is the **generalized autoregressive operator**. One or more of the zeros of the polynomial φ (*i.e.* one or more of the roots of the equation $\varphi(B)=0$) is unity.

The general form of a model to represent the homogeneous nonstationary behaviour of the time series is given by

$$\varphi(B)x(t) \equiv \phi(B)(1-B)^d x(t)$$

or,

$$\varphi(B)x(t) = \theta(B)a_t.$$

Defining now

$$w(t) = (1-B)^d x(t),$$

we get

$$\varphi(B)w(t) = \theta(B)a_t.$$

Introducing the operator

$$\nabla = (1-B)$$

the last equation becomes

$$w(t) = \nabla^d x(t).$$

where,

$$\nabla x(t) \equiv x(t) - x(t-1) = (1-B)x(t).$$

The homogeneous nonstationary behaviour can therefore be represented by a model which calls for the *d*th difference of the process to be stationary. In practice, it is mostly $d = 0$ or 1 but not greater than 2.

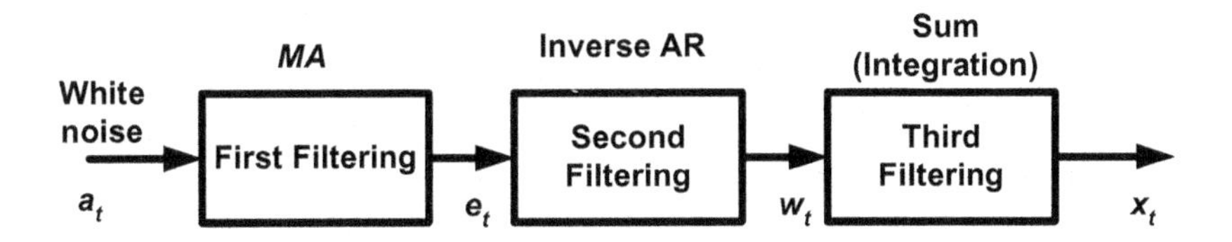

Figure 2.2. Block diagram of an ARIMA model

From the above it follows that the general ARIMA process may be generated from white noise a_t by means of three filtering operation, as shown by the block diagram in Figure 2.2, where the first filter has the input a_t, the transfer function $\theta(B)$, and output

$$e_t = \theta(B)a_t ,$$

where

$$\theta(B) = (1 - \theta_1 B - \theta_2 B^2 - ... - \theta_q B^q)$$

is the moving-average operator.

Table 2.1. Summary of properties of AR, MA and ARMA processes

Sl. No.	Description	AR	MA	ARMA
1.	Model in terms of $\tilde{x}$	$\phi(B)x_t = a_t$	$\theta^{-1}(B)\tilde{x}_t = a_t$	$\theta^{-1}(B)\phi(B)\tilde{x}_t = a_t$
2.	Model in terms of a	$\tilde{x}_t = \phi^{-1}a_t$	$\tilde{x}_t = \theta(B)a_t$	$\tilde{x}_t = \phi^{-1}(B)\theta(B)a_t$
3.	ψ = weights	Infinite series	Finite series	Infinite series
4.	Stationary condition	Roots of $\phi(B) = 0$ lie outside unit circle	Always stationary	Roots of $\phi(B) = 0$ lie outside unit circle

The second filter has the input e_t, transfer function $\phi^{-1}(B)$, and output w_t

$$w_t = \phi^{-1}(B)e_t,$$

and the third filter has the input w_t, the transfer function s^d (*i.e.* it is a nonstationary summation filter) and the output x_t

$$x_t = s^d w_t$$

or

$$x_t = s^d \phi^{-1}(B) \cdot \theta(B) a_t.$$

Table 2.1 summarizes some features of AR, MA, and ARMA model.

2.5.5 CARMAX Model

In systems and control theory the CARMAC model is used for design of minimum variance and predictive control (see Section 2.10.3). For deterministic dynamic systems with an input signal $u(t)$ and a disturbance $e(t)$ the CARMA or CARMAX model is defined as

$$A(z^{-1})y(t) = B(z^{-1})u(t) + C(z^{-1})e(t),$$

where

$$A(z^{-1}) = \sum_{i=0}^{n} a_i z^{-i} \qquad i_0 = 1$$

$$B(z^{-1}) = \sum_{j=1}^{m} b_j z^{-j}$$

$$C(z^{-1}) = \sum_{v=0}^{p} c_v z^{-v} \qquad c_0 = 1.$$

The acronym *CARMAX* stands ***control autoregressive moving-average model with auxiliary inputs***. This has analogy with the Kalman notation of a state-space model of the filter.

2.5.6 Multivariate Time Series Models

The observation values of some time series are multivariate, made up of components that themselves are observations of some time series. Such multivariate values are presented as vector values $x = [x_1, x_2, ... x_n]^T$, and the entire set of multiple values as a matrix made up of individual observation vectors

$$x = \begin{bmatrix} x_{11} & x_{12} & \dots & x_{1n} \\ x_{21} & x_{22} & \dots & x_{2n} \\ \dots & \dots & \dots & \\ x_{n1} & x_{n2} & \dots & x_{nn} \end{bmatrix}.$$

Multivariate time series are processed using ***multivariate analysis***, which is the statistical methodology for processing of multidimensional data.

Model building of multivariate time series is required when the values of one variable of an individual time series are dependent on the values of variables in other related time series. For better modelling and for more accurate analysis, all values concerned should be taken into account simultaneously. For instance, the corresponding joint observations of two mutually dependent variables have to be modeled under consideration of the components of a two-dimensional observation vector $x_i = (y_i, z_i)$, for i=1, 2, 3, ... *etc*. Thus, a ***bivariate time series*** has to be modeled based on two-dimensional observation vectors of the interdependent univate time series. But, before building the model it should be checked whether

- the two time series (represented by y and z values) mutually correlate, in which case only the correlation analysis has to be carried out, or
- the two series are causally related, in which case the time series model should be built.

In practice, the number of time series to be considered simultaneously can be larger than two, so that multivariate time series have to be built using the observation vectors and the related observation matrix. Using this presentation approach, the great majority of basic theory of univariate time series can formally be extended to the multivariate time series. For instance, in an analogous way the equivalent ARMA model for a stationary multivariate time series, with zero mean vector, can be written as

$$x_i = \alpha_1 x_{t-1} + \alpha_2 x_{t-2} + \dots + \alpha_v x_{t-v} + \varepsilon_t - \beta_1 \varepsilon_{t-1} - \beta_2 \varepsilon_{t-2} - \dots - \beta_n \varepsilon_{t-n} - n$$

where x_t and ε_t are n-dimensional column vectors, ε_t being the multivariate white noise, and α_i and β_i are the elements of the corresponding $[n \times n]$ matrix of ARMA model parameters

$$\alpha_j = \{\alpha_{j,kk}\}, \quad \text{for } j = 1, 2, \dots, v$$

$$\beta_j = \{\beta_{j,kk}\}, \quad \text{for} \quad j = 1, 2, \ldots, n.$$

It is usually supposed that the mean value of expectation is

$$E\{\varepsilon_j\} = 0, \quad \text{for all } j$$

and the condition that the covariance matrix of ε_j

$$E\{\varepsilon_j \varepsilon_{j-k}\} = \Sigma_\varepsilon \varepsilon^2, \qquad \text{for } k = 0.$$

Finally

$$x_t = \alpha_1 x_{t-1} + \alpha_2 x_{t-2} + \ldots + \alpha_v x_{t-v},$$

where the values of parameters α_1 to α_v and those of the covariance matrix should be estimated. This is rather mathematically complicated and requires computer support.

For ***dimensionally reduced modelling*** of multivariable time series, the method of ***principal components analysis*** is used (Jolliffe, 1986). The analysis helps to reduce the initial number of correlated variables to a small number of variables, *i.e.* to the ***principal factors*** that still contain (with minimal loss) the essential information of the initial number of variables. This reduces the computational effort needed for further time series data processing.

However, the reduction in the number of initial variables is not a process of simple elimination of some non-relevant variables, as the eliminated variables still have an influence, or "echo", on the remaining variables. This is because the principal components are first determined using a smaller number of linear combinations of the initial variables that are still able to reproduce the entire collection of observed variables within a relatively good accuracy. Applying principal component analysis, the optimal number of linear combinations can be found that are best predictors of the entire set of variables. The prediction accuracy achieved is considered as the best performance measure. It is also to be noted that - after transforming the initial variables to the reduced number of variables using linear combinations - the ***back-transformation*** of the reduced variables to the initial variables is not possible.

Consider now the five observations of each of three variables x_1, x_2, x_3 presented in matrix form

$$x = \begin{bmatrix} x_{11} & x_{21} & x_{31} \\ x_{12} & x_{22} & x_{32} \\ --- & --- & --- \\ x_{15} & x_{25} & x_{35} \end{bmatrix}$$

The components of the corresponding ***mean vector***, made up of the mean of each variable, are calculated as

$$x_{mi} = \frac{1}{5}\sum_{j=1}^{3} x_{ij}\,, \quad \text{where } i = 1, 2, 3, \ldots$$

The elements of the corresponding ***variant-covariant matrix***, or ***dispersion***, made up of the variances of the variables along the main diagonal and the covariances between each pair of variables in the remaining location, are given by

$$s_{ij} = \frac{1}{5}\Sigma_{i,j}(x_i - x_{mi})(x_j - x_{mj})^T,, \quad \text{where } \; i, j = 1, 2, 3.$$

Principal components, which are linear combinations of random variables with some characteristic properties with respect to the variances, play a key role in the analysis of multivariate time series. For instance, the first principal component is the sum of squares of the coefficients having the maximal variance. Furthermore, the principal components are in fact the ***characteristic vectors*** of the covariance matrix, so that they help in the study of the characteristic vectors and ***characteristic roots***.

2.5.7 Linear Time Series Models

Linear models of time series are based on linear relationships between the observed values. Typical examples of linear models are the AR, MA, ARMA, and ARIMA models.

2.5.8 Nonlinear Time Series Models

The difficulty in testing for nonlinearity in a given set of observation values calls for special approaches to building adequate time series models. The observation set of nonlinear time series may contain various shocks of different form and of different intensity. In financial engineering practice, it is common to check the time series nonlinearity using first a linear time series model. If the linear model does not fit the major part of observation data, then a nonlinear model is built and tested. However, the problem then is what nonlinear model should be selected that will best fit the collected data (Casdagli and Eubank, 1992). There are some traditional examples of such models like ***STAR*** (***smooth transition autoregression model***), ***ARCH*** (***autoregressive conditional heteroskedasticity***) and the ***bilinear model***, widely used in econometrics and financial forecasting. Recently, the ***Markov switching model***, *threshold* ***autoregression model***, and ***smooth transition autoregression model*** are also becoming popular.

For STAR models there have been some nonlinear alternatives like

$$x_t = \alpha_0 + \alpha_1 x_{t-1} + f(x_{t-d})(\varphi + \lambda x_{t-1}) + \varepsilon_t,$$

with x_t as a transition variable that can be described by an AR(1) model with the parameters α_0 and α_1 and with the nonlinear component $f(x_{t-d})(\varphi + \lambda x_{t-1})$.

There is also the *ESTAR model*

$$f(x_{t-d}) = 1 - \exp[-k(x_{t-d} - k)^2]$$

with $k > 0$, and the ***LSTAR model***

$$f(x_{t-d}) = \frac{1}{1 + \exp[-k(x_{t-d} - k)^2]}.$$

State-space modelling of nonlinear time series relies on the theory of ***first-order Markov chains*** in the *n*-dimensional state space, where the observation vector is represented by $x_t = (x_t, x_{t-1}, ..., x_{t-n+1})^T$, and the nonlinear time series model is represented by the ***stochastic difference equation***

$$x_t = S(x_t, \varepsilon_{t+1})..$$

Alternatively, the nonlinear state space can be used for modelling the nonlinear time series, relying on transition probability

$$P\{x_{t+1,i+1} < x_{i+1} \perp x_t = x, \; x_{t+1,j} = x_{t+1}, \; j \leq i\}$$

where $x_{t+1,\, i+1}$ denotes the (*i*+1)th component of x_{t+1}.

2.5.9 Chaotic Time Series Models

In the last two decades or so, research in the field of ***chaotic time series analysis*** has steadily grown and it is today an interesting field of work for mathematicians and engineers. Initially, the research interest was in estimating the dimension of the underlying ***attractor*** and the ***Liapunov exponents*** of the chaotic systems that characterize the space-filling properties and the stability of dynamic systems. The attention was later focused on the techniques of chaotic time series modelling and on prediction of future time series values using most frequently the ***nonlinear autoregressive model*** for the state vector *x*(*t*).

$$x(t + d) = F[x(t), x(t - d_1), x(t - d_2), ..., x(t - d_{n-1})]$$

where d is the ***delay factor*** between the individual observations and n is the number of observations considered. Here, the nonlinear time series model is required when the model should hold globally. Otherwise, for local considerations, a ***local linear model*** is preferred.

In reality, the output system signals are corrupted by noise, as well as by chaotic signals generated by dynamic systems. Therefore, for modelling noise-corrupted chaotic signals it should first be established whether the noise present corrupts the systems state vector (like the system noise) in the form

$$y(t+1) = F\,[x(t) + n(t)]$$

or, like the measurement disturbances, whether it only adds to the output signal

$$\hat{y}(t) = y(t) + n(t)\,.$$

The generation of chaotic signals by dynamic systems is based on the phenomena of initial-value sensitivity of the corresponding differential equations. This was first pointed out by Poincare′. For instance, the sequence

$$x_n = 4x_{n-1}(1 - x_{n-1})$$

for any initial value $0 < x_0 < 1$ and for any n = 0, 1, 2, … *etc*., produces the solution

$$x_n = \sin^2[\sin^{-1}(\sqrt{x_0})]\,,$$

which is highly sensitive to the initial value selected, because it determines the value of the arcsin function. A small deviation Δx_0 contributes here the $2^n \Delta x_0$ changes in $\arcsin\sqrt{x_0 + \Delta x_0}$.

2.6 Time-domain Models

Two typical time series modelling approaches in the time domain are to build the

- transfer function model
- state space model

Both models are of fundamental importance in traditional and modern control theory.

2.6.1 Transfer-function Models

Transfer-function models are the extension of regression models in which the transfer function of a dynamic system is integrated into the model. This is used in systems theory for representation of relationships between the systems input and output variables.

Building transfer-function models is based on experimental records of input and output time series. In engineering practice, transfer-function estimation is preferred because it does not require any system disturbance, say by step, pulse, or sinusoidal test signal. Statistical estimation methods largely rely on ***normal operating records*** and are robust against the noise.

Consider the multivariate time series in which supposedly a single output y_t depends on some lagged values x_{t-i}, $i = 0, 1, 2, \ldots$, of the input time series in the following way:

$$y_t = c_0 x_t + c_1 x_{t-1} + c_2 x_{t-2} + \ldots + \xi_t$$

where $\xi(t)$ is the output noise component of y. After introducing the unit delay operator D this becomes

$$y_t = [c_0 + c_1 D + c_2 D^2 + \ldots + \xi_t],$$

which finally results in

$$y_t = [cD]x_t + \xi_t,$$

where $[cD]$ represents the transfer function model of the system. It is supposed that the input series and the noise component $\xi(t)$ are mutually independent.

In systems engineering, the ARMA(n, m) model

$$y(t) + a_1 y(t-1) + \ldots + a_n y(t-n) = e(t) + c_1 e(t) + \ldots + c_m e(t-m)$$

plays a key role in model building. The compact form of the above model is

$$A(z^{-1})y(t) = C(z^{-1})e(t),$$

where the polynomials $A(z^{-1})$ and $C(z^{-1})$ are the respective operators, *i.e.* the polynomials in z^{-1}. The corresponding transfer function of the system is

$$\frac{y(t)}{e(t)} = \frac{C(z^{-1})}{A(z^{-1})}.$$

2.6.2 State-space Models

In systems theory, the widely preferred class of models are the ***state-space models***. The models are made of two sets of equations. One set represents the state-space model of the system

$$x_{t+1} = A_t x_t + B_t u_t + w_t,$$

based on ***state vector*** x_t and the ***system disturbance vector*** w_t, and the other set models the systems output

$$y_t = C_t x + v_t,$$

based on the ***observability vector*** y_t at system output and on the ***system output disturbances vector*** v_t. In the two sets of equations, A_t, B_t, and C_t represent the ***system matrix***, ***control matrix***, and ***observation matrix*** respectively, all at the instant *t*. In addition, the ***covariance matrices*** of the disturbance vectors are supposed to be

$$\text{cov}\{w_t\} = Q_t$$

$$\text{cov}\{v_t\} = R_t$$

The objective of state-space modelling (Aoki, 1990) is to estimate the values of the state vector and to forecast its future values based on observations y_t.

2.7 Frequency-domain Models

In analogy with the signal representation in the frequency domain, also a time series can be represented in the same domain. This is because the time series values are generated by equidistant sampling of signals. Therefore, a time series made up of sampled values can also be presented as a collection of sine and cosine waves with different frequencies and be processed using ***spectral analysis*** (Warner, 1998).

A device that was designed to represent the given time series visually in the frequency domain is the ***periodogram***. It is a simple spectral analysis facility made up of mixtures of sine and cosine components within a frequency spectrum.

To illustrate the calculation of the periodogram, suppose that the number of observations is odd, say $N = 2q+1$, and the ***Fourier series model*** to be fitted using the observation data is given as

$$x_t = a_0 + \sum_{i=1}^{q}(a_i C_{it} + b_i S_{it}) + e_t,$$

where

$$C_{it} = \cos(2\pi f_i t),$$
$$S_{it} = \sin(2\pi f_i t),$$
$$f_i = \frac{i}{N}.$$

If f_i is the ith harmonic of the fundamental frequency $1/N$, then the least square estimates of the coefficients a_0 and (a_i, b_i) will be

$$a_0 = \overline{x},$$
$$a_i = \frac{2}{N}\sum_{t=1}^{N} x_t C_{it},$$
$$b_i = \frac{2}{N}\sum_{t=1}^{N} x_t S_{it}.$$

The periodogram then consists of $q = (N\text{-}1)/2$ values

$$I(f_i) = \frac{N}{2}\sum_{t=1}^{N}(a_i^2 + b_i^2) \qquad i = 1, 2, \ldots, q,$$

called intensities at frequency values f_i.

However, for an even value N, which should be set equal to $2q$, the above equation holds for $i = 1, 2, 3, \ldots, (q\text{-}1)$, and the last coefficients are

$$a_q = \frac{1}{N}\sum_{t=1}^{N}(-1)^t x_t,$$
$$b_q = 0,$$

and correspondingly

$$I(f_q) = I(0.5) = Na_q^2.$$

Because in a given time interval the highest frequency is 0.5 cycles per interval, in the definition of sample spectrum - if f_i is the ith harmonics of the fundamental frequency $(1/N)$ - the definition of the periodogram is modified to

$$I(f) = \frac{N}{2}(a_f^2 + b_f^2)$$

where $0 \le f \le 0.5$. In this case $I(f)$ is referred to as the ***sample spectrum***.

Example

The simplest Fourier model

$$X_t = a_0 + a\cos\omega t + b\sin\omega t + Z_t$$

can, in the case of multidimensional analysis, be represented in matrix form as

$$E(X) = A\theta ,$$

with,

$$X = (X_1, X_2, ..., X_n)^T$$
$$\theta = (a_0, a, b)$$

and

$$A = \begin{bmatrix} 1 & \cos\omega & \sin\omega \\ 1 & \cos 2\omega & \sin 2\omega \\ \ldots & \ldots. & \ldots. \\ 1 & \cos n\omega & \sin n\omega \end{bmatrix}.$$

Minimizing the least-squares

$$\sum_{t=1}^{n} (X_t - a_0 - a\cos\omega t - b\sin\omega t)^2$$

the θ can be estimated using the pseudo inverse relation

$$\hat{\theta} = (A^T A)^{-1} A^T X .$$

In order to use the ***spectral expansion technique*** for forecasting purposes, we need first to observe the given time series carefully to check whether it contains any trend and/or seasonality. This can, for instance, be identified by visual inspection of the graph of the given series.

Trend removal from the time series can be carried out in two ways:

- by taking the first or the second difference of the given time series data
- by fitting a polynomial to it and then by subtracting the fitted polynomial from the given time series data.

The coefficients of the fitted polynomial can be estimated by a least-squares fitting method which minimizes the sum of squares of the difference between actual data and the polynomial data to be fitted.

The first difference is taken as $(x_t - x_{t+1})$ for t ranging from 1 to N-1, and where x_t is the actual time series data at time instant t and N is the total number of observations in the given time series. Similarly, the first difference applied to the resulting first difference will give rise to the second difference. Once the series is de-trended, we have to check for various frequency components present in the residual of the time series. This is accomplished by first transforming the signals from the time domain into the frequency domain using a ***fast Fourier transform***

(FFT), and then by computing the ***power spectral density function*** as a measure of the energy at various frequencies

$$P_{yy} = Y * \tilde{Y}$$

where "*" represents the one-to-one multiplication of vector components or of matrix elements, and $\tilde{Y}$ represents the complex conjugate of Y. The calculated values P_{yy} are then plotted against the frequencies. From the resulting plot, the major frequency components present in the residual time series signal can be identified.

The FFT is a computational technique that substantially reduces the time required to perform Fourier transformations on a digital computer. Introduced in the 1960s by Cooley and Tukey (1965), the transformation has steadily increased its popularity.

2.8 Model Building

The Box and Jenkins methodology (Box and Jenkins, 1976) for building time series models includes a

- ***model identification phase***, in which – apart from some preliminary statistical calculations – the number of model parameters is determined that are needed to ensure that the mathematical model to be built matches the collected time series data with the desired accuracy
- ***model estimation phase***, in which the values of model parameters are estimated by minimizing the sum of squares of residuals
- ***model validation phase***, in which the model accuracy is checked and the possible model improvement is established
- ***model forecasting phase***, in which the model is used to establish the confidence limits of the forecast.

The above methodology, however, is not a straight-forward process, rather it is a chain of iterative actions that Box and Jenkins described using the flow chart shown in Figure 2.3.

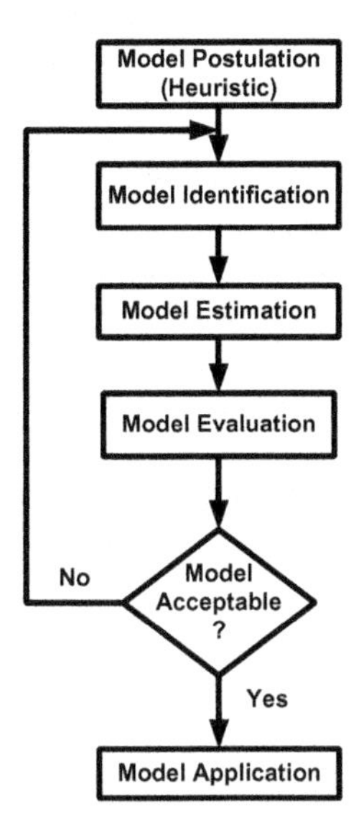

Figure 2.3. Box-Jenkins methodology of model building

The Box-Jenkins model building process assumes that the time series to be modeled is stationary. Otherwise, it should be differenced several times until it becomes stationary. In some cases the time series values should be manipulated so that their mean becomes zero. Further to this, the seasonality of the time series has to be removed, which complicates the related calculations, particularly when building ARIMA models.

2.8.1 Model Identification

Box and Jenkins defined the model identification phase as a rough procedure for laying down the initial model structure that matches good enough with the collected observation data. The essence of the identification process was first demonstrated on the example of an ARMA model, for which the required number of parameters for both the autoregressive and the moving-average parts of the model have been determined. This could be done using the autocorrelation approach, usually by determining the ***sample autocorrelation function*** and the ***sample partial autocorrelation function***.

The sample autocorrelation function is defined as the ratio

$$\hat{\rho}(d) = \frac{\hat{\gamma}(d)}{\hat{\gamma}(0)},$$

where

$$\hat{\gamma}(d) = \frac{1}{n}\sum_{t=1}^{n-|d|}(t + |d| - \overline{x})(x(t) - \overline{x})$$

is the corresponding sample autocovariance function for $-n < d < n$.

In contrast with the autocorrelation function, which is infinite in extent, the ***partial autocorrelation function*** is described in terms of N non-zero autocorrelation functions

$$\rho_i = \sum_{j=1}^{N} \phi_{jN}\rho_{i-j}\,, \quad i = 1, 2, \ldots, N\,;$$

which can be described in a compact form by the ***Yule-Walker equation***

$$\mathrm{P}_N \phi_N = \rho_N\,,$$

using the vectors

$$\rho_N = \left[\rho_1,\ \rho_2,\ \ldots,\ \rho_N\right]^T$$

$$\phi_N = \left[\phi_{N1},\ \phi_{N2},\ \ldots,\ \phi_{NN}\right]^T$$

and the matrix

$$\mathrm{P}_N = \begin{bmatrix} 1 & \rho_1 & \rho_2 & \cdots & \rho_{N-1} \\ \rho_1 & 1 & \rho_1 & \cdots & \rho_{N-2} \\ \cdots & \cdots & \cdots & \cdots & \\ \rho_{N-1} & \rho_{N-2} & \cdots & 1 & \end{bmatrix}$$

Solving the Yule-Walker equation for $N = 1, 2, 3, \ldots$, one gets

$$\phi_{11} = \rho_1, \qquad \phi_{22} = \frac{\begin{vmatrix} 1 & \rho_1 \\ \rho_1 & \rho_2 \end{vmatrix}}{\begin{vmatrix} 1 & \rho_1 \\ \rho_1 & 1 \end{vmatrix}}, \qquad \phi_{33} = \frac{\begin{vmatrix} 1 & \rho_1 & \rho_1 \\ \rho_1 & 1 & \rho_2 \\ \rho_2 & \rho_1 & \rho_3 \end{vmatrix}}{\begin{vmatrix} 1 & \rho_1 & \rho_2 \\ \rho_1 & 1 & \rho_1 \\ \rho_2 & \rho_1 & 1 \end{vmatrix}},$$

etc. The last equality, ϕ_{NN}, represents the ***partial autocorrelation function***. Here, the ***sample autocorrelation function*** for an AR(1) process should have mixed exponentially decaying and damped sinusoidal components. In addition, for higher order autoregressive processes the sample partial autocorrelation function should also be considered, which becomes zero when more model parameters are involved than needed.

The partial autocorrelation function is not helpful for identifying the order of the moving-average process because if the number of model parameters is higher than required, then the autocorrelation process becomes zero. Nevertheless, the fact that both the sample autocorrelation functions and the partial autocorrelation functions are random variables makes the model identification generally difficult, particularly the identification of a mixed ARMA model. Also, developing time series models using sample plots of both autocorrelation functions involves multiple trial-and-error iterations, which is time consuming. Akaike (1974) proposed the information criterion, known as the ***Akaike information criterion*** (AIC):

$$AIC(n) = v \log_e(\sigma ml)^2 + 2n ,$$

with σml = RSS/v and RSS being the residual sum of squares. By minimizing the criterion with respect to n, the model order v can be determined, which helps automate the model identification process. For instance, in the case of two equivalent models being found, with both having acceptable residuals, the one having a lower AIC(n) value can be taken as the better one.

A similar criterion was proposed by Schwarz, known as the ***Bayesian information criterion*** (BIC), defined as

$$BIC(n) = v \log_e(\sigma ml)^2 + n \log_e v.$$

It delivers a lower order model than the AIC, which is an argument for its preference. But also here, in the initial phase of the model identification process, the stationarity, seasonality, *etc.* of the given time series have to be checked and removed by de-trending and de-seasonalization of time series data.

A successful identification phase of model building is to a great extent a matter of knowledge and practical experience, rather than the matter of some given rigid instructions about how to do it. Yet, some recommendations related to the initial parameter estimation of a pure AR process are still available, relying on the use of the Yule-Walker approach. Much more difficult is to model the MA part of an ARMA model, where a system of nonlinear equations has to be solved.

2.8.2 Model Estimation

Once the preliminary time series model has been identified, *i.e.* the number of required model parameters has been determined, the actual model parameter values have to be estimated using the observation data. This is a nonlinear estimation problem that needs some special statistical procedures, like the ***maximum likelihood method*** or ***nonlinear least-squares estimation***. The parameter values estimated at this stage of model building should minimize the sum of squared residuals. The estimated values are usually called ***most likelihood parameter values*** or the ***least-squares parameter values***.

The maximum likelihood method applied to the ARMA(p, q) process with the sampled values arranged as the components of the vector $[y_1, y_2, ..., y_n]^T$ and with the non-zero mean μ starts with the extended model

$$Y_i - \mu = \sum_{i=1}^{p} \alpha_i (Y_{t-i} - \mu) + z_t + \sum_{j=1}^{q} \beta_j z_{t-j}$$

with $p+q+2$ parameters $\alpha_i, \beta_j, \sigma^2 = \mathrm{VAR}(z)$, and $\mu = \varepsilon\{Y_i\}$. A matrix V(α, β) should now be defined so that the relation

$$\mathrm{VAR}(Y) = \sigma^2 V(\alpha, \beta),$$

holds, where

$$\begin{aligned} \alpha &= [\alpha_1, \alpha_2, ..., \alpha_p]^T \\ \beta &= [\beta_1, \beta_2, ..., \beta_q]^T \\ y &= [y_1, y_2, ..., y_n]^T \\ Y &= [Y_1, Y_2, ..., Y_n]^T \end{aligned}$$

with the elements of $V(\alpha, \beta)$ being proportional to the autocorrelation coefficients of $\{Y_i\}$.

Supposing now that z_i values are normally distributed, so will Y also be normally distributed, so that the log-likelihood function will be defined by

$$L(\mu, \sigma^2, \alpha, \beta) = -\frac{1}{2}[n \log \sigma^2 + \log V(\alpha, \beta) + (y - \mu I)^T [V(\alpha, \beta)]^{-1} (y - \mu I) / \sigma^2]$$

where I is the identity vector $I = [1,\ 1,\ ...,\ 1]^T$.

Given initial values of α and β, the maximum estimate of μ and σ^2 are

$$\begin{aligned} \hat{\mu}(\alpha, \beta) &= \{I^T [V(\alpha, \beta)]^{-1} y\} / \{I^T [V(\alpha, \beta)]^{-1} I\} \\ \sigma^2(\alpha, \beta) &= \frac{1}{n}[y - \hat{\mu}(\alpha, \beta) I]^T / \{[V(\alpha, \beta)]^{-1}[y - \hat{\mu}(\alpha, \beta) I] \end{aligned}$$

This, after substituting in the above likelihood equation, gives

$$L_0(\alpha, \beta) = -\frac{1}{2}[n \log \sigma^2(\alpha, \beta) + \log V(\alpha, \beta)].$$

The maximum likelihood estimates $\hat{\alpha}$ and $\hat{\beta}$ can now be determined in the conventional way.

Parameter estimation of autoregressive models can, based on the Yule-Walker approach, be managed in a relatively direct way. Multiplying the autoregressive model

$$X_t = \sum_{i=1}^{p} \alpha_i X_{t-i} + Z_t$$

by X_{t-j}, $j > 0$, and calculating the expectation of each term of the resulting equation, we obtain

$$X_t X_{t-j} = \sum_{i=1}^{p} \alpha_i X_{t-i} X_{t-j} + Z_t X_{t-j}.$$

Taking into account that Z_t and Z_{t-j} are statistically independent, the relation

$$\gamma_j = \sum_{i=1}^{p} \alpha_i \gamma_{j-i},$$

is obtained, which after division by γ_0 becomes the relation of autocorrelation functions ρ

$$\rho_j = \sum_{i=1}^{p} \alpha_i \rho_{j-i}.$$

Subsequently, by substituting the values $j = 1, 2, \ldots, p$ in the last equation a set of linear equations can be built, the matrix form of which will be

$$\hat{\alpha} = R^{-1} \rho,$$

where $\hat{\alpha}^T = [\hat{\alpha}_1, \hat{\alpha}_2, \ldots, \hat{\alpha}_p]$ is the ***parameters vector***, $\rho^T = [\rho_1, \rho_2, \ldots, \rho_p]$ is the ***autocorrelations vector***, and

$$R = \begin{bmatrix} 1, & \rho_1, & \ldots, & \rho_{p-1} \\ \rho_1, & \rho_2, & \ldots, & \rho_{p-2} \\ \ldots & \ldots & \ldots & \ldots \\ \rho_{p-1}, & \rho_{p-2}, & \ldots, 1 & \end{bmatrix}$$

is the corresponding matrix. The parameters of the autoregressive models can now be determined by solving the above matrix equation.

2.8.3 Model Validation and Diagnostic Check

In the last phase of model building the ***model validity*** or the ***model adequacy*** should be verified, *i.e.* it should be checked how good the developed model fits the collected time series data and how close the predicted future values are close to the actual future values of the time series itself. If the model built fits the time series data satisfactorily, then the residuals should behave consistently with the model.

The ***model diagnostic check*** on the other side, includes checking the model sensitivity to the characteristics of the input data. For this, Box and Jenkins proposed the ***overfitting procedure***, which starts with an identified and estimated low-order model and continues with the fitting of more-elaborate models by augmenting the model dimension. If the augmented model is ***overfitted***, then the previous model is taken as the better one. Alternatively, in starting with a high-order model, if the previous model is already overfitted, then the diagnostic check continues with checking of lower order models, the dimensions of which are reduced and repeatedly checked against overfitting.

Anyhow, although the overfitting approach to model diagnostic checks appears to be rather simple, it still presumes the normality of the statistical distribution and is strongly influenced by the ***correlation structure*** of the data. For instance, taking the ARMA(p, q) model for overfitting and repeating the fitting procedure for ARMA(p+1, q) and ARMA(p, q+1), we will get the maximum log-likelihood values, say L_0, L_1, L_2. If the initial ARMA(p, q) model is adequate, then the ***generalized likelihood ratio*** test procedure expects that each of the statistics $(L_1 - L_0)$ and $(L_2 - L_0)$ is ***chi-squared distributed***.

A simplified approach to verifying the minimum number of model parameters really needed to represent the observation data is the check of the mutual non-correlation of residuals. If the residuals correlate, then the number of model parameters should be increased. For verifying this correlation **the *residual diagnostic methods*** are appropriate, and for detection of parameters that are irrelevant for model presentation the ***parameter diagnostics approaches*** are recommended. Both approaches support finding the alternative models that, again, can help in improving the current model.

Residual diagnostics includes statistical calculations of

- *mean percentage error* and the *residual mean, i.e.* the average of all the computed residuals
- *autocorrelation of residuals*
- *closeness-of-fit statistics*.

If the value of the residual mean deviates significantly from zero, then this indicates that the fitted values deviate from the original time series values or that the residuals are not balanced out, in the sense that the positive or the negative residuals are predominant. The *unbalanced residuals* represent a kind of *bias*, so that in this case the *biased forecasts* are determined. The mean percentage error is similar to the residual mean.

For checking the mutual correlation of residuals the ***correlogram of residuals*** is evaluated. The presence of spikes in the correlogram indicates that the residuals might be correlated and that the model developed is not adequate.

The residual values $z_1, z_2, ..., z_n$ of the noise sequence $Z_0, Z_1, ..., Z_n$ of an ARMA process are obtained by substituting in the likelihood function all the estimated values of α and β into each of the related time series $y(t)$, $y(t+1)$, $y(t+2)$, ..., $y(t+n-1)$ and by solving the resulting system of equations. This is generally a difficult issue. It is much easier to extract the residual sequence for the AR(p) and the MA(q) part of the ARMA(p, q) process separately. For instance, in the case of an AR(p) model

$$Y(t) - \mu = \sum_{i=1}^{p} \alpha_i [Y(t-i) + Z(t)],$$

for $t = p+1$, ..., n, the residuals are

$$z(t) = [y(t) - \hat{\mu}] - \sum_{i=1}^{p} \hat{\alpha}_i [y(t-i) - \hat{\mu}],$$

whereby for $t \leq p$ the residuals are not defined. In the case of an MA(q) model

$$Y(t) - \mu = \sum_{i=1}^{q} \beta_i Z(t-i) + Z(t),$$

that can be rewritten as

$$Z(t) = [Y(t) - \mu] - \sum_{i=1}^{q} \beta_i Z(t-i),$$

and the residuals are defined as

$$\begin{aligned}
z_1 &= y_1 - \hat{\mu} \\
z_2 &= y_2 - \hat{\mu} + \beta_1 Z_1 \\
&\dots \quad \dots \quad \dots. \\
z_t &= y_t - \mu + \sum_{i=1}^{q} \beta_i Z_{t-i}
\end{aligned}$$

where the last equality holds for $t > q$.

2.9 Forecasting Methods

Once the time series model has been built, it can be employed in forecasting the future values using an adequate forecasting method. Viewed historically, the term

forecasting is closely linked with the term ***prediction***. The earliest researchers working on methods of determination of future values of empirical functions, based on a set of collected values, coined the term prediction, rather than forecasting. Forecasting is predominantly associated with the problem of time series analysis. The term prediction, however, is still preferably used in systems and control engineering.

2.9.1 Some Forecasting Issues

Forecasting the future values of a time series is defined in the following way:

- given a set of observed values x_1, x_2, x_3, ..., x_n of a time series, the future value x_{n+1}, x_{n+2}, ..., should be estimated
- ***q-steps ahead prediction*** x_{n+q}, calculated at time point n, is denoted by $\hat{x}_n(q)$, where the integer q is called the ***lead time***.

Generally, the forecasting approaches can be classified into

- ***objective forecasts***, made on a subjective basis using judgement, intuition, commercial knowledge and any other relevant information
- ***univariate forecasts***, based entirely on fitting a one-dimensional model to the collected data and on extrapolation of the time series pattern
- ***multivariate forecasts***, based on simultaneous observation of two or more variables and on models of multivariate time series.

In practice, a forecasting approach can include a combination of two of the above approaches. For example, univariate forecasts – after being carried out – can be adjusted subjectively. Or, put in another way, the marketing forecast based on various predictions developed statistically from the past data can be combined with the experience or knowledge of people deeply involved in the market. Finally, the simplest way of more reliable forecasting takes into account the combination of two or more weighted objective forecast estimations to calculate the final forecast value (see Section 2.9.6).

Before selecting a forecasting method it is essential to consider how this is to be used, what forecasting accuracy is expected, what computational resources are available, how many items are to be forecast, how much data are available, and how far ahead forecasts are needed. Furthermore, the forecasting method may somehow depend on the required lead time, although in engineering it is mainly short-term forecasts that are of interest, whereas in management it is mostly lead time of nine months that may be of interest. For example, in stock control, the lead time for which forecasts are required is the time between ordering an item and its delivery, which is usually a few weeks or a few months.

Apart from this, some forecasting methods simply produce point forecasts. But in some cases it is desirable to produce interval forecasts. Some procedures, such as the one from Box-Jenkins, enable one to do this by addressing the upper and lower limits on a subjective basis.

Basically, for all forecasting approaches, plotting the time series data is recommended as the first step of data analysis. This is because much useful information can often be obtained from a visual examination of the plots, which helps in selecting the most appropriate forecasting procedure. In addition to calculating the best forecasts, it is also important to specify the accuracy with which the forecasts are to be determined, so that the risk associated with decisions based upon the forecasts may be calculated.

2.9.2 Forecasting Using Trend Analysis

For trend forecasting, linear or nonlinear regression is mostly used. This is based on ***trend line fitting*** of time series data using a linear, quadratic, or exponential function

$$x_p = ax + b$$

$$x_p = ax^2 + bx + c$$

$$x_p = \exp[ax + b]$$

2.9.3 Forecasting Using Regression Approaches

Regression analysis is a mathematical tool that supports the study of relationships among the observed variables. Its main objective is to estimate and predict the value of one variable by taking into account the values of the possibly related other observed variables. Thus, before using the regression technique for prediction of a specific variable, all variables related to this variable should be identified. For prediction

- *simple regression*
- *multiple regression*
- *nonlinear regression*

can be used.

Forecasting using simple regression is based on the equation

$$Y_i = a_0 + a_1 X_i + \varepsilon_i \,, \quad \text{where} \quad i = 1, 2, \ldots, n;$$

where the mean value of the error ε_i is supposed to be zero, and its variance is one. The unknown values of parameters a_0 and a_1 should be estimated so that

$$\sum_{i=1}^{n} (y_i - a_0 - a_1 x_i)^2$$

is minimized. This can be achieved in a straight-forward way by differentiating the above sum with respect to the parameters a_0 and a_1.

In the majority of practical cases, multiple regression is used as a mutual relation

$$y = a_0 + a_1 x_1 + a_2 x_2 + \ldots + a_n x_n$$

between the observed variables x_i, $i = 1, 2, \ldots, n$, and the resulting variable to be estimated y. Also here, using the above equation and the collected data, the problem is to determine the values of the coefficients a_0, a_1, a_2, ..., a_n that will guarantee the best fitting of the regression line to the experimental data. This is verified through correlation analysis.

The compact form of multiple regression is

$$y = Ax + \varepsilon$$

where

$$y = [y_1, y_2, \ldots, y_n]^T$$
$$x = [x_1, x_2, \ldots, x_n]^T$$
$$\varepsilon = [\varepsilon_1, \varepsilon_2, \ldots, \varepsilon_n]^T$$

are the corresponding vectors and

$$A = \begin{bmatrix} a_{11} & a_{12} & \cdots & a_{1n} \\ a_{21} & a_{22} & \cdots & a_{2n} \\ \cdots & \cdots & \cdots & \cdots \\ a_{n1} & a_{n2} & \cdots & a_{nn} \end{bmatrix}$$

the corresponding parameter matrix.

To apply the *least-squares estimator* to find the best estimation value of $\hat{x}$, we first build the error value

$$\varepsilon(\hat{x}) = (y - A\hat{x})$$

and try to find the $\hat{x}$ value that minimizes the product

$$(y - A\hat{x})^T \, (y - A\hat{x}) \, .$$

Using the least-squares estimation procedure with respect to $\hat{x}$, the equation

$$2A^T A\hat{x} - 2A^T y = 0$$

is obtained, from which the estimated value of $\hat{x}$

$$\hat{x} = [A^T A]^{-1} A^T y,$$

follows.

The generalization of the least-squares estimator consists of minimization of the linear form

$$(y - A\hat{x})B(y - A\hat{x})^T$$

with the diagonal positive definite matrix

$$B = \text{diag}\ [c_1, c_2, \ldots, c_n]$$

playing the role of a weighting matrix. Using the above calculations, the corresponding generalized least-squares formulation

$$\hat{x} = [A^T CA]^{-1} A^T Cy,$$

is achieved.

The mathematical model on which the nonlinear regression relies has the general form

$$y_i = f(x_i, \alpha) + \varepsilon,$$

where y_i is the ith observation of dependent variable y, x_i is the ith observation of x, and f is a selected nonlinear function. In practice, the polynomial

$$y = \sum_{i=0}^{n} \alpha_i x^i + \varepsilon$$

is frequently selected as the nonlinear function.

2.9.4 Forecasting Using the Box-Jenkins Method

Box and Jenkins have developed a general forecasting methodology for time series generated by a stationary autoregressive moving-average process. In the following, the methodology is explained on regressive models described in Section 2.4.

2.9.4.1 Forecasting Using an Autoregressive Model AR(p)

The autoregressive model

$$x_t = \sum_{i=1}^{p} a_i X_{t-i} + \mu$$

can be used to estimate the forecasts for any number of steps ahead. For example, the one-step prediction

$$\hat{x}_{t+1} = \sum_{i=1}^{p} a_i X_{t-t+1} + \mu .$$

2.9.4.2 Forecasting Using a Moving-average Model MA(q)

For the moving average model

$$x_t = \sum_{i=1}^{q} \beta_i e_{t-i} + \mu + e_t$$

the estimated optimal linear forecast of *j* steps ahead is given by

$$\hat{x}_{t+j} = \sum_{i=1}^{q} \beta_i e_{t-i+j} + \mu$$

2.9.4.3 Forecasting Using an ARMA Model

The general form of an ARMA process is written as

$$y(t) = \sum_{i=1}^{p} \alpha_i y(t-i) + \sum_{i=1}^{q} \beta_i y(t-i) + Z(t),$$

with $Z(t)$ as white noise. For simplicity, the transcription

$$\phi(B)y(t) = \theta(B)Z(t),$$

is preferred, where $\phi(B)$ and $\theta(B)$ are the corresponding polynomials.

Considering now the general form of a linear process

$$y(t) = \sum_{i=0}^{\infty} \theta_i Z(t-i) \tag{2.1}$$

the estimated forecast can be built as

$$\hat{y}(t+k) = \sum_{i=0}^{t-1} w_i y(t-i) . \tag{2.2}$$

Combining the last two equations we get the estimated forecast as

$$\hat{y}(t+k) = \sum_{i=0}^{t-1} w_i \sum_{j=0}^{\infty} \theta_j Z(t-i-j).$$

or in final form as

$$\hat{y}(t+k) = \sum_{j=0}^{\infty} W_j Z_{t-j} . \tag{2.3}$$

The next objective is to estimate the mean squared forecast error (MSE) from the difference

$$\text{MSE} = \varepsilon\{[y(t+k) - \hat{y}(t+k)]^2\}$$

$$\text{MSE} = \varepsilon\{[\sum_{i=0}^{\infty} \theta_i Z(t+k-i) - \sum_{j=0}^{\infty} W_j Z(t-j]^2\}$$

or,

$$\text{MSE} = \varepsilon\{[\sum_{i=0}^{k-1} \theta_i Z(t+k-i) - \sum_{i=k}^{\infty} (\theta_i - W_i - k) Z(t-j-i)]^2\}.$$

Assuming that the Z_t are mutually independent with a mean of zero and variance σ^2, the last equation of mean square error becomes

$$MSE = \sigma^2 [\sum_{i=0}^{k-1} \theta_i^2 + \sum_{i=k}^{\infty} (\theta_i - W_{i-k})^2] .$$

From this it follows that the mean square error is minimized by taking

$$(\theta_i - W_{i-k}) = 0,$$

wherefrom it follows that

$$W_i = \theta_{i+k} .$$

This, when introduced into the k-step forecast (or k-step prediction) equation (2.3), gives

$$\hat{y}(t+k) = \sum_{i=1}^{\infty} \theta_{i+k} Z_{t-i}$$

which can also be expressed as

$$\hat{y}(t+k) = \sum_{i=k}^{\infty} \theta_i Z_{t+k-i} . \tag{2.4}$$

2.9.4.4 Forecasting Using an ARIMA Model

The forecasting approaches presented so far refer only to stationary models. In practice, however, many important time series are not stationary, so that they have to be transformed to stationary time series. For instance, the generalization of an ARMA model can be modified to provide a model for a time series that is nonstationary in the mean (see Section 2.4.4). The modified version of an ARMA is known as ARIMA (*i.e.* the ***autoregressive integrated moving average***). The term ***integrated*** indicates the fact that the model is produced by repeated integrating or summing of the ARMA process. For example, by multiple summing the ARMA process we get the ARIMA model

$$y_n = \sum_{i=1}^{p} a_i y_{n-i} + \mu + \sum_{j=1}^{q} \beta_j Z_{n-j}$$

for $n \geq 0$, where

$$x_n = \sum_{i=1}^{n} y_i .$$

Using the last equation we can build

$$x_n - x_{n-1} = y_n$$

which, after applying the *z*-transformation, results in

$$y(z) = (1 - z^{-1}) x(z) ,$$

so that the *z*-transformed ARIMA model is

$$a(z)(1 - z^{-1}) x(z) = \beta(z) Z(z) + \mu .$$

Again, after *d* successive integrations, the last equation is converted to

$$a(z)(1 - z^{-1})^d x(z) = \beta(z) Z(z) + \mu ,$$

This is the ARIMA(p, d, q) model with p and q as the degrees of polynomials $a(z)$ and $\beta(z)$ respectively.

We now consider the ARMA value

$$y(t) = \sum_{i=0}^{\infty} \theta_i Z(t-i) \tag{2.5}$$

and the prediction

$$\hat{y}(t+k) = \sum_{i=k}^{\infty} \theta_i Z_{t+k-i} \tag{2.6}$$

and build the prediction error

$$y(t) = \sum_{i=0}^{\infty} \theta_i Z(t-i) - \hat{y}(t+k) = \sum_{i=k}^{\infty} \theta_i Z_{t+k-i},$$

or definitely

$$y(t) = \sum_{i=0}^{k-1} \beta_i Z_{t+k-i}.$$

2.9.4.5 Forecasting Using a CARIMAX Model

The predictive capability of the CARIMAX model is discussed in detail, along with its application to predictive control, in Section 2.10.6.

2.9.5 Forecasting Using Smoothing

Processing of sampled signals mainly includes

- ***signal smoothing***, *i.e.* optimal estimation of a signal value within a given time interval, based on signal values within the interval
- ***signal filtering***, *i.e.* optimal estimation of actual signal value at the present point based on the past and the present sampled values of the signal
- ***signal prediction***, *i.e.* optimal estimation of future signal values based on the past and the present sampled values

In time series analysis, smoothing is a technique focused on reduction of irregularities or random fluctuations in time series data in order to provide a clean time series pattern out of contaminated observation data. The simplest smoothing technique used is ***moving-average smoothing***, as well as its more advanced modification, *i.e.* ***exponential smoothing***.

2.9.5.1 Forecasting Using a Simple Moving Average

Moving averages are used for prediction of future values based on weighted averages of the past values. They are useful in reducing the random variations present in observation data. For example, the moving average that uses n past observations and the most recent one to calculate the next time series value is

$$x_m(t+1) = \frac{x(t) + x(t-1) + \ldots + x(t-n)}{n}.$$

Some modifications of the moving average are

- ***centred moving average***, a modification of a simple moving average in which the average is placed in the middle of an interval of n periods, *i.e.* at the $n/2$ point, which holds for odd numbers n
- ***weighted moving average***, an averaging algorithm that discriminates the participation of individual observations according to their "age", as shown in the following equation:

$$x_w(t+1) = w_1 x(t) + w_2 x(t-1) + \ldots + w_n x(t-n) .$$

From the equation it is evident that more recent observations could be given higher weights by greater values of weights w than the older ones. However, the sum of all weights used should be equal to one.

The moving average is easy to understand and simple to use, but it gives equal weight to all past data, of which a large number have to be stored and used for forecasts. This also holds for the weighted moving average, for which it is difficult to select the optimal values for individual weights.

Therefore, a more advanced version of the weighted moving average is an alternative like *exponential smoothing*, a version with exponentially decreasing weights as the observation data become older.

2.9.5.2 Forecasting Using Exponential Smoothing

The exponential smoothing approach is particularly convenient for short-time forecasting. Although it also employs weighting factors for past values, the weighting factors here decay exponentially with distance of the past values of the time series from the present time. This enables a compact formulation of the forecasting algorithm in which only a few most recent data are required and less calculations are needed, which is highly relevant to on-line applications in industrial automation, where programmable controllers and signal processors are used.

Smoothing of observation data is basically required when the data are to a certain degree erroneous due to the superposition of some error component $\varepsilon(t)$ and the exact value $x(t)$, *i.e.* when the measured signal $x_e(t)$ is expressed as

$$x_e(t) = x(t) + \varepsilon(t) .$$

In exponential smoothing, the concept of a weighted moving average is used. In using exponentially decaying coefficients not all past values are used for prediction; rather, a reduced number of measured and calculated data are used, represented by the iterative exponential smoothing algorithm

$$x_e(t) = \alpha x(t) + (1-\alpha) x_e(t-1),$$

with the forecast

$$\Phi(t+k) = x_e(t+k) .$$

Here, $x_e(t)$ is the exponentially smoothed value, $x(t)$ is the observed value at the same point of time, α is the smoothing constant value, and $x_e(t-1)$ is the previous exponentially smoothed value.

The value of the smoothing constant α depends on the properties of the given time series. Values between 0.1 and 0.3 are most commonly used because they produce a forecast which depends on a large number of past observations. Values close to one are rarely used because they give forecasts which depend much more on recent observations. For instance, when smoothing constant α = 1, the forecast is equal to the most recent observation.

The term ***exponential*** can be understood from the result of iterative calculation of $x_e(t)$ using $x_e(t-1)$, $x_e(t-2)$, $x_e(t-3)$, *etc.*, which results in

$$\alpha x(t)+(1-\alpha)[\alpha x(t-1)+(1-\alpha)x_e(t-2)]=$$
$$\alpha x(t)+(1-\alpha)\{\alpha x(t-1)+(1-\alpha)[\alpha x(t-2)+(1-\alpha)x_e(t-3)]\}$$

or generally

$$x_e(t)=\alpha x(t)+(1-\alpha)[x(t-1)+(1-\alpha)^2 x(t-2)+\ldots+(1-\alpha)^t x(0)],$$

from where the exponentially decreasing value of weights is evident.

In addition, because the expression in the second term on the right-hand side of the last equation within the bracket is equal to $x_e(t-1)$, it can be rewritten as

$$x_e(t)=\alpha x(t)+(1-\alpha)x_e(t-1)$$

From this equation it follows that in order to estimate the smoothed value $x_e(t)$ of the time series at the time point *t*, we need the current value $x(t)$ and the estimate of the smoothed value $x_e(t-1)$ at the previous time point (*t*-1), supposing that the value of the constant α is time invariant.

Prior to applying exponential smoothing algorithm it should be decided

- how to initialize the exponential smoothing process
- how to select the value of smoothing constant α.

For simplicity, the algorithm is initialized by setting $x_e(2)=x(2)$. With regard the value of exponential smoothing constant α, it can generally be arbitrarily selected within the interval [0, 1]. Its optimal value depends largely on the time series pattern and on the smoothing objectives. Since the value of α determines how strong the older observations are dampened, selection of higher α values dampens the old values more strongly than the selection of lower α values. There is also a direct experimental way to evaluate the optimal value of α in which the values $\alpha =$ 0.1, 0.2, ..., 0.9 are taken and for each value the efficiency of estimation is calculated. In this way the value of α giving the best efficiency is found.

The value of α can also be calculated directly from the past data values using the sum squared prediction errors (SSE) for different values of α. The value of α which minimizes the SSE is taken for forecasting. For instance, given the value of search step $\Delta\alpha = 0.1$, the following algorithm can be used to select the best value for starting with any initial value of α within $0 < \alpha < 1$:

Algorithm 2.1. Algorithm for selection of best smoothing constant

Given a time series $X = \{X_1, X_2, X_3, \ldots, X_n\}$,

for $t = \{1, 2, 3, \ldots, n\}$.

***Set*:**

$$\hat{X}(1,1) = X_1$$
$$e_2 = X_2 - \hat{X}(1,1)$$

***Then*:**

$$\hat{X}(2,1) = X_3 - \hat{X}(1,1)$$
$$e_3 = X_3 - \hat{X}(2,1)$$
$$\ldots \quad \ldots \quad \ldots \quad \ldots \quad \ldots$$
$$e_n = X_n - \hat{X}(n-1,1)$$

***Calculate*:**

$$\text{SSE} = \sum_{i=2}^{n} e_i^2$$

***Repeat*:**

the same procedure for other values of $0 < \alpha < 1$, *say in steps of* 0.1

***Select*:**
the value for which SSE computed is the minimum
***End*:**

Because the surface of SSE near its minimum is quite flat, the choice of α is not very critical and can be found very easily.

The considerable disadvantage of so-called ***single exponential smoothing*** described above is that it does not work efficiently when a remarkable trend component is present in the time series pattern. This can be improved by upgrading the single exponential smoothing algorithm to the ***double exponential smoothing***

algorithm, which simultaneously considers the trend components by processing the equations

$$X_e(t) = \alpha X(t) + (1-\alpha)X_e(t-1)$$
$$X_t(t) = \beta[X_e(t) - X_e(t-1)] + (1-\beta)X_t(t-1)$$

in which the constant β can (under certain boundaries) be freely selected. In this case the resulting estimated forecasting value for ν steps ahead is defined by

$$x_f(t+\nu) = x_e(t) + \nu x_t(t)$$

The double exponential smoothing algorithm can also be extended to deal with the time series containing trend *and* seasonal components. The extended algorithm is the ***triple exponential smoothing*** algorithm or ***Holt-Winter algorithm***, based on simultaneous consideration of the following three equations:

$$x_e(t) = [\alpha / x_s(t-\eta)]x_t + (1-\alpha)[x_e(t-1) - x_t(t-1)]$$
$$x_t(t) = \beta[x_e(t) - x_e(t-1)] + (1-\beta)x_t(t-1)$$
$$x_s(t) = [\gamma / x_e(t) + (1-\gamma)x_s(t-\eta)$$

from which the estimated value for ν steps-ahead forecast is defined as

$$x_f(t+\nu) = [x_e(t) + \nu x_t(t)]x_s(t+\nu-\eta)$$

Also here, the constant value γ, under certain limits, can be freely selected.

For assessment of forecasting results the MAE criterion

$$\text{MAE} = (1/n)\sum_{i=1}^{n}|e_i|$$

can be used, or alternatively the RMSE criterion

$$\text{RMSE} = \sqrt{\sum_{i=1}^{n} e_i^2}\ ,$$

or the MAPE criterion

$$\text{MAPE} = (1/n)\sum_{i=1}^{n}\left|\frac{e_i}{x_f(i)}\right| \cdot 100\%\ .$$

For estimation of the confidence intervals for the forecast $x_f(t+1)$ the criterion

$$\text{CI} = 2\text{RMSE}\sqrt{n(n-p)}$$

is preferred, where p represents the number of parameters estimated in the forecasting method.

2.9.5.3 Forecasting Using Adaptive Smoothing

In ***adaptive smoothing***, which is a more advanced version of exponential smoothing, the smoothing constant α is adjusted on-line according to the actual value of the forecast error. This is presented below on the example of the k-step prediction equation

$$y(t+k) = \frac{C(z^{-1})}{A(z^{-1})}\varepsilon(t+k) \tag{2.7}$$

with the polynomial operators

$$A = \sum_{j=o}^{m} a_j(z^{-j}), \qquad \text{with } a_0 = 0$$

$$C = \sum_{i=0}^{n} c_i z^{-i}, \qquad \text{with } c_0 = 0$$

and $\varepsilon(t)$ as a white noise. The prediction is qualified as ***good*** if it minimizes the cost function

$$V_k = E\{\tilde{y}^2(t+k)\}$$

Introducing the output prediction error

$$\tilde{y}(t+k) = y(t+k) - \hat{y}(t+k|t),$$

where $\hat{y}(t+k|t)$ is the predicted value

$$\hat{y}(t+k|t) = E\{y(t+k)|t\}$$

$$= E\left\{\left[\frac{C(z^{-1})}{A(z^{-1})}\varepsilon(t+k)\right]\middle| t\right\}$$

of the real future value $y(t+k)$ using the minimum mean square error MMSE. Introducing now the ***Diophantine equation***

$$C(z^{-1}) = A(z^{-1})F(z^{-1}) + z^{-k}G(z^{-1}) \tag{2.8}$$

where,

$$F(z^{-1}) = \sum_{i=0}^{k-1} f_i z^{-i} \tag{2.9}$$

$$G(z^{-1}) = \sum_{j=0}^{m-1} g_j z^{-j} \tag{2.10}$$

With $f_0 = 0$ and $g_0 = 0$, and assuming that the noise $\varepsilon(t)$ is time-independent, *i.e.* that the equality

$$E\{\varepsilon(t+v)\} = E\{\varepsilon(t+v)|t\}$$

holds for any positive v, the polynomial ratio C/A in Equation (2.7) can be written as

$$\frac{C(z^{-1})}{A(z^{-1})} = F(z^{-1}) + z^{-k}\frac{G(z^{-1})}{A(z^{-1})}. \tag{2.11}$$

Now, inserting Equation (2.11) into Equation (2.7) we get the expected values

$$E\{y(t+k)|t\} = E\left\{\left[\frac{G(z^{-1})}{A(z^{-1})}\varepsilon(t)\right]\Bigg| t\right\} \tag{2.12}$$

Equation (2.7) can now be rewritten to give the expected values

$$\frac{A(z^{-1})}{C(z^{-1})} y(t+k) = E\{\varepsilon(t+k)|t\} \tag{2.13}$$

for $v \leq 0$. Finally, from the last two equations follows the expected value

$$E\{y(t+k)|t\} = \frac{G(z^{-1})}{C(z^{-1})} y(t).$$

The cost function to be minimized now becomes

$$V_k = E\left\{\frac{G(z^{-1})}{A(k^{-1})}\varepsilon(t) - \hat{y}(t+k|t)\right\} + k_0$$

with

$$k_0 = \sigma_\varepsilon^{\ 2} \Sigma f_i^2, \quad f_0 = 0,$$

where σ_ε^2 is the variance of noise ε.
Finally, the minimum of V_k is

$$\hat{y}(t+k|t) = \frac{G(z^{-1})}{C(z^{-1})} y(t)$$

Taking into account the *k*-step prediction defined by Equation (2.7) and the Diophantine Equation (2.8), this result is now used to find the predictor value $\hat{y}(t+k|t)$ from the relation

$$y(t+k) = \hat{y}(t+k|t) + F(z^{-1})e(t+k),$$

which is equivalent to

$$\tilde{y}(t) = F(z^{-1})e(t).$$

This finally results in

$$\hat{y}(t+k|t) = \frac{G(z^{-1})}{A(z^{-1)})F(z^{-1})} \tilde{y}(t).$$

2.9.5.4 Combined Forecast

Thus far, various traditional methods available for time series forecasting have been presented. It was mentioned that, unfortunately, there are no specific guidelines for selection of a best forecasting method to solve a forecasting problem. Besides, not each available method, applied to the same problem, delivers the forecasting results with the same accuracy. For example, to forecast a nonstationary, non-seasonal time series one can use the autoregressive method, Holt-Winter's exponential smoothing technique, the Box-Jenkins ARMA/ARIMA method, Kalman filtering, *etc*. Different methods will, for a given time series, provide different forecasting results, so that, after comparing the individual forecasting results, a decision has to be made about what prediction method should be ultimately selected for further considerations. This is a difficult task requiring much professional experience. As a way out of the selection dilemma the ***nonlinear combination of forecasts*** has been advocated, as described below.

The need for combined forecast of a time series has been well understood for a long time. Many studies have been done and revealed that not any arbitrary combination of methods is decisive for an improved forecast, but it is essential that the combination is ***nonlinear***. Only the nonlinearity provides a combination with better forecasts than either of the combination components separately, due to a kind of ***synergic effect*** generated. It was also revealed that the forecasting results generally improve as more methods are included in the combination. This is shown in the following example.

Let the forecasts $f_1, f_2, f_3, \ldots, f_k$, of the random variable z be given and let them be linearly combined to give the resulting forecast f_c, defined as

$$f_c = \sum_{i=1}^{k} w_i f_i(z),$$

where w_i, $i = 1, 2, \ldots, k$, are the assigned weights to the individual forecasts. The main problem is how to select the individual weights optimally. The simplest way would be to select an equal weighted combination based on the arithmetic average of the individual forecasts. This has proven to be relatively robust and accurate, which is evident when two unbiased forecasts f_1 and f_2 of a given time series are linearly combined as

$$f_c = kf_1 + (1-k)f_2,$$

which will have a minimum mean square error for suitably chosen k. The corresponding forecast errors for the combination, e_c, is defined using the individual errors e_1 and e_2 as

$$e_c = ke_1 + (1-k)e_2.$$

For the two mutually independent forecast errors the value

$$k = E\left(e_2^2\right) / \left(E\left(e_1^2\right) + E\left(e_2^2\right)\right) \approx \tilde{e}_2^2 / \left(\tilde{e}_1^2 + \tilde{e}_2^2\right)$$

delivers the minimum value of $E\left(e_c^2\right)$, $\tilde{e}^2$ being the local estimate of the expected error squared.

Anyhow, the linear combination of forecasts is not likely to be the appropriate in forecasting practice, as the following example shows, in which k different forecast methods are given, the ith individual forecast having an information set $\{I_i : I_c, I_{si}\}$, I_c being the common part of the information used by all k models and I_{si} the special information for the ith forecast only. Denoting the ith forecast by $f_i = F_i(I_i)$, the linear combination of forecasts can be expressed as

$$F_c = \sum w_i F_i(I_i),$$

where w_i is the weight of the ith forecast. On the other hand, every individual forecasting model given can also be regarded as a subsystem for information processing, while the combination method

$$f_c = F_c(I_1, I_{2,}, \ldots, I_k)$$

is regarded as such a system. It follows that the integration of forecasts is more than their sum, because the performance of the integrated system is more than the sum of the performances of its subsystems. So, the trustworthiness of the linear forecast combination is quite questionable. Rather, more trust should be paid to a nonlinear interrelation between the individual forecasts, such as

$$f_c = \psi[F_1(I_1), F_2(I_2), \dots, F_k(I_k)]$$

where ψ is a nonlinear function. While the given information is processed by individual forecasting models, it is likely that the parts of the entire information can be lost. For instance, it could happen that the information set I_i is not used efficiently, or different forecasts may have different parts of information lost. This is why as many different forecasts should be present in the combination as possible, even when the individual forecast depends on the same set of information. What still remains is how to determine the form of the nonlinear relationship ψ.

2.10 Application Examples

In the following, some examples are given of practical applications of time series analysis and forecasting in business and industry.

2.10.1 Forecasting Nonstationary Processes

As the first example, forecasting of a nonstationary non-seasonal time series is taken, based on collected equidistantly spaced temperature values of an uncontrolled chemical plant (Box and Jenkins, 1976). For forecasting, the ARMA process model and the Holt-Winter exponential smoothing technique are used. It is an experiment based on 226 time series data, approximately fitted by the model

$$z_{t+1} = -0.8z_{t+1-2} + 1.8z_{t+1-1} + a_{t+1}$$

or by

$$\hat{z}_t(1) = -0.8z_{t-1} + 1.8z_t$$

where the time t is the origin at which the forecast $\hat{z}_t(l)$ is made and l is the lead time of forecast, representing the number of time steps ahead the forecast should be made with respect to origin, and a_{t+l} is the random shock. Based on the above model, the forecast has been made with the lead time $l = 1$ at different origins $t = 2, 3, 4, \dots, 225$. Consequently, a total of $m = 224$ data have been generated as a Box and Jenkins forecast series.

Similarly, the Holt-Winter exponential smoothing technique has been applied to generate the second forecasts of the same temperature series

$$\hat{z}_1(t) = c_0 z_t + c_1 z_{t-1} + c_2 z_{t-2} + \ldots$$

where

$$c_i = (1-\alpha)^i, \text{ where } \quad i = 0, 1, 2, \ldots$$

and α is a constant value within the interval $0 < \alpha \leq 1$. This results in

$$\hat{z}_t(1) = \alpha z_t + (1-\alpha)\hat{z}_{t-1}(1).$$

The two forecast series are then arranged as columns 1 and 2 and the actual temperature series as column 3 of an HBXIO matrix

$$\text{HBXIO} = \begin{bmatrix} f_{B1} & f_{H1} & d_1 \\ f_{B2} & f_{H2} & d_2 \\ \ldots & \ldots & \ldots \\ f_{Bm} & f_{Hm} & d_m \end{bmatrix}$$

The sum squared error (SSE) of the generated forecast has been also computed as SSE = $0.5E^T E$, where E is the column vector of errors $e_i = (f_i - d_i)$, with f_i, d_i representing the forecast at ith instant and actual value of the time series at ith instant and E^T is the transposition of E. Consequently, the sum squared error for the Box-Jenkins forecast is 2.0080 and that of the Holt-Winter forecast is 1.1688, computed for the entire forecast series (Palit, 1999).

It is important to note that in the above example of Holt-Winter's smoothing technique the smoothing constant α = 1.6 has been selected because that gave the minimum value of SSE for generated forecasts, which is quite unusual.

2.10.2 Quality Prediction of Crude Oil

In the following example, time series analysis is applied to crude oil physical and chemical qualities prediction (Debska and Ivasczek, 2001). The observation data are collected from oil fields within time period of 5 years and first analyzed statistically for estimation of values of the most relevant chemical physical parameters, such as specific gravity, density, colour, viscosity, relative and kinematic viscosity, drip and set point, *etc.* The statistical methods used for these purposes are: preprocessing and smoothing of data, partial and autocorrelation calculation, seasonality and trend-analysis, decomposition, *etc.* For decomposition of complex time series with cyclic components and for extraction of underlying sine and cosine functions of different frequencies, frequency analysis has been employed, supported by building and analysis of corresponding periodograms for interpretation of the data. Finally, for prediction of crude oil properties the Fourier transformation has been used as a nonlinear, parametric model that can forecast future values by processing the past values.

2.10.3 Production Monitoring and Failure Diagnosis

Production monitoring and failure diagnosis are the major objectives in on-line observation of overall performance of a production plant. In manufacturing, the major attention is paid to the monitoring and diagnosis of numerical machines and of machine tools. In both cases – apart from modern approaches relying on intelligent technology – statistical methods, based on time series analysis, are still used. The main reason is that, for monitoring purposes, an abundant number of observation data are collected on-line to be processed statistically.

Damiano *et al.* (1999) reported on the use of nonlinear time series to form a one-step prediction map for machine monitoring and failure diagnosis in which the sequence of previously collected observation data helped in the estimation of the next time series data point. The map built in this way models efficiently the dynamics of the system generated by a time series. Applying nonlinear time series analysis, the optimum time delay is determined to be used for reconstruction of the ***attractor***, required for creation of the map that approximates the attractor. For reconstruction of a ***multidimensional attractor***, the method of delays was used, where the vector components were created from the given time series using time series values mutually separated by the delay time.

The one-step prediction is now applied to the machinery diagnosis. A baseline time series is built out of data collected from the machine under normal operating conditions and the nonlinear time series analysis used to build the corresponding one-step prediction map. Using the map, the average map error is calculated for the baseline time series. The calculated error and the map built are then employed for machine monitoring by calculating the average absolute map error using the current time series. The calculation results are then compared with the map error for baseline time series and the difference between the two types of map error is finally used to detect the possible changes in the machine being monitored.

2.10.4 Tool Wear Monitoring

In the following example, the most significant problem in flexible manufacturing systems, the problem of ***monitoring of tool wear*** during the cutting and drilling process, is assessed. This monitoring task is needed to maintain constant quality of products and to avoid damage to the workpiece. To achieve this, a set of versatile ***nondestructive*** sensing elements have to be installed for on-line tracing of the status of tool wear during normal operation. The objective is to detect and replace the tool when worn beyond the tolerable limits. In practice, ***acoustic emission sensors*** are regularly used instead of power- or force-based sensors, because of their close relationship between the generation of the emission signal and the wear condition of the tool.

The acoustic emission technique used is an adequate means for monitoring the cutting tool wear condition (Liang and Dornfeld, 1989), because the frequency band emitted by tool wear is much higher than the machine vibration frequency band. Thus, the two frequency bands can be easily separated by a high-pass filter. Also, the frequency signal emitted can be picked up directly from a sensor installed on the tool holder.

Once the measured signal is obtained, the time series data sets can be built by sampling the acoustic emission signal and thereafter its prediction can be made by processing the time series data. The main difficulty here, however, is the high sampling frequency required to build the time series of the emitted acoustic signal having a frequency band of 100 kHz to 1 MHz and requiring a sampling frequency of over 2 MHz.

2.10.5 Minimum Variance Control

Time series analysis and forecasting have been, since the earliest days of engineering, powerful tools for problem solving in signal and system analysis and prediction. Initially, the application of vibration analysis to machines and like objects was the essential field of application, but later this was extended to encompass most various application fields, including systems identification, parameter estimation, and in self-tuning and predictive control.

An excellent representative application example is found in model building, parameter estimation, and predictive control of dynamic systems. For this purpose, the modifications of ARMA and ARIMA models are used, known as CARMA (or CARMAX) and CARIMA (or CARIMAX), where C stands for ***control*** and X for ***auxiliary input signal***.

We would first like to use the CARIMA model

$$A(z^{-1})y(t) = z^{-k}B(z^{-1})u(t) + C(z^{-1})e(t) \tag{2.14}$$

to implement ***minimum variance control***, designed to keep the output of a stochastic system to the set point value. This requires that, for each time instant t, the value of the control signal $u(t)$ should be determined to minimize the output variance

$$J = E\{y^2(t+k)\}$$

Introducing the ***Diophantine equation***

$$C = AF + z^{-k}G \tag{2.15}$$

with the polynomials

$$F = \sum_{i=0}^{k-1} f_i z^{-i}, \qquad f_0 = 0 \tag{2.16}$$

$$G = \sum_{j=0} g_j z^{-n_g}, \tag{2.17}$$

and $n_g = \max(n_a - 1, n_c - k)$, the CARMA equation becomes

$$y(t+k) = \left[\frac{B}{A} u(t) + \frac{G}{A} e(t) + Fe(t+k) \right]. \tag{2.18}$$

Taking $e(t + k)$ from Equation (2.14) as

$$e(t+k) = \frac{A}{C} y(t+k) - z^{-k} \frac{B}{C} u(t+k)$$

and using the Diophantine equation, Equation (2.18) becomes

$$y(t+k) = \frac{BF}{C} u(t) + \frac{G}{C} y(t) + Fe(t+k) \tag{2.19}$$

or

$$y(t+k) = \hat{y}(t+k|t) + Fe(t+k). \tag{2.20}$$

The resulting output variance

$$J = E\{[\hat{y}(t+k|t) + Fe(t+k)]^2\}$$

is now minimized for $\hat{y}(t+k|t) = 0$, to become

$$J_{\min} = F\sigma_e^2.$$

From Equations (2.19) and (2.20) it follows that

$$\frac{BF}{C} u(t) + \frac{G}{C} y(t) = 0$$

or,

$$BFu(t) + Gy(t) = 0$$

which finally results in

$$u(t) = -\frac{G}{BF} y(t).$$

2.10.6 General Predictive Control

General predictive control (Hueseyin and Karasu, 2000) is based on the CARIMA model

$$y(t) = \frac{B(z^{-1})}{A(z^{-1})} u(t-1) + \frac{C(z^{-1})}{\Delta A(z^{-1})} e(t), \tag{2.21}$$

where,

$$\Delta = 1 - z^{-1}.$$

Introducing the term

$$\xi(t) = \frac{e(t)}{C(z^{-1})}$$

and presuming that

$$C(z^{-1}) = 1,$$

the CARIMA model (2.21) takes the form

$$A(z^{-1}) y(t) = B(z^{-1}) u(t-1) + \xi(t)/\Delta. \tag{2.22}$$

Based on this model, the predictive control (Camacho and Bordons, 1999; Clarke *et al.*, 1987) is implemented through the following steps and should be repeated for every sample instant:

- ***prediction of output value*** using the CARIMA model and the observation data collected up to time t
- ***determination of control signal value*** that produces the future output value close to the predicted output value
- ***closing the control loop*** using the above results.

To determine a d-step predictor for $y(t)$, *i.e.* $y(t+d)$ using the Equation (2.22), the Diophantine equation

$$P_d(z^{-1}) A(z^{-1}) \Delta + z^{-d} Q_d(z^{-1}) = 1$$

is introduced in which the polynomials P and Q are uniquely determined given the polynomial A and number of prediction steps d. Now, multiplying the above CARIMA model (2.22) by $P_d \Delta z^d$ and using the Diophantine equation, the predicted value will be

$$y(t+d) = P_d B \Delta u(t+d-1) + Q_d y(t) + P_d \xi(t+d) . \tag{2.23}$$

Next, to determine the predictive control law, the future set-points $w(t+d)$, $d = 1, 2, \ldots$ should be given, or it is supposed that they have a constant value w. The control objectives would then be to find the control law that will drive the system output $y(t+d)$ as close as possible to the set points $w(t+d)$. This value is obtained by minimizing the cost function

$$J(n_1, n_2) = \mathrm{E}\left\{ \sum_{d=n_1}^{n_2} [y(t+d) - w(t+d)]^2 + \sum_{d=1}^{n_2} \lambda(d)[\Delta u(t+d-1)]^2 \right\},$$

which is the expectation value, in which $\lambda(d)$ is a weighting factor of control sequences, and n_1, n_2 are the minimum and maximum cost horizons. But still, the solution found in this way is the ***open-loop feedback-optimal control***. To find the corresponding closed-loop control we will proceed as follows.

The CARIMA Equation (2.23), after ignoring the future noise component $\xi(t+d)$, is written as

$$y(t+d) = G_d \Delta u(t+d-1) + Q_d y(t)$$

where,

$$G_d = P_d B$$

with

$$G_d = g_{d0} + g_{d1} z^{-1} + g_{d2} z^{-2} + \ldots$$

Writing now the above equation for $d = 1, 2, \ldots, n$, the set of generated prediction equations will be

$$y(t+1) = G_1 \Delta u(t) + Q_1 y(t)$$
$$y(t+2) = G_2 \Delta u(t+1) + Q_2 y(t)$$
$$\ldots \quad \ldots \quad \ldots$$
$$y(t+n) = G_n \Delta u(t+n-1) + Q_n y(t)$$

In the above equations the right-hand terms should be taken for further processing. These obviously depend only on the past values, so that they produce the set of equations written in matrix form as

$$y = Gu + f,$$

where,

$$y = [y(t+1), y(t+2)..., y(t+n)]^T$$
$$u = [\Delta u(t), \Delta u(t+1), ..., \Delta u(t+n-1)]^T$$
$$f = [f(t+1), f(t+2), ..., f(t+n)]^T$$

and G is the lower triangular $n \times n$ matrix

$$G = \begin{bmatrix} g_0 \; 0............0 \\ g_1 \; g_0..........0 \\ .... \\ g_{n-1} \; g_{n-2}....g_0 \end{bmatrix}.$$

Introducing now the set-point sequence

$$w = [w(t+1), w(t+2), ..., w(t+n)]^T$$

and minimizing the expected value of

$$J = E\left\{(y-w)^T (y-w) + \lambda \tilde{u}^T u\right\},$$

or, of

$$J = \{(G\tilde{u} + f - w)^T (G\tilde{u} + f - w) + \lambda \tilde{u}^T \tilde{u}\}$$

the projected control increment vector

$$\tilde{u} = (G^T G + \lambda I)^{-1} G^T (w - f)$$

is determined with the unit vector I. From the last result, only the first element of $\tilde{u}$ is taken as the next control value

$$u(t) = u(t-1) + \tilde{g}^T (w - f)$$

where $\tilde{g}^T$ is the first row of

$$(G^T G + \lambda I)^{-1} G^T.$$

References

[1] Akaike H (1974) A new look in the Statistical Model Identification, IEEE Trans. on Automatic Control 19: 716-723.

[2] Aoki M (1990) State Space Modelling of Time Series, Springer-Verlag, Berlin.

[3] Box GEP and Jenkins GM (1976) Time Series Analysis: Forecasting and Control, Revised Edition, Holden-Day, San Francisco.

[4] Brockwell PJ and Davis RA (2002) Introduction to Time Series and Forecasting, 2nd edition, Springer-Verlag, New York

[5] Camacho EF and Bordons C (1999) Modern Predictive Control, Springer, Berlin.

[6] Casdagli M and Eubank S (1992) Nonlinear Modelling and Forecasting, Reading, MA, Addison Wesley.

[7] Clarke DW, Mohtadi C, and Tuffs PS (1987) General Predictive Control: Part 1, The Basic Algorithm., Automatica 23(2): 137-148.

[8] Cooley JW and Tukey (1965) Math. Comp., 19(4): 297-301

[9] Damiano B, Breeding JE, and Tucker RWJr (1999) Machine and Process System Diagnosis Using One-Step Prediction Maps. Proc. MARCON99 Conf., Gatlingburg, Tennessee, May 10-12, Vol. 1, Paper 9.02.

[10] Debska B and Ivasczek G (2001) Prediction of Physical and Chemical qualities Of Crude-Oil Using Statistical Time Series Analysis. Fresenius Journal of Anal. Chem.: 704-708, Springer-Verlag, Berlin

[11] Dillon WR and Goldstein M (1985) Multivariate Analysis: Methods and Applications. Wiley, New York.

[12] Draper N and Smith H (1981) Applied Regression Analysis. 2nd Ed., Wiley, New York.

[13] Hueseyin D and Karasu E (2000) Generalized Predictive Control. IEEE Control Systems Magazine 20(5): 36-47.

[14] Johnson RA and Wichern DW (1988) Applied Multivariate Statistical Analysis. 2nd Ed., Prentice-Hall, Englewood Cliffs, New Jersey.

[15] Jolliffe IT (1986) Principal Component Analysis. Springer-Verlag, New York.

[16] Liang SY and Dornfeld DA (1989) Tool Wear Using Time Series Analysis of Acoustic Emission. ASME J. of Engineering for Industry 111:199-205.

[17] Palit, AK (1999) Artificial Intelligent Approaches to Time Series Forecasting, Ph.D Thesis, University of Bremen, Germany.

[18] Warner RM (1998) Spectral Analysis of Time Serial Data, Guildford Publication.

Selected Reading

[19] Abraham B and Ledolter J (1983) Statistical Methods for Forecasting. Wiley, New York.

[20] Anderson TW (1984) An Introduction to Multivariate Statistical Analysis. 2nd Ed., Wiley, New York.

[21] Brillinger DR (1981) Time Series: Data Analysis and Theory. 2nd Ed., Holden Day, San Francisco.

[22] Brock W and Potter S (1993) Nonlinear Time Series and Macroeconometrics. In: Handbook of Statistics, Manddala, G.S. *et al.* (eds.), North Holland, Amsterdam.

[23] Brockwell PJ and Davis RA (1991) Time Series: Theory and Methods. 2nd Ed., Springer-Verlag, New York.

[24] Cohen L (1995) Time-Frequency Analysis. Prentice-Hall, Englewood Cliffs, New Jersey.

[25] Fuller WA (1976) An Introduction to Statistical Time Series. Wiley, New York.
[26] Kendall MG and Stuart A (1976) An Advanced Theory of Statistics. vol. 3, Griffin, London.
[27] Wellstead PE and Zarrop MB (1991) Self-tuning Systems Control and Signal Processing. Wiley, New York
[28] Zhongjie X (2003) Case Studies in Time Series Analysis. World Scien. Publ. Co.

Part II

Basic Intelligent Computational Technologies

3

Neural Networks Approach

3.1 Introduction

Neural networks are massively parallel, distributed processing systems representing a new computational technology built on the analogy to the human information processing system. That is how we know the neural networks today, but the evolution of artificial neural networks, from the early idea of neuro-physiologist Heb (1949) about the structure and the behaviour of a biological neural system up to the recent model of artificial neural system, was very long. The first cornerstones here were laid down by the neurologists McCulloch and Pitts (1943) who, using formal logic, modelled neural networks using the neurons as binary devices with fixed thresholds interconnected by synapses. Nevertheless, the list of pioneer contributors in this field of work is long. It certainly includes the names of distinguished researchers like Rosenblatt (1958), who extended the idea of the ***computing neuron*** to the ***perceptron*** as an element of a self-organizing computational network capable of learning by feedback and by structural adaptation. Further pioneer work was also done by Widrow and Hoff (1960), who created and implemented the analogue electronic devices known as ADALINE (Adaptive Linear Element) and MADALINE (Multiple ADALINE) to mimic the neurons, or perceptrons. They used the least mean squares algorithm, simply called the ***delta rule***, to train the devices to learn the pattern vectors presented to their inputs. In 1969, Minsky and Papert (1969) portrayed perceptron history in an excellent way but their view, that the multilayer perceptron (MLP) systems had limited learning capabilities similar to the one-layer perceptron system, was later disproved by Rumelhart and McClelland (1986). Rumelhart and McClelland in fact showed that multilayer neural networks have outstanding nonlinear discriminating capabilities and are capable of learning more complex patterns by ***backpropagation learning***. This essentially terminates the most fundamental development phase of perceptron-based neural networks.

After a period of stagnation, the research interest was turned to the possible alternative network variants that have been found in ***self-organizing networks***

(Amari and Maginu,1988), ***resonating neural networks*** (Grossberg, 1988), ***feedforward networks*** (Werbos, 1974), ***associative memory networks*** (Kohonen, 1989), ***counterpropagation networks*** (Hecht-Nielsen, 1987a), ***recurrent networks*** (Elman, 1990), ***radial basis function networks*** (Broomhead and Lowe, 1988), ***probabilistic networks*** (Specht, 1988), *etc*. Nevertheless, up to now, the most comprehensively studied and, in engineering practice, most frequently used neural networks are the multilayer perceptron networks (MLPN) and radial basis function networks (RBFN), which are frequently the subject of further research and applications.

Neural networks have, since the very beginning of their practical application, proven to be a powerful tool for signal analysis, features extraction, data classification, pattern recognition, *etc*. Owing to their capabilities of learning and generalization from observation data, the networks have been widely accepted by engineers and researchers as a tool for processing of experimental data. This is mainly because neural networks reduce enormously the computational efforts needed for problem solving and, owing to their massive parallelity, considerably accelerate the computational process. This was reason enough for intelligent network technology to leave soon the research laboratories and to migrate to industry, business, financial engineering, *etc*. For instance, the neural-network-based approaches developed and the methodologies used have efficiently solved the fundamental problems of time series analysis, forecasting, and prediction using collected observation data and the problems of on-line modelling and control of dynamic systems using sensor data.

Generally speaking, the practical use of neural networks has been recognized mainly because of such distinguished features as

- general nonlinear mapping between a subset of the past time series values and the future time series values
- the capability of capturing essential functional relationships among the data, which is valuable when such relationships are not *a priori* known or are very difficult to describe mathematically and/or when the collected observation data are corrupted by noise
- universal function approximation capability that enables modelling of arbitrary nonlinear continuous functions to any degree of accuracy
- capability of learning and generalization from examples using the data-driven self-adaptive approach.

3.2 Basic Network Architectures

The model of the basic element of a neural network *i.e.* the neuron, as still used today was originally worked out by Widrow and Hoff (1960). They considered the perceptron as an adaptive element bearing a resemblance to the neuron (Figure 3.1). A neuron, as the fundamental building block of a neural information processing system, is made up of (see Figure 3.1)

- a ***cell body*** with an inherent *nucleus*

- **dendrites** that feed the external signals to the cell body
- ***axons*** that carry the signals out of the cell to other cell bodies

This configuration was translated in terms of analogue computational technology as shown in Figure 3.1, where

- the core part of the element, called a perceptron, contains a summing element Σ and a nonlinear element NL
- the multiple signal inputs x_i are connected via adjustable weighting elements w_i with the core part of the element
- the signal output(s) y_d

An additional perceptron input w_0, called the *bias*, is understood as a threshold (switching) element.

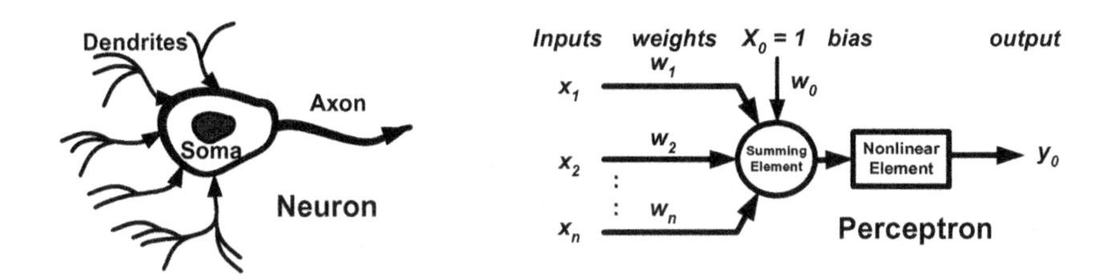

Figure 3.1. Symbolic representation of neuron and perceptron

The output signal is defined as

$$y_0 = f\left(\sum_{i=1}^{n} w_i x_i + w_0\right)$$

and the bias follows the relationship

$$w^{\mathrm{T}} x + w_0 \geq 0$$

meaning that the perceptron *fires*, *i.e.* it is activated and produces an output signal when this condition is met, otherwise not.

Our attention should now be shifted to the question of what nonlinear function should be implemented in the core part of the perceptron as its ***activation function***. The early attempt of Block (1962) to select the ***binary step function*** for this purpose was later modified in favour of a ***sigmoid activation function*** (Figure 3.2).

$$f(x) = \frac{1}{1 + \exp(-x)}.$$

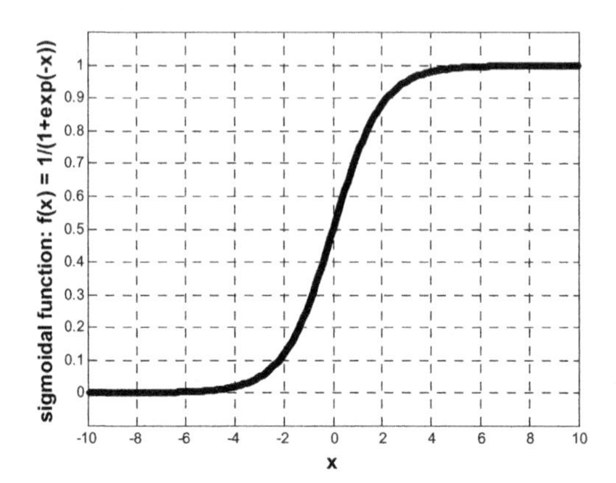

Figure 3.2. Sigmoid activation function

The perceptron basically learns through a training process, based on a set of collected data. During the training, the perceptron adjusts its interconnection weights according to the data presented at its input. For adjusting the perceptron weights, Widrow and Hoff (1960) originally proposed using the ***delta rule***, *i.e.* the recursive gradient-type of learning algorithm (the so-called ***α-LMC Algorithm***) that adds to the current weight value $w(k)$ a compensation term $\eta\varepsilon(k)x(k)$, to build the next weight value

$$w(k+1) = w(k) + \eta\varepsilon(k)x(k),$$

where η is a proportionality term, $\varepsilon(k)$ is the error at the adjusting step k, and $x(k)$ the value of the input signal at the current step k.

Although rather simple, the delta learning rule has, in the majority of cases, demonstrated a high efficiency and a high convergence speed in perceptron training. Even so, a single perceptron alone cannot learn enough to be capable of solving more complex problems because it's radius of computational action is rather restricted by the simplicity of it's structure. This was demonstrated in an example of a perceptron as a pattern classifier. Owing to it's restricted structural capabilities the perceptron can only solve the ***linearly separable problems***. It is thus far away from being a general-purpose processing device. But, the fundamental erroneous belief of Minsky was that even multiple perceptron layer devices cannot build a universal general-purpose processing machine. This was disproved by building the ***multilayer perceptrons*** (MLPs) that, in addition to the perceptron ***input layer*** and ***output layer***, also include so-called ***hidden layers*** inserted between the input and the output layer to form a cascaded network structure with extended connectionist capabilities (see Section 3.3.1). The term hidden layer was selected for the intermediate layer because this layer is only accessible through the input and/or the output layer but not directly. In practice, one hidden layer is usually sufficient to build the network with the extended computational capabilities for solving the majority of practical problems. Only in some rare cases some additional hidden layers could be needed. This also holds in time series analysis and forecasting applications.

Accidentally, the concept of the perceptron emerged at that time when the difficulties in solving complex intelligent problems using classical computing automata of John von Neumann had grown to be insurmountable. It was realized that, for solving such problems, massive, highly parallel, distributed data processing systems are required. Building of such highly sophisticated computational systems was already put on the agenda of some leading research institutions. However, discovery of the perceptron as a simple computing element that can easily be mutually interconnected with other perceptrons to build huge computing networks was viewed as a more promising way for development of the massive parallel computational systems needed at that time. Minsky and Papert (1969) expected that the use of more complex, MLP configurations could help in building the future intelligent, general-purpose computers with learning and cognition capability. This was very soon proven using perceptrons as the basic elements of ADALINE (A) in single-layer perceptrons to build a multi-layer MADALINE architecture (see Figure 3.3).

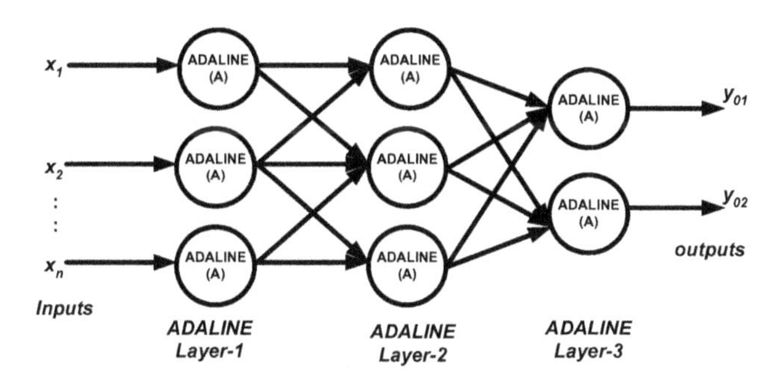

Figure 3.3. ADALINE-based MADALINE

In 1950, Rosenblatt used a single perceptron layer for optical character recognition. It was a multiple input structure fully connected to the perceptron layer with adjustable multiplicative constants w_i called weights. The input signals, before being forwarded to the ***processing elements*** (*i.e.* perceptrons) of the single network layer, are multiplied by the corresponding values of the weighting elements. The outputs of the processing units build a set of signals that determine the number of pattern classes that can be distinguished in the input data sets by the linear separation capability of perceptron layer. For weight adjustment Rosenblatt used the delta rule.

3.3 Networks Used for Forecasting

Hu (1964) was the first to demonstrate - on a practical weather forecasting example - the general forecasting capability of neural networks. Werbos (1974) later experimented with the neural networks as tools for time series forecasting, based on observational data. However, apart from some isolated attempts to solve the forecasting problems using the then still poorly developed neural networks technology, the research work in practical application of neural networks had generally undergone a long period of stagnation. The stagnation was broken and the work on neural network applications enthusiastically resumed after the backpropagation training algorithm was formulated by Rumelhart *et al.* (1986). Experimenting with the backpropagation-trained neural networks, Werbos (1989, 1990) also concluded that the networks even outperform the statistical forecasting methods, such as ***regression analysis*** and the ***Box-Jenkins forecasting approach***. Lapedes and Farber (1988) also successfully used neural networks for modelling and prediction of nonlinear time series.

In the following, typical neural networks used for forecasting and prediction purposes will be described.

3.3.1 Multilayer Perceptron Networks

Although in the meantime the variety of proposed neural network structures has grown, the multilayered perceptron has remained the prevailing one and also the most widespread network structure. This particularly holds for the three-layer network structure in which the ***input layer*** and the ***output layer*** are directly interconnected with the intermediate single ***hidden layer***. The inherent capability of the three-layer network structure to carry out any arbitrary input-output mapping highly qualifies the multilayer perceptron networks for efficient time series forecasting. When trained on examples of observation data, the networks can learn the characteristic features "hidden" in the examples of the collected data and even generalize the knowledge learnt, which will be discussed later in detail.

The multilayer perceptron, because of its cascaded structure, performs the input-output mapping of nonlinearities. For instance, the input-output mapping of a one hidden layer perceptron network can generally be written as

$$y = f_0\left(\sum w_h f_h\left(\sum f_i\left(w_i^T x\right)\right)\right).$$

Relying on the Stone-Weierstrass theorem, which states that any arbitrary function can be approximated with a given accuracy by a sufficiently large-order polynomial, Cybenko (1989) and Hornik *et al.* (1989) proved that a single hidden layer neural network is a ***universal approximator*** because it can approximate an arbitrary continuous function with the desired accuracy provided that the number of perceptrons in it is high enough. This network capability is general, *i.e.* it does not depend on the shape of the perceptron activation function if it is nonlinear.

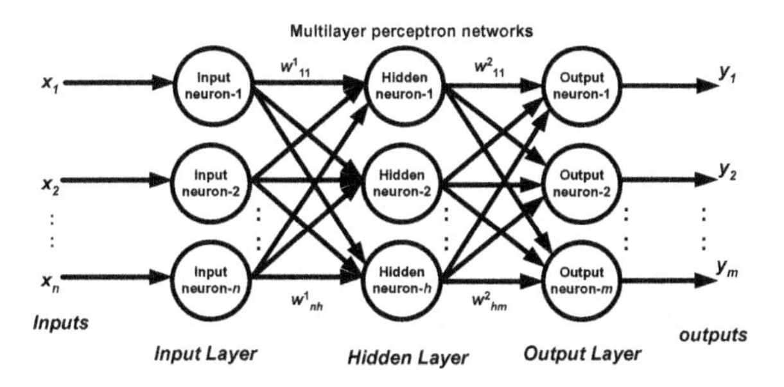

Figure 3.4 Multilayer perceptron architecture

Rumelhart and McClelland (1986, MIT book) suggested for multilayer neural networks the ***backpropagation learning rule***. This has also widely been accepted. Later, various accelerated versions of the rule have been elaborated that speed up the learning process. In the meantime, the multilayer perceptron networks trained to learn using backpropagation algorithm are simply called ***backpropagation networks***.

The learning capability of backpropagation networks is mainly due to the internal mapping of the characteristic signal features in the process of network training onto the hidden layer. The mappings stored in this layer during the training phase of the network can be automatically retrieved during it's application phase for further processing. Although the features-capturing capability of the network can be extended enormously when a second hidden layer is added, the additional training and computational time required in this case, however, advises the network user not to do this, if it is not absolutely required by the complexity of the problem to be solved.

Training of backpropagation networks (without internal feedback) is a process of ***supervised learning***, relying on the ***error-correction learning*** method in which the desired, *i.e.* a given, output pattern is expected to be matched by the final output pattern of the network within a specified accuracy. This is to be achieved by adjusting the network weights according to a parameter tuning algorithm, traditionally performed by a backpropagation algorithm that is considered as a generalization of the delta rule.

3.3.2 Radial Basis Function Networks

The idea of function approximation using ***localized basis functions*** is the result of the research work done by Bashkirov *et al.* (1964) and by Aizerman, Braverman and Rozenoer (1964) on the ***potential function approach*** to pattern recognition. Moody and Darken (1989) used this idea to implement a fast learning neural network structure with locally tuned processing units. Similarly, Broomhead and Lowe (1988) have described an approach to local functional approximation based on adaptive function interpolation. This has found a remarkable resonance within the researchers working on function approximation using ***radial basis functions***, that is considered to be the birth of a new category of neural networks, named ***radial basis function networks***.

The new category of networks was enthusiastically welcomed by the neural network society because the new networks have demonstrated the improved capability of solving pattern separation and classification problems. Backpropagation networks, in spite of their universal approximation capability, fail to be reliable pattern classifiers. This is because during the training phase multilayer perceptron networks build strictly ***separating hyperplanes*** that exactly classify the given examples, so that the new, unknown examples are randomly classified. This is a consequence of using the sigmoidal function as the network activation function with its resemblance to the unit step function, which is a global function. Also, the sigmoidal function, since it belongs to the set of ***monotonic basis functions***, has a slowly decaying behaviour in a large area of it's arguments. Therefore, the networks using this kind of activation function can reach a very good overall approximation quality in the large area of arguments; however, they cannot exactly reproduce the function values at the given points. For this one needs ***locally restricted basis functions***, such as a ***Gaussian function***, ***bell-shaped function***, ***wavelets*** or the ***B-spline functions***.

The locally restricted functions can be centred with the exact values at some selected argument values. The function values around these selected argument positions can decay relatively fast, controlled by the approximation algorithm. Powel (1988) suggested that the locally restricted basis functions should generally have the form

$$F(x) = \sum_{i=1}^{n} w_i \varphi\left(\|x - x_i\|\right),$$

where $\varphi\left(\|x - x_i\|\right)$ is a set of nonlinear functions relying on the Euclidean distance $\|x - x_i\|$. Moody and Darken (1989) selected for their radial basis function networks the exponential activation function

$$F_i = \exp\left(-\frac{\|x_i - c_i\|^2}{\sigma_i^2}\right),$$

which is similar to the Gaussian density function centred at c_i. The function spread σ_i around the centre determines the ratio of the function decay with its distance from the centre.

The common configuration of an RBF network firmly consists of three layers (Figure 3.5): the input layer, the hidden layer, and the output layer. In the neurons of hidden layer the activation functions are placed. The input layer of the network is directly connected with the hidden layer of the network, so that only the connections between the hidden layer and the output layer are weighted. As a consequence, the training procedure here is entirely different from that in the backpropagation networks. The most important issue here is the selection for each neuron in the hidden layer the centre c_i and the spread around the centre σ_i; this is mostly done using the ***k-means clustering algorithm***, which is capable of determining the optimal position of centres. In addition, the value of the ***spread parameter*** σ_i should be selected small enough in order to restrict the basis function spreading, but also large enough to enable a smooth network output through the joint effect with the neighbouring functions.

The network training process mainly includes two training phases:

- ***initialization*** of RBF centres, for instance using ***unsupervised clustering*** methods (Moody and Darken, 1989), ***linear vector quantization*** (Schwenker *et al*, 1994), or ***decision trees*** (Kubat, 1998)
- ***output weight training*** of the RBF using an adaptive algorithm to estimate its appropriate values.

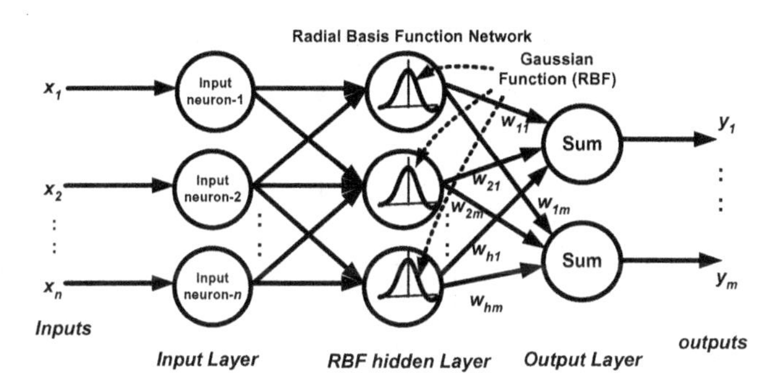

Figure 3.5. Configuration of an RBF network

In some cases, it is recommended to add a third training phase (Schwenker *et al.* 2001) in which the entire network architecture is adjusted using an optimization method.

3.3.3 Recurrent Networks

Research in the area of sequential and time-varying patterns recognition has created the need for time-dependent nonlinear input-output mapping using neural networks. To achieve this extended network capability, the ***time dimension*** has to be introduced into the network topology, for instance by introducing ***short-term memory features***, that would enable network to perform time-dependent mappings. Elman (1990) proposed a kind of ***globally feedforward, locally recurrent network*** using the ***context nodes*** as the principal processing elements of the network. Such nodes have also been the principal processing elements of the network proposed by Jordan (1986) for providing the networks with the dynamic memory. Both Jordan and Elman networks belong to the category of ***simple recurrent networks***.

An Elman network (Figure 3.6) is a ***four-layer network*** made out of input layer, hidden layer, output layer and the ***context layer***, the nodes of which are the one-step delay elements embedded into the local feedback paths. In the network, the neighbouring layers are interconnected by adjustable weights.

Originally, Elman proposed his simple recurrent network for speech processing. Nevertheless, owing to its eminent dynamic characteristics the network was widely accepted for systems identification and control (Sastry *et al.*, 1994). This was followed by applications in function approximation and in time series prediction.

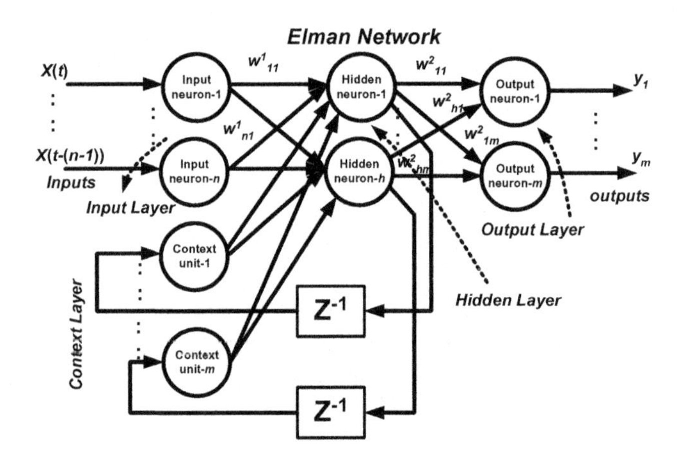

Figure 3.6. Configuration of the Elman network

Independently, Hopfield (1982) reported to the US National Academy of Sciences about neural networks with emergent collective computational abilities. In his report, Hopfield (1984) presented the neurons with graded response and their collective computational properties. He also presented some applications in neurobiology and described an electric circuit that closely reflected the dynamic behaviour of neurons, which is known as the ***Hopfield network*** (see Figure 3.7).

The Hopfield network is a single-layer fully interconnected recurrent network with a symmetric weight matrix having the elements $w_{ij} = w_{ji}$ and zero diagonal elements. As shown in Figure 3.7, the output of each neuron is fed back via a delay unit to the inputs of all neurons of the layer, except to its own input. This provides the network with some ***auto-associative capabilities***: the network can store by learning, following the ***Hebbian law*** or the ***delta rule***, a number of prototype patterns called ***fixed-point attractors*** in the locations determined by the weight matrix. The patterns stored can then be retrieved by associative recalls. On request to recall any of patterns stored, the network repeatedly feeds the output signals back to the neuron inputs until it reaches its stable state.

The recall capability of recurrent networks of retaining the past events and of using them in further computations is the advantage that the feedforward networks do not have. This capability enables the networks to generate time-variable outputs in response to the static inputs.

Because of incorporating internal feedback loops, the critical issue of recurrent networks is their stability, determined by the time behaviour of the network ***energy function***. For a ***binary Hopfield net*** with a symmetric weights matrix this function is defined as

$$E = -\frac{i}{2}\sum_{i=1}^{n}\sum_{j=1}^{n} w_{ij}x_i\dot{x}_j \ .$$

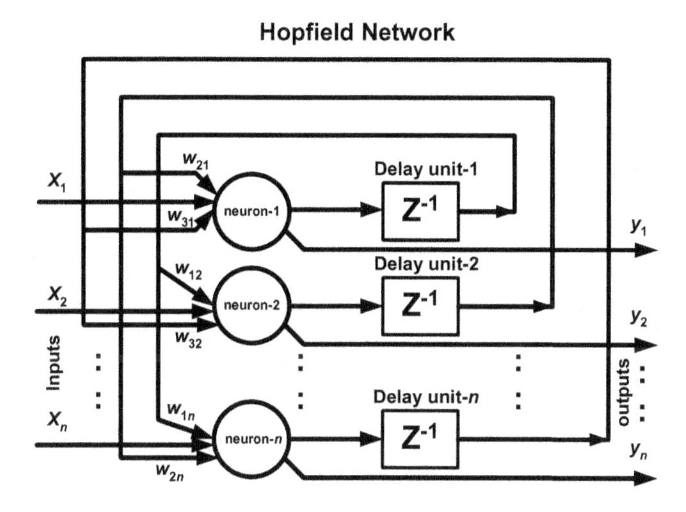

Figure 3.7. Configuration of a Hopfield network

In the case of a stable network this function must decrease with time and ultimately reach its minimum, or it's value remains constant. The minima reached are usually local minima because there are a number of states corresponding to fixed-point actuators or stored patterns to which the network must converge. Each finally reached state of the network has its associated energy defined above.

For the generalized form of binary Hopfield network, in which the sigmoid function

$$f(x) = \frac{1}{1+e^{-x}}$$

is used, the changes in time are continuously described following the equation

$$\kappa \frac{du_j}{dt} = \sum_i w_{ji}y_i - \frac{u_j}{D_j} + U_j ,$$

where κ is a constant positive value, y_i is the output value of the unit i, D_j is the factor controlling the sigmoid decay resistance, and U_j is the external input to the unit j. The resulting energy function in this case is defined by

$$E = -\frac{1}{2}\sum_i \sum_j w_{ij} u_i u_j - \sum_i u_i U_i$$

Network stability, as proven by Hopfield (1982), is generally guaranteed by the symmetric network structure.

For the training of recurrent networks, Rumelhart *et al.* (1986) proposed a general framework similar to that used for training feedforward networks, called ***backpropagation through time***. The algorithm is obtained by unfolding the temporal operation of the network into a layered feedforward growing with each time step. This, however, is not always satisfactory. Williams and Zipser (1988) presented a learning algorithm for continuously running ***fully connected recurrent neural networks*** (Figure 3.9) that adjusts the network weights in real time, *i.e.* during the operational phase of the network. The proposed learning algorithm is known as a ***real-time recurrent learning algorithm***.

There are two basic learning paradigms for recurrent networks:

- ***fixed-point learning***, through which the network reaches the prescribed steady state in which a ***static input pattern*** should be stored
- ***trajectory learning***, through which a network learns to follow a trajectory or a sequence of samples over time, which is valuable for ***temporal pattern recognition***, ***multistep prediction,*** and ***systems control***.

For trajectory learning, both the backpropagation through time and the real-time recurrent learning are appropriate. From the mathematical point of view, using the backpropagation through time we turn the recurrent network - by unfolding the temporal operation - into a layered feedforward network, the structure of which at every time step grows by one layer.

Almeida (1987) and Pineda (1987) have presented a method to train the recurrent networks of any architecture by backpropagation. Under the assumption that the network outputs strictly depend only on present and not on the past input values, Almeida derived the ***generalized backpropagation rule*** for this type of network, and addressed the problem of network stability using the energy function formulated by Hopfield (1982). Pineda (1987), however, directly addressed the problem of generalization of the backpropagation training algorithm and it's extension to recurrent neural networks. Hertz *et al.* (1991), based on the results of this work, have worked out a backpropagation algorithm for networks, the activation function of which obeys the ***evolutionary law***

$$\tau \frac{dv_i}{dt} = -v_i + g\left(\sum_j w_{ij} v_j + x_i\right),$$

that was formulated by Cohen and Grossberg (1983). In the above equation, τ is the time constant and x_i is the external input to the unit i. Solving this equation and defining the network equilibrium state for the unit k of the network

$$h_k = \sum_j w_{kj} v_j + x_k ,$$

the network should relax and ultimately reach the value y_k. Thereafter, the weights are updated using the gradient descent method by

$$\Delta w_{lk} = \alpha v_l g(h_k) y_k ,$$

where v_l and h_k are the equilibrium values of unit l and the equilibrium net input to the unit k respectively, and y_k is the equilibrium value of the ***matrix inverse unit***.

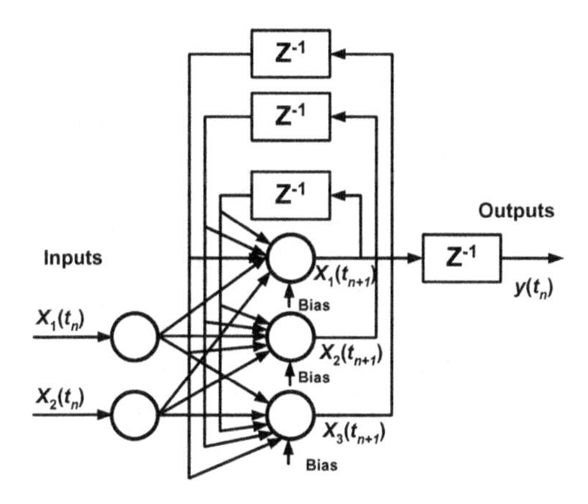

Figure 3.8. Fully connected recurrent neural network

A particular type of recurrent networks that do not obey the restrictions of the Hopfield networks are the ***dynamic recurrent networks***, proposed for representation of systems whose internal state changes with time. They are particularly appropriate for modelling of nonlinear dynamic systems, generally defined by the ***state-space equations***

$$\boldsymbol{X}(k+1) = \boldsymbol{f}(\boldsymbol{x}(k), \boldsymbol{u}(k))$$
$$\boldsymbol{Y}(k) = \mathbf{C}\boldsymbol{x}(k).$$

3.3.4 Counterpropagation Networks

A ***counterpropagation network,*** as proposed by Hecht-Nielsen (1987a, 1988), is a combination of a Kohonen's ***self-organizing map*** of Grossberg's learning. The combination of two neuro-concepts provides the new network with properties that are not available in either one of them. For instance, the network can for a given set of input-output vector pairs $(x_1, y_1), (x_2, y_2), ..., (x_n, y_n)$ learn the functional relationship $y = f(x)$ between the input vector $x = (x_1, x_2, ..., x_n)$ and the output vector $y = (y_1, y_2, ..., y_n)$. If the inverse of the function $f(x)$ exists, then the network can also generate the inverse functional relationship

$$x = f^{-1}(y).$$

When adequately trained, the counterpropagation network can serve as a ***bi-directional associative memory***, useful for pattern mapping and classification, analysis of statistical data, data compression and, above all, for function approximation.

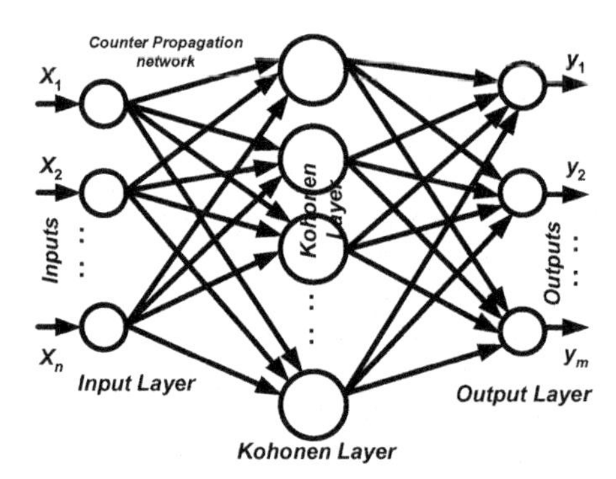

Figure 3.9. Configuration of a counterpropagation network

The overall configuration of a counterpropagation network is presented in Figure 3.9. It is a three-layer network configuration that includes the input layer, the Kohonen ***competitive layer*** as hidden layer, and the ***Grossberg output layer***. The hidden layer performs the key mapping operations in a competitive ***winner-takes-all fashion***. As a consequence, each given particular input vector $(x_{1p}, x_{2p}, ..., x_{np})$ activates only a single neuron in the Kohonen layer, leaving all other neurons of the layer inactive (see Figure 3.10). Once the competition process is terminated, a set of weights connecting the activated neuron with the neurons of the output layer defines the output of the activated neuron (say p) as the sum of products

$$y_p = \sum_{i=1}^{n} w_{ji} x_i ,$$

where n is the number of input layer neurons connected with the activated neuron. Using the set of weights learnt and stored, the network is capable of recognizing the pattern once learnt and the patterns in its neighbourhoods because similar inputs will activate the same Kohonen neuron.

After locating the Kohonen neuron, we turn to the Grossberg layer, *i.e.* the output layer of the network, and train it. To produce the desired mapping of the pattern at the network output using the output of the activated Kohonen neuron, all we need is to connect this neuron with each neuron in the Grossberg layer using the corresponding weights. As a result, a star connection between the Kohonen neuron and the network output, known as ***Grossberg's outstar***, builds the output vector $(y_{1p}, y_{2p}, \ldots, y_{mp})$, as shown in Figure 3.10.

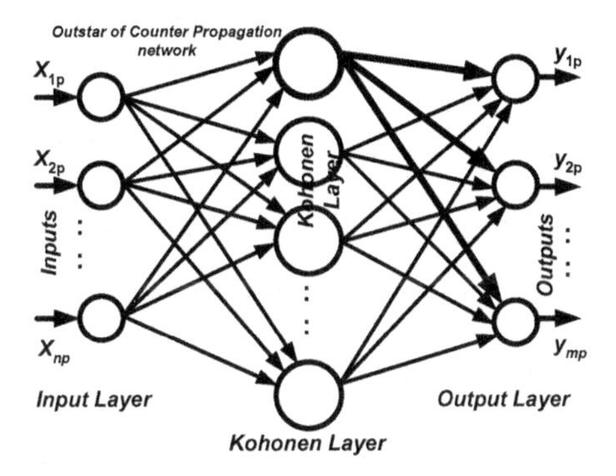

Figure 3.10. Outstar of counterpropagation network

The input vectors of a counterpropagation network should generally be normalized, *i.e.* they should satisfy the relation

$$\|x\| = 1 .$$

The normalization can be carried out by decreasing or increasing the vector length to be on the unit sphere using the relation

$$\bar{x} = \frac{x}{\|x\|} .$$

The question that remains is how to initialize the weight vectors before the network training starts. The preference of taking the randomized weight vectors has not always given reliable learning results. It has in some cases even created serious solution problems. The way out was found in using the *convex combination method* by taking for all the weight vectors the same value $1/\sqrt{n}$, where n is the dimension of weight vectors.

3.3.5 Probabilistic Neural Networks

The idea of probabilistic neural networks was born in the late 1980s at Lockheed Palo Alto Research Centre, where the problem of special patterns classification into submarine/non-submarine classes was to be solved. Specht (1988) suggested using a newly elaborated special kind of neural network, the *probabilistic neural networks.* To solve the classification problem, the new type of network had to operate in parallel with a *polynomial ADALINE* (Specht, 1990).

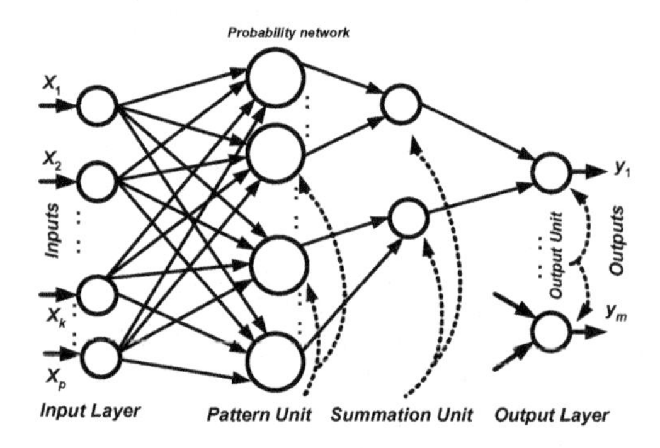

Figure 3.11. Architecture of a probability network

Supposing that $P_1, P_2, ..., P_m$ are the ***a priori probabilities*** for the vector **x** to belong to a corresponding category, and denoting by L_i the merit of classification loss for the category i, the Bayesian decision rules $P_i L_i p_i$, for i = 1, 2,…, m, can help determine the largest product value. In case that, say, $P_i L_i p_i \geq P_j L_j p_j$ holds, the input vector **x** is assigned to the category i. In this case the ***decision boundary*** for the above decision, that can be a nonlinear ***decision surface*** of arbitrary complexity, is defined by

$$p_i = \frac{P_j L_j p_j}{L_i P_i}.$$

The structure of probabilistic networks is similar to that of backpropagation networks, but the two types of network have different activation functions. In probabilistic networks the sigmoid function is replaced by a class of exponential functions (Specht, 1988). Also, the probabilistic networks require only a single training pass, in order that - with the growing number of training examples - the decision surfaces finally reach the Bayes-optimal decision boundaries (Specht, 1990). This is achieved by modelling the well-known Bayesian classifier that follows the strategy of minimization of the expected classification risk. The strategy can be explained in terms of an n-dimensional input vector $\mathbf{x}$ belonging to one of m possible classes with the ***probability density functions***

$$p_1(x), p_2(x), ..., p_m(x) .$$

The architecture of a probabilistic network, shown in Figure 3.11, consists of an input layer followed by three computational layers. It has a striking similarity with a multilayer perceptron network. The network is capable of discriminating two pattern categories represented through the positive and negative output signals. To extend the network capability of multiplying discrimination, additional network outputs and the corresponding number of summation units are required.

The input layer of a probabilistic network is simply a distribution layer that provides the normalized input signal values to all classifying networks that make up a multiple classes classifier. The subsequent layer consists of a number of ***pattern units***, fully connected to the input layer through adjustable weights that correspond to the number of categories to be classified. Each pattern unit forms the product of the input vector $\mathbf{x}$ with the weight vector $\mathbf{w}$. The product value, before being led to the corresponding ***summation unit***, undergoes the initial nonlinear operation

$$F(xw_i) = e^{\frac{(xw_i - 1)}{\sigma^2}} .$$

However, since both the input pattern and the weighting vectors are normalized to the unit length, the last relation is to be rewritten as

$$F(xw_i) = e^{\frac{\sum_{j=1}^{n}(x_j - w_{ij})^2}{2\sigma^2}} .$$

The summation units finally add the signals coming from the pattern units corresponding to the category selected for the current training pattern.

3.4 Network Training Methods

We now turn our attention to some training aspects of neural networks, particularly to the aspects of training process acceleration and training process results. Our primary interests are the ***supervised learning algorithms***, the most frequently used in real applications, such as the ***backpropagation training algorithm***, also known as the ***generalized delta rule***.

The backpropagation algorithm was initially developed by Paul Werbos in 1971 but it remained almost unknown until it was "rediscovered" by Parker in 1982. The algorithm, however, became widely popular after being clearly formulated by Rumelhart *et al.* (1986), which was a triggering moment for intensive use of multilayer perceptron networks in many simulated engineering applications. The real-life application had at that time to be "postponed" due to the lack of a suitable neuro-technology. In the 1990s Rumelhart put much effort into popularizing the training algorithm among the neural network scientific community. Presently, the backpropagation algorithm is also used (in slightly modified form) for training of other categories of neural networks.

In the following, we will confine our discussion mainly to multilayer perceptron networks. As mentioned earlier, this kind of networks, based on given training samples or input-output patterns, implements nonlinear mapping of functions that is applicable to function approximation, pattern classification, signal analysis, *etc*. In the process of training, the network learns through adaptation of ***synaptic weights*** in such a way that the discrepancy between the given pattern and the corresponding actual pattern at network output is minimized. Because the synaptic adaptation mostly follows the ***gradient descent law*** of parameter tuning, the backpropagation training algorithm is considered as the search algorithm of ***unconstrained minimization*** of a suitably constructed error function at network output.

In order to illustrate the basic concept of the backpropagation algorithm, let us consider its application to the training of a single neuron located in the output layer of a multilayer perceptron (see Figure 3.12). In addition, let us suppose that as the nonlinear activation function the hyperbolic tangent function

$$y = f\left(u_j\right) = \tanh(\gamma u_j) = \frac{1-\exp\left(-\gamma u_j\right)}{1+\exp\left(-\gamma u_j\right)} \tag{3.1}$$

is chosen, where

$$u_j = \sum_{i=1}^{n} w_i x_i + \theta_j, \qquad \gamma > 0. \tag{3.2}$$

Furthermore, x_i is the ith input with corresponding interconnecting weight w_i to the neuron and θ_j is the bias input to the same neuron. Typically, all neurons in a particular layer of the multilayer perceptron have the same activation function. The aim of the learning algorithm is to minimize the instantaneous squared error function of the network output

$$S_j = 0.5\left(d_j - y_j\right)^2 = 0.5\left(e_j\right)^2, \tag{3.3}$$

defined as the square of the difference $(d_j - y_j)$ between the desired output signal and the actual output signal of the network, by modifying the synaptic weights w_i. The minimization process in parameter tuning steps Δw_i is based on the steepest descent gradient rule

$$\Delta w_i = -\eta \frac{\partial S_j}{\partial w_i} \tag{3.4}$$

where η is a positive learning parameter determining the speed of convergence to the minimum.

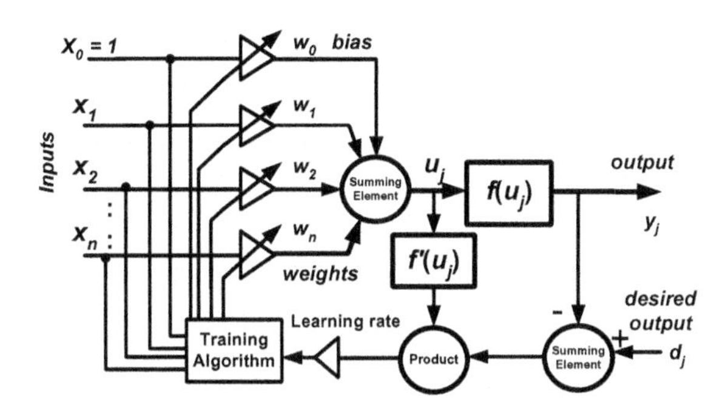

Figure 3.12. Backpropagation training implementation for a single neuron

Now, taking into account that from (3.3) follows:

$$e_j = \left(d_j - y_j\right) = \left(d_j - f\left(u_j\right)\right), \tag{3.5}$$

where

$$u_j = \sum_{i=0}^{n} w_i x_i \,.$$

By applying the chain rule

$$\Delta w_i = -\eta \frac{\partial S_j}{\partial e_j} \cdot \frac{\partial e_j}{\partial w_i} \tag{3.6}$$

to Equation (3.5) we get

$$\Delta w_i = -\eta e_j \cdot \frac{\partial e_j}{\partial w_i} = -\eta e_j \cdot \frac{\partial e_j}{\partial u_j} \cdot \frac{\partial u_j}{\partial w_i} \tag{3.7}$$

This can further be transformed to

$$\Delta w = \eta e_j \cdot \frac{\partial f(u_j)}{\partial u_j} \cdot x_i = \eta e_j \cdot f'(u_j) \cdot x_i = \eta \delta_j \cdot x_i$$

where δ_j can be expressed as

$$\delta_j = e_j f'(u_j) = -\frac{\partial S_j}{\partial u_j}. \tag{3.8}$$

The derivation $f'(u_j)$ of the selected activation function (3.1) is

$$f'(u_j) = \frac{\partial f(u_j)}{\partial u_j} = \gamma\left[1 - \tanh^2(\gamma u_j)\right] = \gamma\left[1 - y_j^2\right], \tag{3.9}$$

and the corresponding weight updates (3.7)

$$\Delta w_i = \eta\gamma e_j \cdot \left(1 - y_j^2\right) \cdot x_i, \tag{3.10}$$

with $\eta\gamma > 0$.

Note that the weight update stabilizes if y_j approaches –1 or +1, since the partial derivative $\partial y_j / \partial u_j$, equal to $\gamma\left(1 - y_j^2\right)$, reaches its maximum for $y_j = 0$ and its minima for ± 1. However, if the sigmoidal activation function is used and if it is unipolar, described by

$$y_j = f(u_j) = \frac{1}{1 + \exp(-\gamma y_j)}, \tag{3.11}$$

then

$$f'(u_j) = \frac{\partial f(u_j)}{\partial u_j} = \gamma y_j (1 - y_j). \tag{3.12}$$

Therefore, the weight increment takes the form

$$\Delta w_i = \eta\gamma e_j \cdot y_j (1 - y_j) \cdot x_i. \tag{3.13}$$

It should also be noted that in this case the partial derivative $\partial y_j / \partial u_j$ reaches its maximum for $y_j = 0.5$ and, since $0 \leq y_j \leq 1$, it approaches its minimum as the output y_j approaches the value zero or the value one.

The synaptic weights are usually changed incrementally and the neuron gradually converges to a set of weights which solve the specific problem. Therefore, the implementation of the backpropagation algorithm requires an accurate realization of the ***sigmoid activation function*** and of its derivative.

The backpropagation algorithm described can also be extended to train multilayer perceptron networks.

3.4.1 Accelerated Backpropagation Algorithm

The backpropagation algorithm generally suffers from a relatively slow convergence and with the possibility of being trapped at a local minimum. Also, it can be accompanied by possible oscillation around the located minimum value. This may restrict its practical application in many cases. Therefore, such unwanted drawbacks of the algorithm have to be removed, or at least reduced. For instance, the speed of algorithm convergence can be accelerated:

- by selection of the best initial weights instead of taking the ones that are generated at random
- through adequate preprocessing of training data, *e.g.* by employing the feature extraction algorithms or some data projection methods
- by improving the optimization algorithm to be used.

Numerous heuristic optimization algorithms have been proposed for speed acceleration; unfortunately, they are generally computationally involved and time exhausting. In the following, only two of the most efficient are briefly reviewed:

- adaptation of learning rate
- using a momentum term.

It is usually assumed that the learning rate of the algorithm is fixed and uniform for all weights during the training iterations. In order to prevent parasitic oscillations and to ensure the convergence to the global minimum, the learning rate must be kept as small as possible. However, a very small value of learning rate slows down the convergence speed of algorithm considerably. On the other hand, a large value of the learning rate results in an unstable learning process. Therefore, the learning rate has to be optimally set between the two extreme values of learning rate, *e.g.* by using the ***adaptive learning rate***, and in this way the training time can be considerably reduced. Similarly, the speed up of convergence can be achieved by extending the training algorithm by a ***momentum term*** (Kröse and Smagt, 1996). In this case the learning rate can be kept at each iteration step as large as possible within the admitted values, while maintaining the learning process stable.

One of the simplest heuristic approaches of learning rate tuning is to increase the learning rate slightly (typically by 5%) in an iteration step if the new value of the output error (sum squared error) function S is smaller than the previous iteration step. On the other hand, if the new value of the error function exceeds the value of the previous one, then the learning rate should be decreased by approximately 30%, and in the latter case the new weight updates and the error function are discarded, *i.e.* in this case we set weight update as

$$\Delta w_{ij}(k+1) = 0,$$

and that leads to weights in $(k + 1)$th iteration as identical as $(k - 1)$th, *i.e.*

$$w_{ij}(k+1) = w_{ij}(k-1)).$$

After starting with a small learning rate, the approach will behave as follows:

$$\begin{aligned} \eta^{(k)} &= a\eta^{(k-1)}, \textit{ for } S(w(k)) < S(w(k-1)), \\ \eta^{(k)} &= b\eta^{(k-1)}, \textit{ for } S(w(k)) \geq k_0 S(w(k-1)), \\ \eta^{(k)} &= \eta^{(k-1)}, \quad \text{otherwise} \end{aligned} \tag{3.14}$$

with a = 1.05, b = 0.7 and k_0 = 1.04 being typical values (Vogl *et al.* 1988; Cichocki and Unbehauen, 1993).

In some training applications not all the training patterns are available before the learning starts. In such situations an on-line approach has to be used. Schmidhuber (1989) proposed the simple global updates of the learning rate for each training pattern as

$$\Delta w_{ij}(k) = -\eta^{(k)} \frac{\partial S_p}{\partial w_{ij}}, \tag{3.15}$$

with

$$\eta^{(k)} = \min\left\{ \frac{S_p - S_0}{\left\|\nabla S_p\right\|_2^2}, \eta_{(\max)} \right\}, \tag{3.16}$$

where the index $\eta_{(\max)}$ indicates the maximum learning rate (typically $\eta_{(\max)} = 20$) and S_0 is a small offset error function (typically $0.01 \leq S_0 \leq 0.1$).

Various suggestions have been made for practical use of both adaptable learning rate and the momentum term, with the best known being the conjugate gradient algorithm (Johansson *et al.*, 1992). Alternatively, the second-order derivative-based Levenberg-Marquardt algorithm (Hagan and Menhaj, 1994), proposed for accelerated minimization of the cost function, is preferably used for accelerated neural networks training. The key idea of the algorithm is to use a

search vector P_k to calculate the parameter value W_{k+1}, based on a current value W_k as

$$W_{k+1} = W_k + \alpha_k p_k, \tag{3.17}$$

where α_k is a scalar value. The search vector P_k is to be chosen so that the relation $V(W_{k+1}) < V(W_k)$ holds, where $V(W)$ is the performance index of the network, generally a sum square error function.

Now, considering the Taylor series expansion of $V(W_{k+1})$ at point W_k

$$V(W_{k+1}) = V(W_k + \alpha_k P_k) \approx V(W_k) + \alpha_k \nabla V(W_k)^T P_k. \tag{3.18}$$

it is obvious that, in order for the cost function V to decrease and for a positive value of α_k, the second term of (3.18) must be negative. This will be the case if the steepest descent condition

$$W_{k+1} = W_k - \alpha_k \nabla(W_k) \tag{3.19}$$

is met. However, the steepest descent method, as discussed earlier, when used in its original form, exhibits some drawbacks that need to be eliminated for its practical use. To overcome this, the approximation of the objective function in the immediate neighbourhood of a strong minimum by a quadratic function with positive definite Hessian matrix or by using Newton's method for pursuing the minimization problem is preferred.

Let us now consider the Taylor series expansion

$$V(W_{k+1}) \approx V(W_k) + \nabla V(W_k)^T \Delta W_k + \frac{1}{2} \cdot \Delta W_k^T \cdot \nabla^2 V(W_k)^T \Delta W_k \tag{3.20}$$

where $\nabla^2 V(W_k)$ is the Hessian matrix and $\Delta W_k = \alpha_k P_k$. If the gradient of the truncated Taylor series expansion (3.20) is taken with respect to ΔW_k and set to zero (since we are looking for the minimum of the cost function), it follows that

$$\Delta W_k = -\left[\nabla^2 V(W_k)\right]^{-1} \nabla V(W_k). \tag{3.21}$$

This reduces the Newton method to

$$W_{k+1} = W_k - \left[\nabla^2 V(W_k)\right]^{-1} \nabla V(W_k). \tag{3.22}$$

Direct practical use of this method, however, is hampered by the need for Hessian matrix calculation, whose elements are the second derivatives of the performance index with respect to the parameter vector. To overcome this obstacle, the first and the second derivatives of the performance index

$$V(W_k) = \sum_{i=1}^{N} e_i^2(w_k) = e^T(w_k)e(w_k) \tag{3.23}$$

are built and expressed as

$$\nabla V(w_k) = J^T(w_k)e(w_k) \tag{3.24}$$

and

$$\nabla^2 V(w_k) = J^T(w_k)J(w_k) + \sum_{i=1}^{N} e_i(w_k)\nabla^2 e_i(w_k), \tag{3.25}$$

where $J(w_k)$ is the Jacobian matrix and

$$e(w_k) = T - Y(w_k), \tag{3.26}$$

with the target vector T and the actual output of the neural network $Y(w_k)$.

The Gauss-Newton modification of the method assumes that the second term in the right-hand side expression of (3.25) is zero. Therefore, applying the former assumption (3.22) yields the Gauss-Newton method as

$$W_{k+1} = W_k - \left[J^T(w_k)J(w_k)\right]^{-1} J^T(w_k)e(w_k), \tag{3.27}$$

An additional difficulty appears here with when the Hessian matrix is not positive definite, *i.e.* its inverse does not exist. In this case the modification of the Hessian matrix

$$G = \nabla^2 V(w_k) + \mu I \tag{3.28}$$

should be considered. Suppose that the eigen-values and the eigen-vectors of $\nabla^2 V(W_k)$ are the sets $\{\lambda_i\}$ and $\{z_i\}$ respectively. Multiplying both sides of (3.28) by z_i we have

$$Gz_i = \nabla^2 V(w_k)z_i + \mu I z_i = \lambda_i z_i + \mu z_i \tag{3.29}$$

$$Gz_i = (\lambda_i + \mu)z_i \tag{3.30}$$

Therefore, the eigen-values and eigen-vectors of G are $\{\lambda_i + \mu\}$ and $\{z_i\}$ respectively. G can be made positive definite by increasing μ until $\lambda_i + \mu > 0$ for all i.

Therefore, the Levenberg-Marquardt modification to Gauss-Newton method is

$$W_{k+1} = W_k - \left[J^T(w_k)J(w_k) + \mu I\right]^{-1} J^T(w_k)e(w_k) \qquad (3.31)$$

whereby the parameter μ is multiplied by some factor β whenever a step would result in an increased value of $V(w_k)$. When a step reduces this value, μ is divided by β. Notice that when μ is large the algorithm becomes steepest descent with the step size approximately $1/\mu$. On the other hand, for small μ the algorithm becomes Gauss-Newtonian.

Obviously, the calculation of the Jacobian matrix is the key step in applying this algorithm. At first, all the adjustable parameters of the network should be arranged in one column vector w_k. For a neural network mapping problem the terms in the Jacobian matrix can be computed by simple modification to the backpropagation algorithm (Hagan and Menhaj, 1994). In the standard backpropagation version, partial derivatives of the performance function with respect to the adjustable parameters are needed, while in Levenberg-Marquardt algorithm the derivative of the error is needed for the Jacobian matrix. This means that the Jacobian matrix can be calculated using the sensitivity term of the performance index derived in the standard backpropagation algorithm with one modification at the final layer, *i.e.* by dropping the error term (Hagan and Menhaj, 1994). The Jacobian matrix computation for a neuro-fuzzy network is described in Chapter 6.

The algorithm described above can easily be extended to train the multilayer perceptron networks.

3.5 Forecasting Methodology

Forecasting methodology is generally understood as a collection of approaches, methods, and tools for collection of time series data to be used for forecast or prediction of future values of the time series, based on past values. The forecasting methodology includes the following operational steps:

- ***data preparation*** for forecasting, *i.e.* acquisition, preprocessing, normalization, and structuring of data, determination of training and test data sets, and the like
- ***network architecture determination***, *i.e.* selection of the type of network to be used for forecasting, determination of number of network input and output nodes, number of layers, the number of neurons within the layers, determination of interconnections between the neurons, selection of neuron activation functions, *etc.*

- design of ***network training strategy***, *i.e.* selection of training algorithm, performance index, and the training monitoring approach
- ***overall evaluation*** of forecasting results using fresh observation data sets.

3.5.1 Data Preparation for Forecasting

Data used for analysis and forecasting of time series are generally collected by observations or by measurements. In engineering, of major interest is the analysis of data obtained by sampling of corresponding sensor signals and forecasting their future behaviour. Therefore, our attention will be primarily focused on forecasting of experimental data taken from ***sensing elements*** placed within the experimental setups or within the plant automation devices. Here, depending on the nature of signals provided by sensors, two main critical issues are:

- the number of data needed for representative characterization of the observed signal in view of its linearity, stationarity, drift, *etc.*
- the sampling period required for recording the entire frequency spectrum of the sampled signal, but that will still considerably limit the noise frequency spectrum.

In practice, the ***preprocessing*** of acquired data, because of the presence of noise, drift, and sensor inaccuracy, represents a trial-and-error procedure. In the preprocessing phase it should also be made clear whether data filtering, smoothing, *etc.* are needed, or whether mathematical transformation of data will facilitate the learning process of the network within its training and/or reduce the network training time.

Data normalization is a process of final data preparation for their direct use for network training. It includes the normalization of preprocessed data from their natural range to the network's operating range, so that the normalized data are strictly shaped to meet the requirements of the network input layer and are adapted to the nonlinearities of the neurons, so that their outputs should not cross the saturation limits.

In practice, the simplest normalization

$$x_{ni} = \frac{x_i}{x_{\max}}$$

and the linear normalization

$$x_{ni} = \frac{x_i - x_{\min}}{x_{\max} - x_{\min}}$$

are most frequently used. Moreover, instead of linear normalization, ***nonlinear scaling*** or ***logarithmic scaling*** of input signals is used to moderate the possible nonlinearity problems during the network training. For instance, logarithmic transformation can squeeze the scale in the region of large data values, and exponential scaling can expand the scale in the region of small data values, *etc*. But by far the most critical data preparation issue here is the risk of possible loss of critical information present within the acquired data.

Structuring of data is needed when preparing the mutually related input and output data pairs to be used in supervised learning and/or when preparing multivariate data in general. In the case of training the networks for forecasting purposes, the next value x_{t+1} of the univariate time series is related to the past values of the time series up to the present value x_t. In the next training step the value x_{t+2} is related to the past values of the time series up to the value x_{t+1}, *etc.*

Before structuring the data of a multivariate time series for training of a network forecaster, the fact should be recalled that this kind of time series is a set of simultaneously built multiple time series with the values of each individual time series being related to the corresponding values of other time series. This is because the multivariate time series are built by simultaneous observation of two or more processes, so that the resulting observation across all the individual samplings at a certain time builds an ***observation vector***

$$x_i = [x_{i1} x_{i2} \ldots\ldots x_{in}].$$

Thus, the resulting multiple time series in fact represents a set of observation vectors x_i, $i = 1, 2, \ldots, m$, building up the ***observation matrix***

$$X = \begin{bmatrix} x_{11} \; x_{12} \ldots\ldots x_{1n} \\ x_{21} \; x_{22} \ldots\ldots x_{2n} \\ \ldots \; \ldots \; \ldots \; \ldots \\ x_{m1} \; x_{m2} \ldots\ldots x_{mn} \end{bmatrix},$$

in which the time series of individual processes are represented through the corresponding matrix columns.

A ***training set*** is used to teach the network to behave as a forecaster and the ***test set*** is used, after the training, to test its forecasting capability. Both data sets are to be built from the entire collected data set. Unfortunately, no selection guide is available for splitting the prepared data set into two subsets. The recommendations range from a 90% to 10% ratio, up to a 50% to 50% ratio. Haykin (1995) advocated that the numbers of patterns N in the training set required to classify the test examples with an error of ε should approximately be

$$N = \frac{W}{\varepsilon},$$

where W is the number of weights in the network.

Yet, whatever ratio is selected, attention should be paid to ensuring that the training data set is large enough to cover all the dominant characteristic features required for reliable network training as a forecaster. The remaining data set can then be used for testing the trained network on the data samples never used in the training. For this reason, it is recommended that the non-training data set should be large enough to enable building of not only the test data set but also the ***validation data set*** to be used in the overall network evaluation.

3.5.2 Determination of Network Architecture

This is the core task in building the neural network structure optimally adapted to the specific problem the network should optimally solve. In our case it would be the optimal predictor or the optimal forecaster. This task, although being very challenging, is also the most difficult to execute because it requires from the designer much skill and practical experience. Since being a nontrivial task with a multiplicity of possible solutions, there are opinions that this work is more a kind of art than an expert's routine. The issues addressed in the following present the activities to be carried out when developing the network architecture. They include the

- determination of input nodes required
- determination of output nodes
- selection of number of hidden layers
- selection of hidden neurons
- determination of node interconnection pattern
- selection of activity function of neurons.

Determination of the required ***number of input nodes*** is a relatively easy task, because it depends predominantly on the number of independent variables presented in the data set prepared. As a rule, each independent variable should be represented by its own input node. In the case of input data prepared for forecasting, the number of input nodes is directly determined by the number of ***lagged values*** to be used for forecasting of the next value

$$x(t+1) = f[x(t), x(t-1), x(t-2), \dots, x(t-n)],$$

as represented in Figure 3.13.

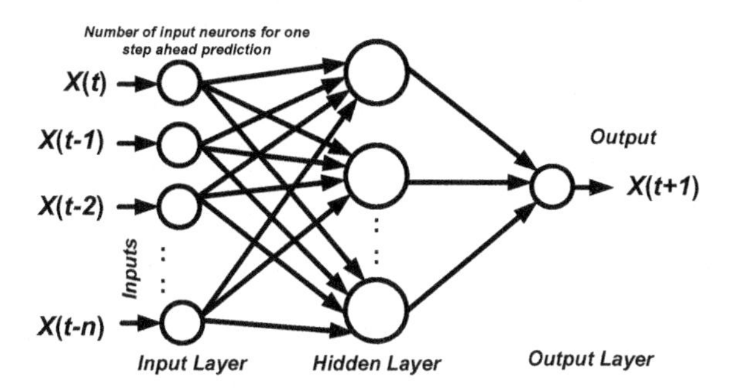

Figure 3.13. Number of input neurons for one-step-ahead forecasting

In practice, the single-step-ahead forecaster is most frequently selected because it is relatively simple and guarantees the most accurate forecasting results. Otherwise, when building a multistep predictor, the determination of the required number of input nodes is a trade-off process in the sense that (following the general inclination) this number should be selected as small as possible but so that it still guarantees good forecasting results, and as large as needed for the extraction of all relevant characteristic features and the ***autocorrelation structure*** embedded in the training data. To solve this problem optimally, some experimental runs could be of considerable use.

The ***number of output nodes***, again, is also a problem-oriented task. In the one-step-ahead forecasting it is apparent that only one output node is sufficient as the forecasting node. Correspondingly, in the case of multistep-ahead forecasting, the number of output nodes should correspond to the forecasting horizon, *i.e.* to the number of forecasts to be simultaneously presented at the network output. Alternatively, a single output node can be used and all the future forecasts required determined in the iterative steps.

In most forecasting applications, only one ***hidden layer*** is used, although some aberrations are exceptionally needed. The sufficiency of a single layer is covered by the ***Kolmogorov's superposition theorem***, which states that any continuous function $f(x)$ – which can also be an n-dimensional vector function $f(x_1, x_2, \ldots, x_n)$ – defined on a closed n-dimensional cube, say $[0,1]^n$, can be represented as

$$f(x_1, x_2, \ldots, x_n) = \sum_{i=1}^{2n+1} \psi_i \left(\sum_{j=1}^{n} \varphi_{ji}(x_j) \right),$$

where ψ_i and φ_{ji} are continuous, single-variable functions. The functions ψ_i depend on the function to be approximated f and the functions φ_{ji} are monotonously increasing functions fixed for a given n.

The theorem, as originally formulated by Kolmogorov, is an existence theorem that does not suggest any particular function to be used for approximation of a given mapping, so that its relevancy to neural networks was not directly evident. There were even opposite views to the relevance: one opposing the relevancy (Girosi and Poggio.1989) and another in favour of it. However, it was the refinement of the theorem by Sprecher (1965) that motivated Hecht-Nielsen (1987b) to point out this reliance. He also proposed that the kth processing elements of the hidden layer should have the activation function

$$z_k = \sum_{i=1}^{n} \lambda^k \varphi(x_i + \varepsilon k) + k \,,$$

where the real constant λ and the monotonously increasing real continuous function □ depend on n, but are independent of f. Furthermore, the rational constant ε should satisfy the conditions of the ***Sprecher theorem*** $0 < \varepsilon < \delta$, $\delta > 0$. The activation function of the output layer units should be

$$y_j = \sum_{k=1}^{2n+1} g_j(z_k) \,,$$

where g_j are the real and continuous functions depending on φ and ε. Consequently, as it was shown (Hecht-Nielsen, 1987b), the Kolmogorov's theorem can be implemented exactly by a three-layer feedforward neural network having n input elements in the input layer, $(2n+1)$ processing elements in the hidden layer, and m processing elements in the output layer. This confirms the statement that even a single hidden-layer network is sufficient to reveal all the characteristic features present on the input nodes of the network. Introducing additional hidden layers increases the feature extraction capability of the network at the cost of the significantly extended training and operational time of the forecaster.

Lippmann (1987), in his celebrated paper on neurocomputing, stated clearly that a three-layer perceptron can form arbitrarily complex decision regions and can separate meshed classes, which means that no more than three network layers are needed in perceptron-like feedforward nets. This particularly holds for the networks with one output, as required for one-step-ahead forecasting. Cybenko (1989), finally underlined that the networks never need more than two hidden layers to solve most complex problems. Also, the investigation of neural network capabilities related to their internal structure has proven that two-hidden-layer networks are more prone to fall into bad local minima. DeVilliers and Barnard (1992) even pointed out that both the one- and two-hidden-layer networks perform similarly in all other respects. This can be understood from the comparison of complexity degree of two investigated networks measured by the ***Vapmik-Chervonenkis dimension***, as was done by Baum and Hausler (1989).

We now turn to the problem of the ***number of hidden neurons*** placed within the hidden layer. To determine the optimal number of hidden neurons there is no straight-forward methodology, but some rules of thumb and some suggestions how to do this have been proposed. For instance, in single-hidden-layer networks, it is recommended to take the number of hidden-layer neurons in the neighbourhood of 75% of the number of network inputs, or say between 0.5 and 3 times the number of network inputs. The ***geometric pyramid rule***, on the other hand, suggests assigning

$$N_h = \alpha\sqrt{N_i \times N_o} \; ,$$

hidden neurons to a single hidden layer, where N_i is the number of network inputs, N_o the number of its outputs, and α is multiplication factor the value of which, depending on the complexity of the problem to be solved, should be selected in the range $0.5 < \alpha < 2$. Baum and Haussler (1989) suggested the number of neurons in the hidden layer be determined as

$$N_h \le \frac{N_{tr} \times E_{tol}}{N_{dp} + N_o} \, ,$$

where N_{tr} is the number of training examples, E_{tol} is the error tolerance, N_{dp} is the number of data points per training example, and N_o is the number of output neurons.

Anyhow, the determination of the optimal number of hidden neurons involves trial-and-error experimentation: starting with a number of neurons within the layer to be decided – based on final accuracy of each learning process – to increase or decrease the number of hidden neurons and to start a new learning process. In this way the redundant hidden neurons can be deleted and the neurons needed for optimal performance of the layer added. Here, both starting with a relatively large or small number of neurons is possible, but starting with a large number of neurons bears the risk of long-time computation and of getting trapped in local minima.

Khorasani and Weng (1994) have presented an approach to structural adaptation of feedforward neural networks by neuron pruning, *i.e.* by addition and deletion of hidden neurons based on the activity status of individual neurons during the learning, measured by the variance of the neuron output signal and by the strength of the backpropagated error. This is a proper indication of neuron activity that helps decide which low-activity redundant neurons are to be deleted.

There is also a reliable way to determine the number of hidden neurons using the ***Akaike's information criterion*** (AIC), originally defined as

$$\text{AIC} = (-2)\ \ln(\text{Maximum likelihood}) + 2(\text{number of adjusted parameters}).$$

The criterion statistically evaluates the goodness of a model by combining the evaluated mean squares error for training data and the number of parameters to be estimated. Seen otherwise, AIC combines a measure of fit and the penalty term to account for model complexity. Its potential application suitability for neural networks model building was recognized by Kurita (1990) and Fogel (1991), who reformulated the original form of the criterion (for statistically independent, normally distributed output errors with zero mean and with constant variance) as

$$\text{AIC} = Nk\ln(\sigma^2) + 2K\ ,$$

where N is the number of training data, k is the number of output units of the network, σ^2 is the ***maximum likelihood estimate*** of the mean square error for training data and K is the number of model parameters.

The application principle of the AIC is that, if two models have the same mean square error for a training data set, then the smaller sized model should be selected. Alternatively, from a set of possible models, the model with the smallest value of AIC is to be selected (Ishikawa and Moriyama, 1996; Anders and Korn, 1999). This, however, requests a set of models to be built and their parameter estimated before this application principle is used.

Unfortunately, direct application of the AIC to neural networks is rather circumstantial. It is, however, facilitated when using the ***network information criterion*** (NIC) of Stone (1977)

$$\text{NIC} = -\frac{1}{T}\ln\left(L(\widehat{w})\right) + \frac{\text{tr}[BA^{-1}]}{T},$$

which is a generalization of the AIC. The first term in the above expression represents the estimated maximum logarithmic likelihood. The matrices A and B are defined as

$$A \equiv -E[\nabla^2 \ln L_t]$$
$$B \equiv E[\nabla \ln L_t \nabla \ln L_t].$$

If the classes of models investigated include the true model, then it holds asymptotically that $A = B$ and

$$tr[BA^{-1}] = tr[I] = K,$$

where K is, again, the number of model parameters. In this case the NIC takes the form

$$\text{NIC} = -\frac{1}{T}\ln L(\widehat{w}) + \frac{K}{T}.$$

This is similar to the AIC, which in this transcription becomes

$$\text{AIC} = -\frac{2}{T}\ln L(\widehat{w}) + \frac{2K}{T}.$$

Murata *et al.* (1994) used this generalization to determine the number of hidden units required to mimic the system based on input-output examples only. Attention was paid to avoiding possible network ***overfitting*** by taking a small number of redundant hidden neurons. A large number of hidden layer neurons could, for the given training example, deliver better learning results but, due to the increased network complexity, for some fresh examples could deliver worse results.

What the interconnections of network nodes concerns, ***full interconnection*** is recommended for initial network configuration, in which the output of each neuron of a layer is connected with the input of each neuron of the subsequent layer. However, in some applications, deviations from full interconnection have also been successful.

For ***activation function selection***, there is generally no rich choice left. For backpropagation networks, mostly the

- ***sigmoid function***

$$y = \frac{1}{1 + \mathrm{e}^{-x}}$$

  is selected as an activation function in numerous applications, including time series forecasting. But in some applications the

- ***hyperbolic tangent function***

$$y = \frac{\mathrm{e}^{x} - \mathrm{e}^{-x}}{\mathrm{e}^{x} + \mathrm{e}^{-x}},$$

  has also been used successfully, for instance when solving the problems that rely on learning of deviations from average behaviour (Klimasauskas, 1991)

- ***step*** and ***ramp function*** are some additional alternatives favourable for processing binary variables.

In any case, to avoid functional destruction of the neuron, the function selected should be limited at its output, usually between the values -1 and $+1$. Although there are no guidelines for selecting the activation functions in individual network layers and for distributing them within the layers, it is still best to build homogeneous individual layers and for the hidden neurons possibly to use the ***sigmoid activation function***. But still, some researchers have successfully used the hyperbolic tangent as an activation function of hidden-layer neurons. Very seldom heterogeneous network layers have been used. For time series forecasting, the general experience has shown that for output neurons the linear activation function delivers the best results. Some theoretical evidence for this has also been given (Rumelhart *et al.*, 1986). It was shown that only for forecasting of time series with trend, output neurons with a nonlinear activation function are required.

3.5.3 Network Training Strategy

Network training is a process in which the network learns to recognize the patterns inherent to the training signals. In network training for time series forecasting all relevant characteristic features embedded in the training data that reflect the autocorrelation structure of the time series should be revealed and learnt. The training is usually carried out in off-line mode using an unconstrained nonlinear minimization algorithm, most frequently a gradient descent method, for tuning the interconnection weights of the network. The objective is to achieve the optimal network behaviour across the training set.

Network learning can generally be executed in ***supervised mode*** (Hopfield model) or in ***unsupervised mode*** (Kohonen model). For supervised learning the network is provided by data examples that include the desired output. For unsupervised learning the desired output values are not required because the network finds the adequate output values itself.

The objective of training is to find the set of most suitable values of interconnecting weights through their tuning during the network training. By doing so, the network should still attain the highest ***generalization attribute***. This, however, can be aggravated if, instead of the global minimum, only a local minimum has been found. So, particular precautions should be provided to avoid pitting into one of the local minima. Such and similar issues seriously affect the training success, so that some careful considerations are required when preparing the ***experiment design*** for network training. This includes some decisions to be made concerning the network initialization for training, selection of the appropriate training algorithm, monitoring the training process using an appropriate performance index, formulation of training stopping criteria, *etc*.

Network initialization is a decision that is to be made before the weights tuning process starts. This is a difficult decision, because the training speed and the total training time required are strongly influenced by this decision. To circumvent this, various suggestions have been made, the most popular being that, in order to prevent neuron saturation and other unpleasant phenomena, some small, randomly distributed parameter values should initially be taken. However, setting all weights initially at the same small value should be avoided because it could possibly hamper the tuning process to start and/or to learn. This definitely does not hold for unsupervised training, like it holds for training of a Kohonen layer of a counterpropagation network, where the competition process take place. Here, the unique value $1/\sqrt{N}$ is initially taken for all weights, N being the number of network inputs. This is required because by starting the competition process it is advantageous that all competitors have the same initial parameter values for every training run.

Hebb (1949) has proposed the simplest training algorithm for neural networks, known as the ***Hebb learning rule***. A neurophysiologist himself, he enunciated the learning principle of natural neurons: if two interconnected neurons at the same time fire, then the strength (weight) of the synapse connecting them increases. Extended to artificial neural networks, this principle states that the common weight

w_{ij} connecting the output of the perceptron i and the input of the perceptron j will increase by an amount

$$\Delta w_{ij} = \eta x_j y_i ,$$

where x_j is the output of the perceptron j, y_i the output of the perceptron i, and η is a measure controlling the learning step size (Figure 3.14). Accordingly, the Hebbian learning updating the weights, or the ***Hebbian learning rule***, can be expressed as

$$w_{ij}(t+1) = w_{ij}(t) + \eta x_j(t) y_i(t) .$$

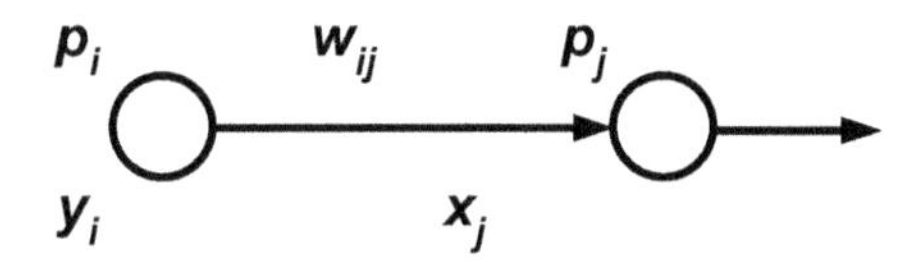

Figure 3.14. Interconnected perceptrons

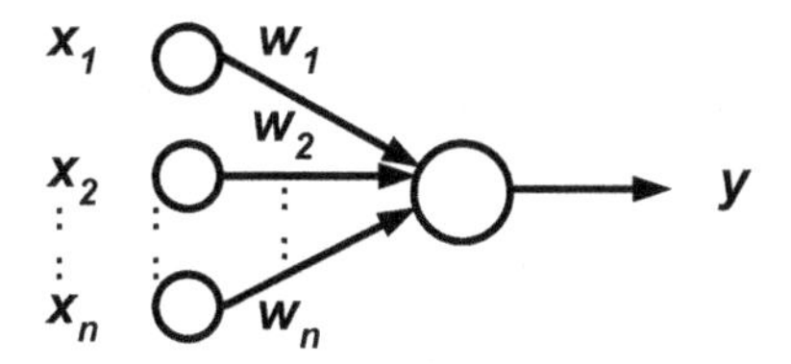

Figure 3.15. Multiple interconnected perceptron

The rule can be generalized and applied to a multiple-input perceptron as

$$w(t+1) = w(t) + \eta x^T wx ,$$

where the relation

$$y = \sum_{j=1}^{n} w_j x_j = w^T x = x^T w$$

is taken into account (Figure 3.15).

Nevertheless, the direct application of the Hebbian rule bears the risk of an endless increase of weight values, which could saturate the output neurons. As a remedy, an increase in the normalization of weights at every iteration step is necessary. Oja (1982) proposed using for this the normalization relationship

$$w_i(t+1) = \frac{w_i(t) + \eta x\ (t)y(t)}{\sqrt{\sum_i [w_i(t) + \eta x_i(t)y(t)]^2}},$$

derived through modification of the Hebbian rule itself. The modification normalizes the weight vector size to the value 1 by decreasing the values of all other weight vectors if one of its components increases, in this way keeping the total length of the vector constant.

The above rule modification can, for a small value of η and after power expansion, be approximated as

$$w_i(t+1) \cong w_i(t) + \eta y(t)[x_i - y(t)w_i(t)],$$

which is known as Oja's rule.

Yet, the fact that the application of the Hebbian rule is considerably limited to single-layer neural networks, the original version of the ***backpropagation algorithm*** is favoured for training of multilayer networks. The training is performed off-line in a supervisory learning mode, which is convenient because, in practice, a large number of data are available that have to be processed prior to their application for training. Besides, for forecasting purposes the pairs of related input and output data also have to be built and processed. Finally, the supervisory mode of learning facilitates the implementation of monitoring of training performance and the determination of the training stopping point.

When applying the backpropagation algorithm, which is a typical gradient steepest descent method, decisions have to be made concerning the

- ***learning rate***, *i.e.* the step size or the magnitude of weight updating
- ***momentum***, which is required for escaping the trapping in local minima.

An appropriate selection of learning rate is particularly important because the steepest descent method suffers from slow convergence and weak robustness. Convergence acceleration by taking a larger learning rate bears the danger of network oscillatory behaviour around the minimum. To avoid this, and still to take a larger learning rate, addition of a momentum parameter was recommended (Rumelhart *et al.*, 1986). By doing this, the original learning step according to the delta rule

$$w(t+1) = w(t) + \eta \varepsilon_p(t)x_p(t)$$

is extended by the momentum term to result in

$$w_{ij}(t+1) = w_{ij}(t) + \eta \delta_i(t)x_j(t) + \alpha[w_{ij}(t) - w_{ij}(t-1)],$$

where α is the ***momentum constant***, with the value $0.5 < \alpha < 0.9$. The added term represents the memorized value of the last increment so that the next weight change keeps approximately the same direction as the last one. This stabilizes the learning convergence.

An alternative way for speeding up and stabilizing the convergence was found in adaptive step size implementation. Silva and Almeida (1990) recommend the following weight update strategy

$$w_{ij}(t) = w_{ij}(t-1) + \eta_{ij}(t)\nabla_{ij}C(t),$$

where $\nabla_{ij}(t)C(t)$ are the gradient components of individual iteration steps

$$\nabla_{ij}C(t) = \sum_{v=1}^{N} \frac{\partial J(\theta_v)}{\partial w_{ij}},$$

with N as the number of training set samples. In the above updating relation, $\eta_{ij}(t)$ is taken as

$$\eta_{ij}(t) = c_1\eta_{ij}(t-1) \quad \text{if} \quad \nabla_{ij}C(t)\nabla_{ij}C(t-1) > 0$$

$$\eta_{ij} = \frac{1}{c_1}\eta_{ij}(t-1) \quad \text{if} \quad \nabla_{ij}C(t)\nabla_{ij}C(t-1) < 0,$$

where c_1 is a positive constant.

To circumvent the problem of avoiding the numerous flat and steep regions of the error surface Yu *et al.* (1995) advocated the ***dynamic learning rate*** to be imbedded into the backpropagation algorithm, based on information delivered by the first and the second derivatives of the objective function with respect to the learning rate. The clue to the proposed strategy is that it avoids the calculation of the values of the second derivative in weight space, using the information collected from the training instead. To bypass the calculation of the pseudo-inverse Hessian matrix that is inherent in second-order optimization methods, the conjugate gradient method is used.

The overwhelming number of upgraded learning algorithms are mainly focused on learning velocity increase and search stability improvement by adding a term containing the derivatives in weight space. But, some improvements of both objectives, namely of learning velocity and of convergence stabilization, are also achievable by manipulating the parameters of the neuron transfer function. Such an updating proposal was made for supervised pattern learning that adaptively manipulates the learning rate by updating neuron internal nonlinearity (Zhou *et al.*, 1991). Using some simulated data sets, it was shown that the updating law proposed increases the learning speed and is very suitable for identification of nonlinear dynamic systems.

3.5.4 Training, Stopping and Evaluation

Originally, the simple principle was accepted that the network should be trained until it has learnt it's task. This is certainly difficult to find out, because there is no direct approach how to do this. The general statement that a high enough number of iterations, or training steps, is good enough, in the sense that the network has learnt well enough to be a qualified expert in a specific domain, say in forecasting, does not hold. Thus far, at least theoretically, reaching the global minimum of the objective function is accepted as the ***training efficiency merit***, so that by approaching this minimum the error function will steadily decrease until the minimum has been reached. Finding out that there is no further decrease of the error function would then be an indication to stop the training process.

In practice, to find the global minimum, network training can require a number of repeated training trials with various initial weight values. After each training run the training results have to be evaluated and compared with the results achieved in the previous runs, this in order to select the best run. Some researchers have here centred their attention on the problem of *a priori* determination of a maximum number of training runs required for the training. Iyer and Rhinehart (2000) have developed an analytical procedure for determining the desirable lower number of training runs, sufficient - within a certain level of confidence - that the best one is within them. The procedure is based on the ***weakest-link-in-the-chain analysis*** described by Bethea and Rhinehart (1991).

The authors use the cumulative distribution function for the weakest link in a set of N training, with runs starting with the random initial weight values

$$F_w(a) = 1 - [1 - F_x(a)]^N .$$

This, rearranged as

$$F_x(a) = 1 - [1 - F_w(a)]^{\frac{1}{N}} ,$$

represents the probability that any single optimization has an error value $x \leq a$. The two relations, simultaneously taken, define the required number of random starts as

$$N = \frac{\ln[1 - F_w(a)]}{\ln[1 - F_x(a)]} .$$

For example, if, at the confidence of 99% level, the best of random starts should result in one of the best 20% values for the sum of squared errors, then the required number of random starts will be

$$N = \frac{\ln(1 - 0.99)}{\ln(1 - 20)} \cong 20 .$$

A more recent approach to solving the problems of appropriate training termination departs from some ***stopping*** criteria. For instance, based on the automated stopping criterion of Natarajan and Rhinehart (1997), Iyer and Rhinehart (2000) take as the stopping criterion the ***performance-to-cost ratio*** of the network. Assuming that the entire cost of a validation set consisting of N_v data points is $C_v = CN_v$, where C is the cost of single data points, and assuming that the cost of training and test data sets are CN_t and CN_c respectively, then the corresponding performance-to-cost ratio is

$$\rho = \frac{1}{E_{ce}C(N_t + N_c + N_v)},$$

where E_{ce} is the cumulative error on the test set for a trained network. Setting this result in relation to the total costs for training termination has reached the minimum RMS error without the validation cost will become

$$\sigma = \frac{1}{\mathrm{E}C(N_T + N_C)},$$

so their ratio

$$\xi = \frac{\rho}{\sigma} = \vartheta \frac{N_T + N_C}{N_t + N_c + N_v},$$

with

$$\vartheta = \frac{\mathrm{E}}{E_{ce}}.$$

However, even when using the predetermined number of training steps, there will generally be no guarantee that the network parameters will be adequately tuned. The optimal stopping strategy is to stop training after the network has learnt all about the problem class it has to solve. This happens when the training stopping is effected at the point where the network has reached the maximal ***generalization***. For the practising expert, this means that the stopping should be triggered exactly at the point where the network output error has reached its minimal value, This is known as ***early stopping***. If the training is continued beyond this point, then the result could be the ***network overtraining*** or ***network overfitting.***

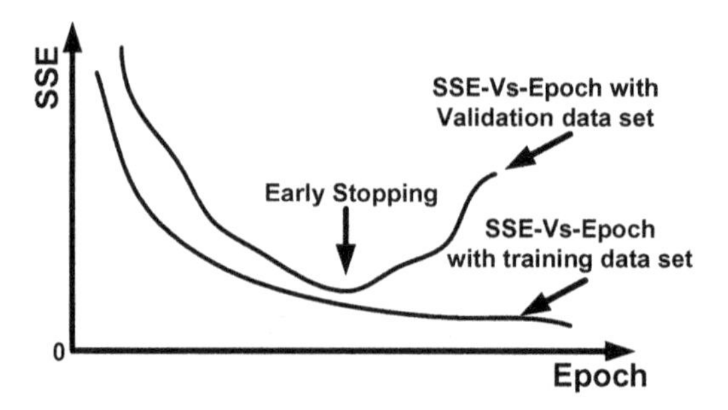

Figure 3.16. Early stopping of training

But still, the dilemma remains: in order to stop the training process, how do we realize that the network has learnt all the required knowledge from the training data and has reached its maximum generalization? Then, from learning theory we know that after reaching the point of maximum generalization, the network – although learning more and more from the training set - will start impairing the related test set performance (Figure 3.16) due to its overtraining (Vapnik, 1995). To prevent this, the method of ***early stopping with cross-validation*** has been suggested by Prechelt (1998).

Cross-validation is a traditional statistical procedure for random partitioning of collected data into a ***training set*** and a ***test set***, and for further partitioning of the training set into the ***estimation set*** and the ***validation set***. It is obvious that, if only a restricted data set is available, the partition of the entire set reduces the size of the training set. This, again, makes the location of the early stopping point difficult. For managing this problem, a predicate or a ***stopping criterion*** should be found that can indicate when to stop the training.

Prechelt (1998), using the error function (or the objective function) E, training error E_{tr} (as the average error per example across the training set), and the test and validation errors E_t and E_{v} respectively, has defined three possible stopping criteria:

- Stop as soon as the generalization loss exceeds a threshold value ε, *i.e.* when $g_{loss}(t) > \varepsilon$, where the error function $g_{loss}(t)$ is based on the lowest validation set error E_{opt} and the validation error E_{v}.
- Stop as soon as the quotient

$$\frac{g_{loss}(t)}{P_{tr}(t)} > \varepsilon,$$

where $P_{tr}(t)$ is the ***training progress*** defined by

$$P_{\mathrm{tr}}(t) = 1000\frac{\sum_{t'}^{t} E_{\mathrm{tr}}(t')}{k \min_{t'}^{t} E_{\mathrm{tr}}(t')},$$

with $t' = t - k + 1$, and the **training strip length** k.

- Stop when the generalization error increased in v successive strips.

Prechelt (1998), in order to interrogate the validity of the criteria, conducted 1296 training runs, producing 18144 stopping criteria. In the experiments, 270 of the records from 125 different runs reached automatically the 3000 epoch limit without using stopping criteria.

We will now consider the problem of ***network overtraining*** or ***network overfitting*** in more detail. Both the problem of overfitting and the opposite problem of ***underfitting*** arise as a consequence of improper training stopping. Therefore, both of them should be prevented because each of them lowers the generalization capability of the trained network. For example, if a network to be trained is less complex than the task to be learnt, then the network - after being trained - can suffer from underfitting and can, therefore, poorly identify the features within a large training data set. On the contrary, a too complex network can, after being trained, suffer from overfitting and can, therefore, extract the features within the training set along with the superposed noise. As a consequence, a complex network can produce predictions that are not acceptable.

Network complexity is primarily related to the number of weights. The term is used in connection with the model selection for prediction in the sense that the prediction accuracy of a network determines its complexity. This is the starting point of network model selection: how many and of what size of weights (and how many hidden units) should the model have in order to implement the wanted prediction accuracy without (or at least with a low) overfitting?

From the statistical point of view, the underfitting and overfitting are related to the statistical ***bias*** and the statistical ***variance*** they produce. They strongly influence the generalization capability of the trained network as follows:

- the ***statistical bias*** is related to the degree of target function fitting and restricts the network complexity, but does not care about the trained network generalization
- ***statistical variance***, which is the deviation of network learning efficiency within the set of training data, cares about the generalisation of the trained network.

For instance, underfitting produces a very high bias at network outputs, whereas overfitting produces a large variance. The difficulty of their simultaneous reduction or their balancing in the process of learning, which is essential for achieving the highest possible degree of generalization, is known as the ***bias-variance dilemma***. The dilemma is to be understood as follows: the bias of a neural network with a high fitting performance across the given training set of data is very low, but its variance is very high. By reducing the variance the network data fitting performance of the network will decrease. As a consequence, a trade-off between the low bias and the low variance is necessary, as demonstrated in Figure 3.17 on the example of ***polynomial curve fitting*** of a set of given data points.

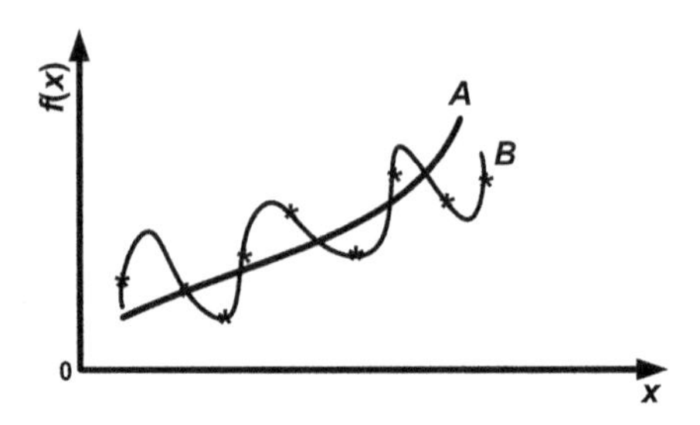

Figure 3.17. Polynomial curve fitting of data

A polynomial of degree n can exactly fit a set of $(n + 1)$ data points, say training samples. If the degree of the polynomial is lower, then the fitting will not be exact because the polynomial (as a regression curve A) cannot pass through all data points (Figure 3.17). The fitting will be erroneous and will suffer from ***bias error***, formulated as the minimized value of the mean square error. In the opposite case, if the degree of the polynomial is higher than the degree required for exact fitting of the given training data set, the excess number of it's degrees will lead to oscillations because of missing constraints (curve B in Figure 3.17). The polynomial approximation will, therefore, suffer from ***variance error***. Consequently, a polynomial of the optimal degree should be chosen for data fitting that will provide a low bias error as well as a low variance error, in order to resolve the ***bias-variance dilemma.***

Translated in terms of neural network training, polynomial fitting is seen as an optimal nonlinear regression problem (German et *al.*, 1992). This means that, in order to fit a given data set optimally using neural network, we need a corresponding model implemented as a structured neural network with a number of interconnected neurons in hidden layer. If the size of the selected network (or the order of its model) is too low, then the network will not be able to fit the data optimally and the data fitting will be accompanied by a bias error that will gradually decrease with increasing network size until it reaches its minimal value. Increasing the network size beyond this point, the network will also start learning the noise present in the training data, because there will be more internal parameters than are required to fit the given data. With this, also the variance error of the network will increase. The cross-point of the bias and the variance error curve will guarantee the lowest bias error and the lowest variance error for fitting the given data set. The corresponding network size (*i.e.* the corresponding number of neurons) will solve the given data fitting problem optimally. At this point the network training should be stopped, which is known as ***early stopping*** or ***stopping with cross-validation***. The network trained in this way will guarantee the ***best generalization***.

For probabilistic consideration of polynomial fitting, the expected value of the minimum square error across the set of training data

$$MSE_D = E_D\{[p(x) - f(x)]^2\}$$

is taken, where the training points are represented by the function $f(x)$ and the fitting polynomial or the actual network output by $p(x)$. Expanding the MSE_D formally as

$$MSE_D = E_D\{[p(x) - E_D\{p(x)\} + E_D\{p(x)\} - f(x)]^2\}$$

and rearranging its expansion as

$$MSE_D = E_D\{[p(x) - E_D\{p(x)\}]^2\} + E_D\{E_D\{p(x)\} - f(x)]^2\},$$

one gets the sum of the statistical variance

$$VAR_D = E_D\{[p(x) - E_D\{p(x)\}]^2\}$$

and the statistical bias

$$BIAS_D = E_D\{E_D\{p(x)\} - f(x)]^2\}.$$

In summary, the optimal network size is essential for optimal problem solving because a relatively small network will not be able to fit the given data accurately and thus will not be able to learn the most important features incorporated in the data. For this reason, the network size should be increased. On the other hand, because a large-sized network tends to learn not only the characteristic features of the given data, but also the accompanying noise and other non-relevant components' idiosyncrasies hidden in the data, its size should be reduced. In both cases, a network size reduction and/or an increase in optimal network size should be found that ensures the optimal network performance. In practice, this is usually achieved by balanced ***network growing*** and/or by ***network pruning***.

Network growing is a process of successive addition of new neurons and their related interconnections to the initial small-sized network until the optimal network performance is reached. This is a common way of designing optimal-sized radial basis function networks.

Network pruning, again, is a process of successive elimination of less relevant interconnections between the neurons within the large-sized network until the further elimination essentially worsens the network performance. A survey of algorithms to be used for network pruning was given by Reed (1993), who distinguished two major pruning methods:

- ***sensitivity calculation methods***, based on the sensitivity of the error function of the trained network with respect to the removal of individual weight connections as the indication of their pruning
- ***penalty term methods***, based on modification of the error function of a trained network by a penalty term.

Mozer and Smolensky (1988) used ρ as a measure of relevancy, defined as the difference between the error after removing a unit and the error before removing a unit. Karinin (1990), however, considers the error sensitivity with respect to removal of individual connections and removes the low-sensitivity connections. Le Cun *et al.* (1990), again, proposed the ***optimal brain damage*** procedure under the condition that the Hessian matrix H is diagonal and estimated the ***saliency of the weights*** and the second derivative of the error with respect to the weights. Hassibi *et al.* (1992) removed the diagonallity restriction of the Hessian matrix and considered the general case of an arbitrary form of Hessian matrix, which they termed the ***optimal brain surgeon***. Both approaches are based on consideration of sensitivity of weights perturbation on the error function E using the Taylor series

$$\delta E = \left(\frac{\partial E}{\partial w}\right)\delta w + \frac{1}{2}\delta w^T H \delta w + \left(\|\delta w\|^3\right),$$

where

$$\delta E = E(w + \delta w)$$

and

$$H = \left(\frac{\partial^2 E}{\partial w^2}\right)$$

is the corresponding Hessian matrix.

Now, knowing that for a network trained to the local minimum in error, the partial derivative

$$\frac{\partial E}{\partial w} = 0$$

holds. Neglecting all higher order terms in the corresponding Taylor series and eliminating a specific weight, say w_{ij}, measures should be undertaken to minimize the increase in error δE, taking into account the condition of weight elimination as given by

$$\delta w_{ij} + w_{ij} = 0\,.$$

The condition of weight elimination in vectorial form is given by

$$e_{ij}^T \delta w + w_{ij} = 0\,,$$

where e_{ij}^T is the unit vector in the weight space and w_{ij} is the weight connecting the *i*th input of the *j*th hidden unit.

To solve the minimization problem, we form the corresponding Lagrangian

$$L = \frac{1}{2}\delta w^T H \delta w + \lambda(e_{ij}^T \delta w + w_{ij}),$$

where λ is the Lagrange multiplier. The derivative of the Lagrangian with respect to δw and the equation

$$e_{ij}^T \delta w + w_{ij} = 0,$$

define the optimal weight change

$$\delta w = \frac{w_{ij}}{[H^{-1}]_{ij}} H^{-1} e_{ij}.$$

Correspondingly, the related optimal value of Lagrangian *L* for the weight w_{ij} is

$$L_{ij} = \frac{1}{2}\frac{w_{ij}^2}{[H^{-1}]_{ij}},$$

where $[H^{-1}]_{ij}$ is the *i*th element of the inverse Hessian matrix *H*. The L_{ij} value of the Lagrangian determined in this way represents the increase of mean square error caused by the removal of the weight w_{ij}, known as ***saliency of the weight*** w_{ij}. It is obvious that, because the saliency depends on the square value of w_{ij}, the small values of weights have a low influence on the mean square error. However, because the saliency is inversely proportional to $[H^{-1}]_{ij}$, small values of $[H^{-1}]_{ij}$ can also have a strong influence on the mean square error.

Although ***pruning*** methods, such as ***optimal brain damage***, and ***optimal brain surgeon***, rely on the weight ranking with respect to saliency, *i.e.* on changes in training error caused by pruning an individual weight, there is still an essential difference between them: the optimal brain damage procedure does not require retraining of the network after removing a weight element, whereas the optimal brain surgeon procedure requires this.

The disadvantage of both methods is that, if no ***stopping criterion*** is built, the removal of the least significant weights can lead to network overfitting. As an efficient stopping criterion, the calculation of the test error using Akaike's (1970) ***final prediction error*** (FPE) estimation and its modification is used to cover the estimation of average generalization error in regularized networks (Moody, 1991).

In practice, to apply the above procedures, the second derivative (Buntine and Weigend, 1994) of the inverse of Hessian matrix (Hassibi *et al.*,1992) has to be calculated anew for every weight to be eliminated. Stahlberger and Riedmiller (1996) proposed a fast network pruning method, called Uni-OBS, that still relies on the optimal brain surgeon procedure but it requires only a single calculation of the inverse Hessian matrix to eliminate a group of weights. This certainly simplifies the calculation of net pruning. For accelerated calculations of matrix multiplication, some fast computational algorithms are required or some algebraic transformations that also accelerate the calculation process. An amendment of the Uni-OBS method, called G-OBS (***generalised optimal brain surgeon***), can simultaneously eliminate, say m, weights in one step with slight increase in error given as

$$\delta E = \frac{1}{2}\delta w^T H \delta w,$$

The related elimination condition is given by

$$(w+\delta w)^T S_m,$$

S_m being the selection matrix that determines the m weights to be removed simultaneously. Using the above weights elimination conditions and the corresponding Lagrange method, we get for the resulting error the relation

$$\delta w = -H^{-1}S(S^T HS)^{-1}S^T w$$

and

$$\delta E = \frac{1}{2}w^T S(S^T HS)^{-1}S^T w.$$

For acceleration of the pruning process, Levin *et al.* (1994) proposed a method for elimination of excess weights.

Another way was followed by Jollife (1986). To improve the network generalization capability, he used the method of ***principal component analysis***. This is a valuable mathematical tool for reducing a system's dimensionality by eliminating it's redundant variables. This method transforms the variables to a basis in which the system covariance is diagonal and the projection is in the low variance directions. To detect the variables that have a low significant influence on the error function, a ***salience measure*** is used, which demonstrates the relationships between the proposed methods and the optimal damage and optimal surgeon procedures of network pruning. The pruning consists in removing the ***eigen-nodes*** with low saliency to reduce the effective number of network parameters. In contrast to the optimal brain damage and optimal brain surgeon procedures, which reduce the rank by eliminating actual weights, the proposed method reduces the rank of weights in each layer by deletion of the smallest salient eigen-nodes. Finally, the proposed method does not require network training.

A network pruning approach is preferably used in designing networks with a high ***generalization capability***, *i.e.* networks that are not only good enough to solve the prediction or classification problems present in the training set, but also some similar problems using some fresh, never seen and not previously known training sets of data. This is achieved through a trade-off between the intention that the trained network should be capable of learning a broad spectrum of similar problem categories, which would require a large-sized network, and the requirement that the network should be as simple as possible, in order to avoid the ***overtraining***.

In practical application of a trained network, there is a fundamental recommendation, *i.e.* where several trained networks have approximately the same final performances, the structurally simplest network should be selected as the best generalized one. This recommendation reflects ***Occam's razor philosophy***, which recommends that a scientific model should favour simplicity.

Many training strategies have been interrogated for network simplification at lower training cost. Such strategies have been discovered within the framework of minimization of the error function extended by a penalty term. To this category of strategies belong:

- the ***weight decay*** approach (Hinton, 1989), a subset of regularization approaches based on minimization of the weight tuning rule augmented by a complexity penalty term

  $$\Delta w_{ij}(t+1) = \pi\delta_i x_j - \lambda w_{ij}$$

  that penalizes the large weight values.

- the ***weight elimination*** approach (Weigend *et al.*, 1991), based on minimization of network training cost function to which a term is added that accounts for the number of parameters:

  $$\Delta w_{ij}(t+1) = \eta\delta_i x_i - \lambda \frac{w_{ij}(t)}{[1+w^2{}_{ij}(t)]^2},$$

  where λ represents the weight decay constant, δ_i is the local error, x_j is the local activation, and η is the learning rate.

In contrast to weight decay, which shrinks large values of weights more than small ones, the weight elimination shrinks predominantly the small weight values and is to a certain degree similar to the pruning process. Hansen and Rasmussen (1994) have demonstrated that network pruning may result when the weight decay parameter is determined by data. The added term punishes the large weight values and forces them to obtain small absolute values and simultaneously retains the other values unchanged. This, however, is favourable in preventing worsening of the network generalization capability. Therefore, care should be taken in selecting the decay constant λ, because an inappropriate value can deteriorate the generalization capability of the weight decay process. As a remedy, Weigend *et al.* (1991) recommend updating the λ value on-line during the network training in iterative steps.

Adding the penalty function in the weight decay and optimizing the augmented performance index corresponds to the ***regularization method*** in which the penalty term is added to the cost function to act as a restriction to the subsequent optimization problem. In approximation theory, the added term penalizes the curvature of the original solution, seeking for a smoother solution of the optimization problem.

The regularization method is generally used to solve ***ill-posed problems***. In the theory of learning, the problems of learning smooth mappings from examples are mostly ill-posed problems. For their solution Tikhonov (1963) proposed optimization of the cost function I extended by a term J, which also represents a cost function. Thus, the resulting cost function to be optimized becomes

$$I_{res} = I + \lambda J,$$

where λ represents the ***regularization parameter***, which determines the ***degree of regularization*** in the sense of balancing the ***degree of smoothness*** of the solution and its ***closeness*** to the training data. The regularization helps in stabilizing the solution of the ill-posed problem because the added term, representing the penalty to the original optimization problem, smoothens the cost function (Morozov, 1984).

The regularization approach determines the so-called ***Tikhonov functional***

$$I_{res}(f) = \sum_{i=1}^{n}(y_i - f(x_i))^2 + \lambda\|Pf\|^2,$$

the first term of which represents the closeness to the data, and in the second term f is the input-output function, P is a linear differential constraint operator, and $\|*\|^2$ is a norm on the function space to which Pf belongs. This operator also embodies the *a priori* knowledge about the problem solution.

To solve the regularization problem we proceed with the minimization of extended cost function I_{res}, using the resulting partial derivatives with respect to f in order to build the Euler-Lagrange equation

$$\hat{P}Pf(x) = \frac{1}{\lambda}\sum_{i=1}^{n}(y_i - f(x))\delta(x - x_i),$$

in which the operator P and its adjoint operator $\hat{P}$ build the differential operator $\hat{P}P$. Therefore, the above Euler-Lagrange equation is a partial difference equation. Its solution can, therefore, be expressed as the integral transformation of the right-hand side of the equation, with the kernel defined by Green's function of the differential operator $\hat{P}P$

$$\hat{P}PG(x,x_i) = \delta(x - x_i) \,.$$

Bearing in mind the definition of Green's function and taking into account the presence of the delta function on the right-hand side of the equation, the integral transformation will generate a discrete sum of terms, so that the function f can be defined as

$$f(x) = \frac{1}{\lambda}\sum_{i=1}^{n}(y_i - f(x_i))G(x,x_i) \,,$$

where $G(x,x_i)$ is ***Green's function*** centred at x_i. The last equation represents the solution of the regularization problem as a linear combination of n Green's functions with the expansion centre x_i and expansion coefficients $(y_i - f(x_i))$. Consequently, the solution of the regularization problem lies in the n-dimensional subspace of the space of smooth functions, with the n Green's functions as its basis (Poggio and Girosi, 1990). Furthermore, the basis function depends on stabilizer P, that represents the *a priori* knowledge of the problem domain as a kind of constraint.

Introducing the definition of the expansion weights as

$$w_i = \frac{y_i - f(x_i)}{\lambda} \,,$$

the above solution equation becomes

$$f(x) = \sum_{i=1}^{n} w_i G(x,x_i) \,.$$

Now, to determine the expansion weights w_i, the last two equations have to be written in matrix form as

$$w = \frac{1}{\lambda}(y - f)$$

and

$$f = Gw$$

which result in

$$(G+\lambda I)w = y.$$

Here, I represents the n-dimensional identity matrix and G is the corresponding **Green's matrix**

$$G = \begin{bmatrix} G(x_1,x_1) & G(x_1,x_2) \ldots G(x_1,x_n) \\ G(x_2,x_1) & G(x_2,x_1) \ldots G(x_2,x_n) \\ \ldots & \ldots \\ G(x_n,x_1) & G(x_n,x_2) \ldots G(x_n,x_n) \end{bmatrix},$$

which is a symmetric matrix with the property

$$G(x_i,x_j) = G(x_j,x_i)$$

because the identity matrix I is also symmetric.

From the solution equation

$$f(x) = \sum_{i=1}^{n} w_i G(x,x_i)$$

the corresponding ***regularization network*** (Figure 3.18) can be structured. The input layer of the network has an equivalent number of units to the dimension of the input vector, *i.e.* to the number of independent variables of the problem to be solved. The subsequent hidden layer, fully connected with the input layer with the fixed value weights, has the same number of nonlinear units as the number of data points and the activation function in the form of a Green's function with the output $G(x,x_i)$. It does not participate in the training process. Finally, the output layer, also fully connected to the hidden layer, contains one or more linear units with the weights w_i that correspond to the unknown coefficients of the above solution equation.

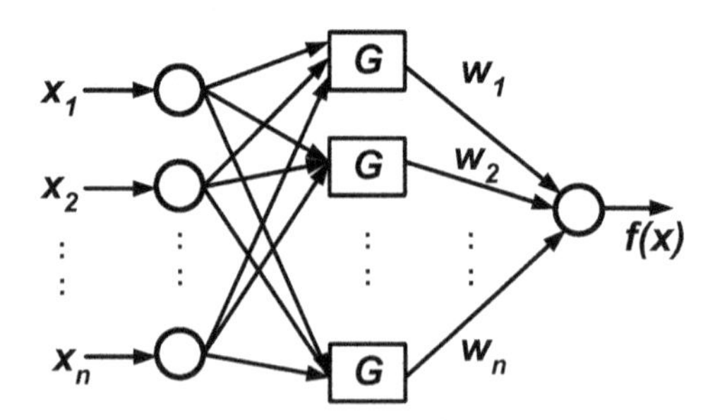

Figure 3.18. Regularization network

Obviously, the structure of the regularization network is mainly determined by the problem to be solved, with the exception of the weights between the input layer and the hidden layer, which are fixed. The main attributes of the network are:

- the regularization network is an optimal network because it minimizes the performance index that defines the proximity of the elaborated solution to the real solution defined by the training data
- the regularization network represents the ***best approximator*** (Girosi and Poggio, 1990) in the sense that for a given function there always exists a number of coefficients that approximate the given function better than any other set of coefficients and – by properly defining the stabilizer – guarantee that the regularization network has the desirable degree of smoothness
- the regularization network is a ***universal approximator*** that, given a sufficiently large number of hidden neurons, can approximate any continuous multivariate function arbitrarily well on a compact domain, a property that is based on the classical ***Weierstrass theorem***.
- when it is used for simplification of linear networks, particularly of basis function networks, this corresponds to the ***ridge regression method***.

The above objectives can, at least in principle, be reached by "extensive" network training. Although this might lead to network overfitting, this can be prevented by training stopping with cross-validation and by network structure reduction, for which various approaches have been suggested.

3.6 Forecasting Using Neural Networks

Unlike the traditional approaches to time series analysis and forecasting, neural networks need a reduced quantity of information to forecast the future time series data. Based on the available time series data, network internal parameters are tuned using an appropriate tuning algorithm. This can, if necessary, also include the modification of the initially chosen network architecture to better match the architecture required by the problem at hand. The related issues have been discussed extensively in this chapter, so that our attention will be focused on the comparison of the traditional approach to time series forecasting and on the approach using neural networks. This will be followed by pointing out the benefits of forecasting by merging both kinds of approaches and by building a nonlinear combination of forecasts. Finally, some issues related to the forecasting of multivariable time series using neural networks will be presented.

3.6.1 Neural Networks versus Traditional Forecasting

Comparison of forecasting performance of traditional statistical methods and of ***neuro forecasters*** has, since the early 1990s, attracted the attention of many researchers. Their reports have, however, been inconsistent because they were based on experimental investigations using various network configurations with various performance quality. Added to this came that the experiments used different time series data. For instance, forecasting collected linear data using nonlinear mapping of neural networks cannot give better results than the forecasting using linear statistical algorithms. In the reverse case, when dealing with considerably nonlinear time series data, forecasting using nonlinear neural networks could definitely deliver better results than the traditional algorithms. Consequently, when dealing with mixed linear/nonlinear time series data a combination of the traditional and the neural approach could be optimal.

Lapedes and Farber (1988) were the first to report that simple neural networks can outperform traditional methods by up to many orders of magnitude. This was radically investigated by Sharda and Patil (1990) on a set of 75 different time series with the objective to compare the forecasting accuracy of the Box-Jenkins method and of a neuro forecaster. Using a subset of 14 time series of Sharda and Patil, Tang *et al.* (1991) extended the comparative analysis to some additional aspects and identified a number of facts that make neural networks or traditional approaches deliver better forecasting results. They found by experiments that, generally:

- for time series with long memory, both approaches deliver similar results
- for time series with short memory, neural networks outperform the traditional Box-Jenkins approach in some experiments by more than 100%
- for time series of various complexity, the optimally tuned neural network topologies are of higher efficiency than the corresponding traditional algorithms.

As typical examples for experimental study

- international airline passenger data
- domestic car sales data in the US and
- foreign car sales data in the US

were used.

For experiments, the most typical traditional forecasting approach, the ARMA model of Box-Jenkins approach

$$\phi_p(B)\phi_p(B^L)(1-B^L)^D(1-B)^d y_t = \theta_q(B)\theta_Q(B^L)a_t + \delta$$

was used with the autoregressive operator ϕ, moving-average operator θ, and the back shift operator B. In the model equation, a_t, y_t, and δ represent the white noise, the time series data, and a constant value respectively.

To simplify matters, in all experiments with neuro forecasters, one-hidden-layer networks and networks without a hidden layer were used alternatively. The experimental results showed that hidden-layer networks have a better forecasting performance.

Hill *et al.* (1996) compared six traditional methods with the neuro forecaster on 111 different time series and found that neuro forecasters are significantly better than the statistical methods taken into consideration. However, Foster *et al.* (1992) came to the opposite conclusion. After extensive analysis of forecasting accuracy of neuro and traditional forecasters, they concluded that linear regression and the simple average of the exponential smoothing method are superior to a ***neuro forecaster***. Denton (1995), again, demonstrated that, under standard statistical conditions, there is only a slight difference in prediction accuracy between the regression models and neural models. Some additional results of comparative analysis have been communicated by Nelson *et al.* (1994), Gorr *et al.* (1994), Srinivasan *et al.* (1994), and Hann and Streurer (1996).

3.6.2. Combining Neural Networks and Traditional Approaches

Application of ***hybrid***, *i.e.* combined neural networks and traditional approaches, to time series forecasting was a challenging attempt to increase forecasting accuracy beyond the limits that either one of the two approaches used alone would be able to reach. In the following, we will consider the advantages of combining the neural and ARIMA model approach in time series forecasting. Voort *et al.* (1996) used for this combination the Kohonen self-organizing map as the neural network part for short-term traffic-flow forecasting. Sue *et al.* (1997) used this type of hybrid combination to forecast a time series of reliability data and showed that the hybrid model produced better forecasts than either the ARIMA model or the neural network by itself could produce. Tseng *et al.* (2002) investigated the combination of a seasonal time series model SARIMA and a backpropagation network, resulting in a SARIMABP hybrid combination. They found that the combination outperforms the SARIMA model used alone and the backpropagation model with the de-seasonalized or differentiated data.

For experimental purposes, the time series $z_i, i = 1, 2, 3, ..., k$, is generated by a SARIMA $(p, d, q)(P, D, Q)$ process with mean μ and modeled by

$$\varphi(B)\Phi(B^S)(1-B)^d(1-B^S)^D(z_t-\mu)=\theta(B)\Theta(B^S)a_t,$$

where S is the periodicity, d and D are the number of regular and seasonal differences respectively, B is the polynomial degree, and a_t is the estimated residual at time t. The experimental results show that the SARIMABP method benefits from the forecasting capability of the SARIMA and from the capability of backpropagation to reduce the residuals further, which guarantees a lower forecasting error. As forecasting accuracy evaluation criteria, the mean square error (MSE), mean absolute error (MAE), and mean absolute percentage error (MAPE) have been used.

For a real-life application example, time series data of the total production revenues of the Taiwanese machinery industry were taken for various periods of time. For instance, a five-year data set has been used as the input of the ARIMA $(0,1,1)(1,1,1)_{12}$ model

$$(1+0.309B^{12})(1-B)(1-B^{12})z_t=(1-0.7159B^{12})a_t$$

and for a three-year data set as the input of the ARIMA $(0,1,1)(0,1,0)_{12}$ model

$$(1-B)(1-B^{12})z_t = (1-0.88126B)a_t .$$

In both cases the experiments were carried out with two, three, and seven neurons in the network hidden layer.

Hybrid ARIMA-neural network methodology was also the subject of an experimental study by Zhang (2003), whose objective was to identify whether the given time series data were generated by a linear or a nonlinear process. This is essential for making a decision on whether, in a given case, the use of a linear (*i.e.* the traditional) or a nonlinear (*i.e.* a neural network) approach will be more appropriate. Here, the combined approach could ease the problem solution. After all, because real-world time series are seldom purely linear or nonlinear, it is favourable to use a hybrid approach.

In experimental practice, the assumption is made that a time series to be processed is composed of a linear autocorrelation structure L_t and a nonlinear component N_t :

$$z_t = L_t + N_t .$$

The linear component of the time series can be processed using an ARIMA model, and the residuals

$$e_t = z_t - \hat{L}_t ,$$

containing only the nonlinear relationships, can be processed by neural networks. This can be done using a residual model, *e.g.*

$$e_t = f(e_{t-1}, e_{t-2}, ..., e_{t-n}) + \varepsilon_t ,$$

which corresponds to a neural network with n input nodes and the nonlinearity function $f(.)$. In the above residual model, ε_t represents the random error. The benefits of the proposed hybrid methodology approach have been confirmed on three real-life examples from different application areas.

A remarkable contribution was reported by Wedding and Chios (1996), who combined the Box-Jenkins model and an RBF network.

3.6.3 Nonlinear Combination of Forecasts Using Neural Networks

Because a large number of time series forecasting methods are available, it makes sense for the application expert to select the best one among them in each particular case. Thus, it becomes interesting to combine a group of forecast methods and to examine the forecasting accuracy of the combination. The issue was discussed in Section 2.8.6 from the traditional point of view. It was shown that the best forecasting results are achievable when the combination of traditional forecasting methods is nonlinear. In the meantime, various combination techniques have been suggested and examined using different intelligent technologies, primarily with neural networks.

In engineering practice, choosing the "best" forecasting method means choosing a method that is the best in the given circumstances. For instance (McNees, 1985), experience has shown that no forecasting model retains its accuracy for all values of variables all the time. Also, it has been experimentally proven that if for a forecasting method the short run is good, then there is no guarantee that the long run will also be good. Therefore, it is worthwhile seeking for an adequate combination for each application situation. This is because the combination of methods incorporates different cognition capabilities and can, in a specific case, produce better forecasts than either of methods within the combination itself. Moreover, experimental investigations confirm (Winkler and Markridakis, 1983) that the resulting accuracy of combined forecasts increases with the increase in the number of forecasting methods involved. Mahmoud (1984) also came to a similar conclusion, that the accuracy of the combined forecast improves as more methods are included in the combination.

In forecasting non-stationary, non-seasonal time series one can evaluate the forecast values subsequently generated by a Box-Jenkins ARMA or ARIMA model, Holt-Winter's exponential smoothing, extrapolation of trend curve, Kalman filtering, *etc.* and mutually compare the results achieved. Out of the possible forecasting methods the analyst may prefer to use his own favourite methods that will produce different forecasts of a given time series. Moreover, using a particular method (say, ARMA/ARIMA) different analysts may come up with a different order of the models required for forecasting and, again, with different forecast results. Therefore, forecast models developed using different methods and by different analysts will rarely be identical. This may be very confusing to someone who wants to take a decision on the basis of various forecasts suggested by various analysts.

From the above, it follows that it is inadvisable to prefer one particular forecasting method over another, because no single forecasting method will in every situation produce forecasts of the same accuracy. Rather, it is more advisable to take a combination of a few forecasts generated by different methods. This was even clearly formulated by Bates and Granger (1969).

A number of advanced approaches have been suggested for nonlinear combination of forecasts using neural networks (Shi and Liu, 1993; Harald and Kamastra, 1997). The problem is defined here starting with the availability of k different forecasts f_1, f_2, f_3, ..., f_k, of some random variable z, that should be combined into a single forecast f_c. The straight away step would be to form a linear combination of forecasts

$$f_c(z) = \sum w_i f_i(z)$$

where w_i is the assigned weight of ith forecast f_i.

The simplest approach to determine the weights w_i of the combination would be to take equal weights for each term. This has proven to be relatively robust and accurate. But still, in practice, the linear combination of forecasts is not likely to be the optimal combination like the nonlinear combinations are. This can be demonstrated on the following example.

Suppose that k different forecast models are available and the ith individual forecast has an information set $\{I_i : I_c, I_i\}$, where I_c is the common part of information used by all k models and I_i is the specific information for the ith forecast only. Denoting the ith forecast by $f_i = F_i(I_i)$, we can express the linear combination of forecasts as

$$F_c = \Sigma w_i F_i(I_i),$$

where w_i is the weight of the ith forecast. On the other hand, every individual forecasting model can also be regarded as a subsystem for information processing, while the combination model $f_c = F_c(I_1, I_2, ..., I_k)$ is regarded as such a system. It follows that the integration of forecasts is more than their sum, *i.e.* the performance of the integrated system is more than the sum of its subsystems. So, the trustworthiness of the linear forecast combination is quite questionable. More trust should be paid to a nonlinear interrelation between the individual forecasts, such as

$$f_c = \psi[F_1(I_1), F_2(I_2), F_3(I_3), ..., F_k(I_k)],$$

where ψ is a nonlinear function. While the given information is processed by individual forecasting models, it is likely that parts of the entire information can be lost, which means that, say, the information set I_i is not being used efficiently. Furthermore, different forecasts may have different parts of information lost. This is why it is preferable that as many different forecasts as possible should be present in the combination, even when the individual forecasts depend on the same set of information.

As a forecasting example (Palit and Popovic, 2000), a 2-6-6-1 feedforward network, *i.e.* a network with two inputs, and two hidden layers with each layer containing six neurons and one output, is used, as shown in Figure 3.19b. The network is trained using the Levenberg-Marquardt algorithm, which guarantees much faster learning speed than the standard backpropagation method, and hence requires less training time. The algorithm also uses the gradient descent method, based on Jacobian matrix, according to which the update is

$$\Delta w = -\left[J^T(w)J(w) + \mu I \right]^{-1} J^T(w)e(x)$$

or

$$w(k+1) = w(k) + \Delta w(k)$$
$$w(k+1) = w(k) - \left[J^T(w)J(w) + \mu I \right]^{-1} J^T(w)e(w)$$

where $J(w)$ is the Jacobian matrix with respect to network-adjustable parameters w (all weights and the biases) of dimension $(q \times N_p)$, and q being the number of training sets, N_p being the number of adjustable parameters in the network, and I is the identity matrix of dimension $(N_p \times N_p)$.

Table 3.1. Nonlinear combination of two forecasts of a temperature series using an artificial neural network (ANN: Neural networks combined forecast; BJ: Box-Jenkins forecast, HW: Holt-Winters exponential smoothing)

Serial No.	Forecast	Data sets from HBXIO matrix	SSE	RMSE
1.	BJ	151 to 224 (column-1)	0.4516	0.112
2	HW	151 to 224 (column-2)	0.3174	0.0933
3	ANN (2-6-6-1)	1 to 150 (training)		
4	ANN (2-6-6-1)	151 to 224	0.1306	0.0594
5	ANN (2-2-6-1)	151 to 224	0.2425	0.0810

The parameter μ is multiplied by some factor μ_{inc} whenever an iteration step increases the network performance index (*i.e.* sum squared error) and it is divided by μ_{dec} whenever a step reduces the network performance index. Usually the factor $\mu_{inc} = \mu_{dec}$ and in our case it is selected as 10.

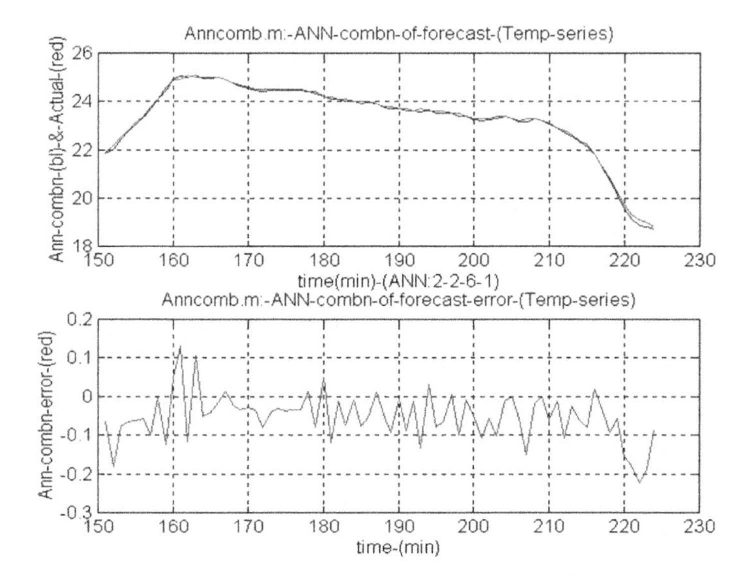

Figure 3.19(a). The combination of forecasts using a 2-2-6-1 artificial neural network

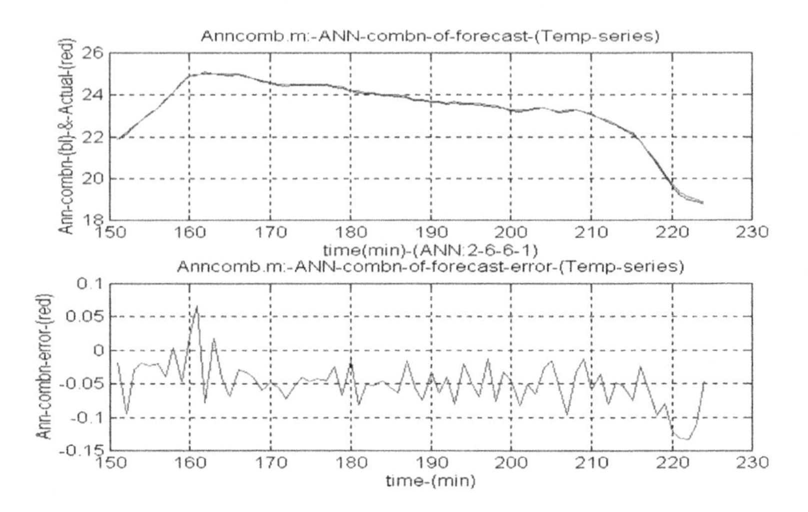

Figure 3.19(b). The combination of forecasts using a 2-6-6-1 artificial neural network

In our practical example, the first 150 input-output samples were used to train the network. Thereafter, the values of the interconnecting weights and biases are saved for network performance testing using the remaining 151 to 224 samples of data. From the experimental results shown in Figure 3.19(a) and Figure 3.19(b) and Table 3.1, it is obvious that the network output very closely matches the actual time series, indicating that a nonlinear combination of the forecasts is better than the individual forecasts.

3.6.4 Forecasting of Multivariate Time Series

Chakraborty *et al.* (1992) conducted experimental investigations on forecasting of multivariate time series using neural networks. They focused their attention on the statement that, in the case of substantial cross-correlation of individual variables of multivariable time series data, the forecasting accuracy of each variable can be improved when simultaneously changing the values of other variables within the time series is taken into account. This has been observed in ***multivariate statistical analysis*** when, based on observation data, identifying the interdependencies of variables involved in a multivariate system. To prove this, Chakraborty *et al.* (1992) analyzed the one-step and multistep prediction behaviour of a ***trivariate time series*** $x_t = [x_{1t}, x_{2t}, x_{3t}]$ in the interval of $t = 1–100$ samplings using

- ***separate modelling*** of each component of the multivariable time series, interpreted as mutually independent univariate time series
- ***combined modelling***, by simultaneous consideration of all three variables
- ***statistical modelling***, using the statistical model developed by Tiao and Tsay (1989).

The analysis of separate modelling was carried out using alternatively 2-2-1, 4-4-1, 6-6-1, and 8-8-1 networks and by evaluating the results for each time series component using the mean square error as the performance indicator. The analysis has shown that a combined modelling approach is superior to separate modelling, and that both of them are superior to statistical modelling. In addition, the experiments with the 2-2-1 backpropagation networks have delivered, in one-step and multistep cases, the best forecasting accuracy, which shows that the 4-4-1 and 6-6-1 networks are oversized for this purpose.

The experimental investigations presented above deliver forecasting results that depend considerably on the art of experiment design used for this purpose. For this reason the results are not coherent and are sensitive to the application field. We are still short of a general theoretical formulation of this phenomenon, but some encouraging trials have been made in this direction (reported by Yang, 2000), related to methods of combining forecasting procedures for forecasting continuous random univariate time series.

References

[1] Aizerman MA, Braverman EM, and Rozenoer LI (1964) Theoretical foundation of potential function method in pattern recognition. Automation and Remote Control 25: 917–936.

[2] Akaike H (1970) Statistical predictor identification, Annals of the Institute of Statistical Maths., 22: 202–217.

[3] Almeida LB (1987) A learning rule for asynchronous perceptrons with feedback in a combinatorial Environment. IEEE 1st International Conf. on Neural Networks, San Diego, CA II:609–618.

[4] Amari S and Maginu K (1988) Statistical neurodynamics of associative memory, Neural Networks 1: 63–73.

[5] Anders U and Korn O (1999) Model selection in neural networks. Neural Networks 12: 309–323.

[6] Bashkirov OA, Braverman EM, and Muchnik IB (1964) Potential function algorithms for pattern recognition learning machines. Automation and Remote Control 25:692–695.

[7] Bates JM and Granger CWJ (1969) The combination of forecasts, Operation Research Quart. 20: 451–461.

[8] Baum EB and Haussler D (1989) What Size Net Gives Valid Generalisation? Neural Computation 1:151–160.

[9] Bethea RM and Rhinehard RR (1991) Applied Engineering Statistics. Marcel Dekker, New York.

[10] Block HD (1962) The Perceptron: a model of brain functioning. Review of Modern Physics, 34:123–135.

[11] Broomhead DS and Lowe D (1988) Multivariable functional interpolation and adaptive networks. Complex Systems 2: 321–355.

[12] Butine WL and Weigend AS (1994) Computing Second Derivatives in Feedforward Networks: A Review. IEEE Trans. on Neural Networks 3: 480–488.

[13] Chakraborty K, Mehrotra K, Mohan ChK, Ranka S (1992) Forecasting the behavior of Multivariate Time Series Using Neural Networks. Neural Networks 5: 961–970.

[14] Cichocki A and Unbehauen R (1993) Neural Networks for Optimization and Signal Processing. Wiley, Chichester, West Sussex, UK.

[15] Cohen MA and Grossberg S (1983) Absolute Stability of global pattern formation and parallel memory storage by competitive neural networks. IEEE Trans. on Systems, Man, and Cybernetics 13: 815–826.
[16] Cybenko G (1989) Approximation by superpositions of a sigmoidal function. Mathematical Control Signals Systems 2:303–314.
[17] Denton JW (1995) How good are neural networks for causal forecasting? J. of Business Forecasting 14(2):17–20.
[18] Elman JL (1990) Finding structure in time. Cognitive Science 14: 179–211.
[19] Fogel DB (1991) An Information Criterion for Optimal Neural Network Selection. IEEE Trans. On Neural Networks 2: 490–497.
[20] Forster WR, Collopy F, Ungar LH (1992) Neural network forecasting of short, noisy time series. Computers and Chemical Engineering 16(2): 293–297.
[21] German SE, Bienenstock, and Doursat R (1992) Neural networks and the bias/variance dilemma. Neural Computation 1: 1–58.
[22] Girosi F and Poggio T (1989) Representation Properties of Networks: Kolmogorov's Theorem is Irrelevant. Neural Computation 1: 465–469.
[23] Girosi F and Poggio T (1990) Networks and the best approximation properties. Biological Cybernetics:169–176.
[24] Gorr WL, Nagin D, Szczypula J (1994) Comparative study of artificial neural network and statistical models predicting student grade point averages. Intl. J. of Forecasting 10: 17–34.
[25] Grossberg S (1988) Competitive Learning: From interactive activation to adaptive resonance, Neural Networks and Neural Intelligence, Grossberg S. (Eds.), MIT Press, Cambridge, MA.
[26] Hagan MT and Menhaj MB (1994) Training feedforward networks with the Marquardt algorithm, IEEE Trans. on Neural Networks, vol. 5(6): 989–993.
[27] Hann TH and Steurer E. (1996) Much ado about nothing? Exchange rate forecasting: Neural networks vs. linear using monthly and weekly data. Neurocomputing 10: 323–339.
[28] Hansen IK and Rasmussen CE (1994) Pruning from adaptive regularization. Neural Computation 6: 1223–1232.
[29] Harald PG and Kamastra M (1997) Evolving artificial neural networks to combine the financial forecasts, IEEE Trans. on Evolutionary Computation, vol. 1(1): 40–51.
[30] Hassibi B, Stork DG, and Wolff GJ (1992) Optimal brain surgeon and general network pruning. IEEE Intl Conf on Neural Networks, San Francisco 1:293–299.
[31] Haykin S (1994) Neural Networks: a comprehensive foundation. McMillan, USA
[32] Hebb DO (1949) The organisation of behaviour. Wiley, New York.
[33] Hecht-Nielsen R (1987a) Counterpropagation Networks. Applied Optics 26(23): 4979–4984.
[34] Hecht-Nielsen R (1987b) Kolmogorov's Mapping Neural Network Existence Theorem, IEEE Conf. On Neural Networks; San Diego, CA. III: 11–14.
[35] Hecht-Nielsen R (1988) Application of counterpropagation networks, Neural Networks 1: 131–139.
[36] Hertz J, Krogh A, and Palmer RG (1991) Introduction to theory of neural computation, Addison-Wesley, Reading, MA.
[37] Hill T, O'Connor M, Remus W. (1996) Neural network models for time series Models forecasts. Management Sciences 42(7): 1082–1092.
[38] Hinton GE (1989) Connectionist learning procedures, Artificial Intelligence, 40: 185–243.
[39] Hopfield JJ (1982) Neural Networks and physical systems with emergent collective computational abilities. Proc. of the Nat. Acad. of Sciences, USA, 79: 2554–2558.

[40] Hopfield JJ (1984) Neurons with graded response have collective computational properties like those of two-state neurons. Proc. of the Nat. Acad. of Sciences, USA 81: 3088–3092.

[41] Hornik K, Stinchcombe M, White H (1989) Multilayer feedforward networks are Universal approximators. Neural Networks 2(5): 359–366.

[42] Hu MJC (1964) Application of the ADALINE system to weather forecasting. Master Thesis, Technical Report 6775–1, Stanford El. Lab., Stanford, CA.

[43] Ishikawa M. and Moriyama T (1996) Prediction of time series by a structural learning of neural networks. Fuzzy Sets and Systems 82: 167–176.

[44] Iyer MS and Rhinehart RR (2000) A novel method

[45] Iyer MS and Rhinehart RR (2000) A Novel Method To Stop Neural Network Training. 2000 American Control Conference, paper WM17–3

[46] Johanson EM, Dowla EU, and Goodman DM (1990) Backpropagation learning for multi-layer feedforward neural networks using the conjugate gradient method, Report UCRL-JC–104850, Lawrence Livermore National Laboratory, CA.

[47] Jollife IT (1986) Principal Components Analysis. Springer-Verlag.

[48] Jordan M (1986) Attractor dynamics and parallelism in a connectionist sequential machine. Proc. of the Eight Annual Conference on Cognitive Science Society :532–546.

[49] Karnin ED (1990) A simple procedure for Pruning back-propagation trained neural networks. IEEE Trans on Neural Networks 2: 188–197.

[50] Khorasani K and Weng W (1994) Structure Adaptation in Feedforward Neural Networks. Proc. IEEE Internat. Conf. on Neural Networks, III: 1403–1408.

[51] Klimasauskas CC (1991) Applying Neural Networks. Part 3: Training a Neural Network. PC-AI, May/June: 20–24. B,

[52] Kohonen T (1989) Self-Organisation and Associative Memory. 3rd Edition, Springer, Berlin, NY.

[53] Kröse B and Smagt P (1996) An introduction to neural networks, The University of Amsterdam, Eighth edition, November, http://www.fwi.uva.nl/research/neuro.

[54] Kubat M (1998) Decision trees can initialise radial-basis-function networks. IEEE Trans. on Neural Networks. 9: 813–821.

[55] Kurita T (1990) A method to determine the number of hidden units of three-layered neural networks by information criteria, Trans. of Inst. of Electronics, Information and Commun. Engineers, J73-D-II–11: 1872–1878 (in Japanese).

[56] Lapedes A and Farber R (1988) Nonlinear signal processing using neural networks: Prediction and system modelling. Technical Report LA-UR-87-2662, Los Alamos National Laboratory, Los Alamos, NM.

[57] Le Cun Y, Denker JS, and Solla SA (1990) Optimal Brain Damage. In: Touretzky S (Ed.). Advances in Neural Information Processing Systems 2, San Mateo, CA, Morgan Kaufman.

[58] Levin AU, Leen TK, and Moody JE (1994) Fast pruning using principle components, In: Advances in Neural Information Processing Systems 6, Covan JD, Tesauro G and Alspector J, Editors: 35–42, Morgan Kaufman Publi. Inc., San Mateo, CA.

[59] Lippmann RP (1987) An introduction to computing with neural nets. IEEE ASSP Magazine (April): 4–22

[60] Mahmoud E (1984) Accuracy in forecasting: A survey. J. of Forecasting 3:139–159.

[61] McClelland JL and Rumelhart DE (1988) Exploration in Parallel Distributed Processing. Cambridge, MA, MIT Press.

[62] McCulloch WS, Pitts W, (1943) A logical Calculus of the ideas Immanent in nervous activity. Bulletin of Mathematical Biophysics 5:115–133.

[63] McNees SK (1985) Which forecast should you use. New England Economic Review, July/August: 36–42.

[64] Minsky ML and Papert S, (1969) Perceptrons. MIT Press, Cambridge MA.

[65] Moody JE (1991) Note on Generalization, Regularization and Architecture Selection in Nonlinear Systems, Proc. of the IEEE-SP Workshop : 1–10.

[66] Moody JE and Darken CJ (1989) Fast learning in networks of locally-tuned processing units. Neural Computation 1: 281–294.

[67] Morozov VA (1984) Methods for Solving Incorrectly Posed Problems. Springer-Verlag, Berlin.

[68] Mozer MC and Smolensky P (1990) Skeletonization: A technique for trimming the fat from a network via relevance assessment. In: Advances in Neural Information Processing 1, Touretzky DS (Ed.) : 107–115.

[69] Murata N, Yoshizawa S, and Amari S (1994) Network Information criterion – Determining the number of Hidden Units for an Artificial Neural model. IEEE Trans. On Neural Networks 6: 865–871.

[70] Natarajan S and Rhinehart RR (1997) Automated Stopping Criteria For Neural Network Training. Proc. of the 1997 American Control Conf., paper #TP09–4.

[71] Nelson M, Hill T, O'Connor M (1994) Can a neural network be applied to time series forecasting and learn seasonal patterns: An empirical investigation. Proc. of the 20th Annual Hawaii Intl. Conf on System Sciences: 649–655.

[72] Oja E (1982) A simplified neuron model as a principal component analyzer. Journal of Mathematical Biology 15: 267–273.

[73] Palit AK and Popovic D (2000) Nonlinear combination of forecasts using artificial neural network, fuzzy logic and neuro-fuzzy approaches, FUZZ-IEEE, 2: 566–571.

[74] Pineda FJ (1987) Generalisation of back-propagation to recurrent neural networks. Physical Review Letters 59: 2229–2232.

[75] Poggio T and Girosi F (1990) Networks for Approximation and Learning. Proc. IEEE 78:1481–1497.

[76] Powel MID (1988) Radial basis function approximation to polynomials, Numerical Analysis Proceedings, Dundee, U.K.: 223–241.

[77] Prechelt L (1998) Early Stopping – but when? In: Orr GB and Moeller K-R (Eds.), Neural Networks: Tricks of the Trade. Springer, Berlin: 55–69.

[78] Reed R (1993) Pruning Algorithms – A Survey. IEEE Trans. on Neural Networks 4: 740–747.

[79] Rosenblatt F, (1958) The Perceptron: A probabilistic model for information storage and organisation of the brain. Psych. Review 65: 386–408.

[80] Rumelhart DE and McClelland (1986) Parallel Distributed Processing: Explorations in the Microstructure of Cognition MIT Press, Cambridge, MA.

[81] Rumelhart DE, Hinton GE, and Williams RJ (1986) Learning internal representation by back-propagation errors. In: Rumelhart DE, McClelland JL, the PDP Research Group(Eds.), Parallel Distributed Processing: Explorations in the Microstructure of Cognition. MIT Press, MA.

[82] Sastry PS, Santharam G, and Unikrishnan KP (1994) Memory neuron networks for identification and control of dynamic systems. IEEE Trans. on Neural

[83] Schmidhuber J (1989) Accelerated learning in backpropagation net, In: Connectionism in Perspective, Elsevier, North Holland, Amsterdam, pp. 439–445.

[84] Sharda R and Patil RB (1990) Neural Networks as Forecasting Experts: An Empirical Test, Proc. of the IJCNN Meeting, Washington: 491–494.

[85] Shi S and Liu B (1993) Nonlinear combination of forecasts with neural networks. Proc. of Intl. Joint Conf. on Neural Networks '93 (IJCNN '93), Nagoya, Japan, 952–962.

[86] Silva FM and Almeida LB (1990) Speeding-up backpropagation, In: Advances of Neural Computers, Eds. Eckmiller R, Elsevier Science Publish. BV., North Holland, pp. 151–158.

[87] Specht DF (1988) Probabilistic neural networks for classification, or associative memory, Proc. of IEEE Intern. Conf. on Neural Networks, San Diego, 1: 525–532.
[88] Specht DF (1990) Probabilistic neural networks and the polynomial ADALINE as complementary techniques for classifications. IEEE Trans. on Neural Networks, 1: 111–121.
[89] Sprecher DA (1965) On the Structure of Continuous Functions of Several Variables. Trans. Amer. Math. Soc. 115:340–355.
[90] Srinivasan D, Liew AC, Chang CS (1994) A neural network short-term load forecaster. Electric Power Systems Research 28: 227–234.
[91] Stahlberger A and Riedmuller M (1996) Fast network pruning and feature extraction using the Unit-OBS algorithm. Advances in Neural Information Processing systems (NIPS'96), Denver.
[92] Stone M (1977) An asymptotic equivalence of choice of model by cross-validation and Akaike's criterion cross validation. J. of the Royal Statistical Soc. B36:44–47.
[93] Sue CT, Tong LI, and Leou CM (1997) Combination of time series and neural network for reliability forecasting modelling. J. Chin. Inst. Ind. Eng. 14(4): 419–429.
[94] Tang Z, Almeida de Ch, and Fishwick, PA (1991) Time series forecasting using neural networks vs. Box-Jenkins methodology. Simulation 57(5): 303–310.
[95] Tiao GC and Tsay RS (1989) Model specification in multivariate time series. J. of the Royal Statistical Society B 51: 157–213.
[96] Tikhonov AN (1963) On solving incorrectly posed problems and methods of regularisation. Docklady Akademii Nauk USSR 151: 501–504.
[97] Tseng F-M, Yu H-Ch, and Tzeng G-H (2002) Combining neural network model with seasonal time series ARIMA model. Technological Forecasting.
[98] Vapnik V (1995) The Nature of Statistical Learning Theory, Springer-Verlag, NY.
[99] Villiers de J and Bernard E (1992) Backpropagation Neural Nets with one and Two Hidden Layers. IEEE Trans. On Neural Networks : 136–141.
[100] Vogl TP, Mangis JK, Rigler AK, Zink WT and Allcon DL (1988) Accelerating the convergence of backpropagation method, Biological Cybernetics, vol. 59: 257–263.
[101] Voort VD, Dougherty M, and Watson M. (1996) Combining Kohonen Maps with ARIMA time series models to forecast traffic flow. Transp. Res. Circ. (Emerg. Technol.) 4C(5): 307–318.
[102] Wedding II DK and Cios KJ (1996) Time series forecasting by combining RBF networks certainty factors, and the Box-Jenkins model. Neurocomputing 10: 149–168.
[103] Weigend AS, Rumelhart DE, and Huberman BA (1991) Generalisation by weight-elimination with application to forecasting. Adv. In Neural Information Processing Systems, Morgan Kaufmann, San Mateo, CA 3: 875–882.
[104] Werbos P (1990) Backpropagation through time what it does and how to do it, Proc. of IEEE, 78(10):1550–1560.
[105] Werbos PJ (1974) Beyond Regression: New Tool for Prediction and analysis in the Behavioural sciences. Ph.D. Thesis, Harvard University, Cambridge, MA.
[106] Werbos PJ (1989) Backpropagation and neural control: A review and prospectus. Internat. Joint Conf. of Neural Networks, Washington, 1: 209–216.
[107] Widrow B and Hoff ME (1960) Adaptive Switching Circuits. In: Anderson J and Rosenfeld E. (eds.) Neurocomputing. MIT Press, Cambridge, MA, 126–134.
[108] Williams RJ and Zipser D (1989) A learning algorithm for continually running fully recurrent neural networks. Neural Computation 1: 270–280.
[109] Winkler R and Makridakis S (1983) The combination of forecasts, Journal of the Royal Statistical Society, Series A: 150–157.
[110] Yang Y (2000) Combining different procedures for adaptive regression, J. of Multivar. Analysis, 74: 135–161.

[111] Yu X-H, Chen G-A, and Cheng S-X (1995) Dynamic Learning Rate Optimization of the Backpropagation Algorithm. IEEE Trans. on Neural Networks 3: 669– 677.
[112] Zhang PG (2003) Time series forecasting using a hybrid ARIMA and neural network models. Neurocomputing 50:159–175.
[113] Zhou S, Popovic D, and Schulz-Ekloff G (1991) An Improved Learning Law for Backpropagation Networks. IEEE Int. Conf. on Neural Networks, San Francisco: 573–579.

Selected Reading

[114] Anderson JA (1972) A Simple Neural Network Generating an Interactive Memory, Mathematical Biosciences 14: 197–220.
[115] Cybenko G (1988) Continuous valued neural networks with two hidden layers are sufficient. Technical Report, Taft University.
[116] Kohonen T (1972) Correlation Matrix Memories. IEEE Transactions on Computers 21: 353–359.
[117] Kolmogorov AI (1957) On Representation of Continuous Function of Many Variables by Superposition of Continuous Functions of One Variable and Addition. Dokl. Akad. Nauk USSR 114:953–956.
[118] Kurkova V (1991) Kolmogorov's Theorem is Relevant, Neural Computation, 3: 617–622.
[119] Kurkova V (1992) Kolmogorov's Theorem and Multilayer Neural Networks, Neural Networks 5: 501–506.
[120] Moody JE (1992) The Effective Number of Parameters: An Analysis of Generalization and Regularisation in Nonlinear learning Systems. In: Advances in Neural Information Processing 4 (Moody JE, Hanson SJ, and Lippmann RP (Eds.), Morgan Kaufman Publ., San Mateo, CA.
[121] Schwenkler F, Kestler H, Palm G (2001) Three learning phases for radial-basis-function networks. Neural Networks 14: 439–458.
[122] Schwenkler F, Kestler H, Palm G, and Höher M (1994) Similarities of LVQ and RBF learning, Proc. IEEE International Conference SMC: 646–651.
[123] Xiaosong D, Popovic D, and Schulz-Ekloff G (1995) Oscillation-Resisting in the Learning of Backpropagation Neural Networks. 3rd IFAC/IFIP Workshop on Algorithms and Architectures for Real-Time Control, 31 May – 2 June, Ostend, Belgium.

4

Fuzzy Logic Approach

4.1 Introduction

The term "fuzzy" was introduced by Zadeh (1965) in his paper on fuzzy sets, where a new mathematical discipline, ***fuzzy logic***, based on the ***theory of fuzzy sets***, was presented. The proposed logic was aimed at supporting of presentation and consideration of inexact or imprecise concepts by fuzzy sets. The imprecision is to be understood as grouping of set members into classes, the boundaries of which are not sharply defined. It was expected that the theory of fuzzy sets should become a novel methodology suitable enough to help formulate and solve complex problems in engineering and science that are difficult to handle using "precise" ***crisp logic***, such as ***binary logic***, where the variables can be either ***true*** or ***false***. The theory of fuzzy sets allows the concept of partial belongingness of an object or a variable in a fuzzy set and, therefore, allows a gradual transition from a full membership to a totally non-membership. Thereby, in fuzzy logic an object or a variable within a domain may partially belong to several fuzzy sets in the same domain simultaneously and, thus, it provides a framework for a ***multivalued logic***. This is essential for capturing the vagueness in a natural linguistic description of any system. Moreover, the underlying fuzzy logic incorporates a variety of rules with the premises containing fuzzy propositions generally defined using ***linguistic terms***, such as ***low*** and ***high*** (temperature, pressure, flow, frequency, voltage, *etc*.), ***old***, ***older***, ***very old*** (person, engine, sensor, measured value, *etc*.). The related linguistic rules are of the IF-THEN art.

The linguistic rules enable the use of both numerical information represented by numerical values, obtained from the various sensors, or given as set point values, and linguistic information represented by words such as high, medium, low, or fast, moderate or slow, *etc*., obtained from an experienced plant operator or a human expert. They replace the traditional approach to modelling of dynamic systems based on differential equations, and the like, that is unsuccessful in modelling of nonlinear and complex systems. Moreover, traditionally modeled engineering systems cannot directly integrate human expert's linguistic knowledge.

It has frequently been reported that the design approaches of fuzzy-logic-based systems have been found to be very robust when embedded in control and signal-processing systems. However, the development of fuzzy logic systems, based on human expert's knowledge, is not an easy attempt, primarily because it is very difficult to extract the complete and consistent human expert's knowledge correctly by interviewing him or her.

The objective of this chapter is to develop some suitable fuzzy logic systems capable of efficiently modelling time series data and forecasting their values. Because the efficient functioning of fuzzy logic systems depends primarily on fuzzy rules used for modelling, and because the automated generation of such rules is rather difficult, various data-driven algorithms for automated rule generation are presented.

4.2 Fuzzy Sets and Membership Functions

The ***membership function*** is the key idea introduced in fuzzy set theory to measure the degree to which the fuzzy set elements meet the specific properties, *i.e.* to measure the ***degree of belongingness*** of an element in a specific fuzzy set. Consequently, the propositions used need not be true or false, but can be to any degree partially true.

Using a membership function μ, we can define a fuzzy set F on a ***universe of discourse*** U as

$$\mu_F(x): U \rightarrow [0, 1],$$

which is nothing but a mapping from the universe of discourse U into the unit interval [0, 1] and $\mu_F(x)$ represents the extent (degree/grade) to which x belongs to fuzzy set F. The concept of membership functions allows any element within the universe of discourse to have partial membership to a specific fuzzy set and also to have partial membership to other fuzzy sets. In order to demonstrate the idea of membership functions, two examples are given, one each for a crisp set and a fuzzy set.

Let C be a ***crisp set*** and x be any element of the set C such that $x \in X$, where X is the universe of discourse (domain), then the degree of membership of x in crisp set C will be 1 and 0 respectively if the element x belongs to C completely (full member) or it does not belong to it at all. Mathematically, this is stated as

$$\mu_C(x) = \begin{cases} 1; & \text{if } x \in C \\ 0; & \text{if } x \notin C \end{cases}$$

Let us now consider that F be a ***fuzzy set*** and x be any element of the fuzzy set F such that $x \in X$, where X is the universe of discourse (domain), then the degree of membership of x in fuzzy set F will be 1 and 0 respectively if the element x belongs to F completely (full member) or it does not belong to it at all. However, if

x belongs to F partially, then the degree of membership of x in fuzzy set F can have any intermediate value, such as 0.5, 0.9, *etc*., within 0 and 1. Mathematically, this is stated as

$$\mu_F(x) = \begin{cases} 1, & \text{if } x \in F \text{ (completely)} \\ (0,1), & \text{if } x \in F \text{ (partially)} \\ 0, & \text{if } x \notin F \text{ (totally non-member)} \end{cases}$$

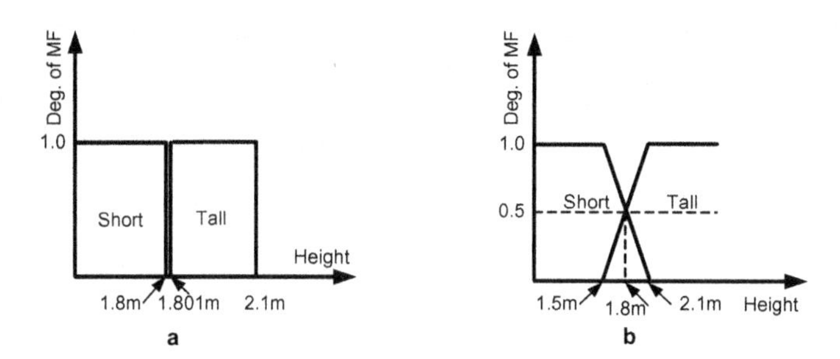

Figure 4.1. (a) Crisp (ordinary) set; (b) fuzzy set

Figure 4.1(a) shows an example of two crisp sets, "short" and "tall", where it is shown that even if the height of a person is 1.7999 m then that person definitely belongs to the "short" category only. This is because the crisp set "short" includes heights up to 1.8 m. In contrast, if the height of the same person had been just 1.8011 m, as per the same Figure 4.1(a), then the person would belong to the category "tall", as in this case the height is 0.0001 m greater than 1.801 m and that categorizes the person into the crisp set "tall". This is obviously quite impractical.

Similarly, Figure 4.1(b) shows the example of two fuzzy sets, "short" and "tall", where it is shown that if the height of a person is less than or equal to 1.5 m, then the person belongs to the category "short", whereas if the height is say 1.8 m then the person belongs to the category "short" with a degree of membership 0.5 and at the same time the person is considered as "tall" with a degree of membership equal to 0.5.

In order to explain the importance of fuzzy sets or membership functions, Boyle's law, as a practical example, is considered, that states that the pressure (P) of a given mass of gas varies inversely proportional to the volume (V) of the gas, provided the gas temperature (T) remains constant.

Using the fuzzy linguistic rules, Boyle's law can be stated as:

Rule-1: IF pressure is **high** and temperature is **constant** THEN volume is **low**

Rule-2: IF pressure is **medium** and temperature is **constant** THEN volume is **medium**

Rule-3: IF pressure is **low** and temperature is **constant** THEN volume is **high**.

The same fact can be written mathematically as PV = constant.

The above three IF-THEN rules are sufficient to model Boyle's observations and is in fact very similar to the way we understand a system or describe our observations and experience about any system in day-to-day life. In the above three rules, pressure, temperature and volume are the *linguistic variables,* whereas (*fuzzy sets*) high, medium, low, *etc.* are the *linguistic terms* or *linguistic labels,* generally represented by triangular or trapezoidal or even by Gaussian ***membership functions (fuzzy sets)***.

For example, in the above rules, say 0.9 to 1.5 bar represents high pressure, 0.4 to 1.0 bar represents medium pressure and 0 to 0.5 bar represents low pressure, *etc.* Note that, here, instead of exact and specific values of pressure we used a range to specify high, low and medium, and also note that ranges are partially overlapping.

So, from the above example it is clear that fuzzy logic (IF-THEN linguistic rules) is a very convenient mathematical tool to describe our observations or experiences about any system for system modelling with the application of fuzzy sets.

4.3 Fuzzy Logic Systems

Fuzzy logic systems have a direct relationship with fuzzy concepts, such as fuzzy sets, linguistic variables, and fuzzy logic. Fuzzy systems are unique in the sense that they can simultaneously process numerical data and linguistic knowledge. From the mathematical point of view, a fuzzy logic system is a nonlinear mapping of an input feature (data) vector into a scalar output.

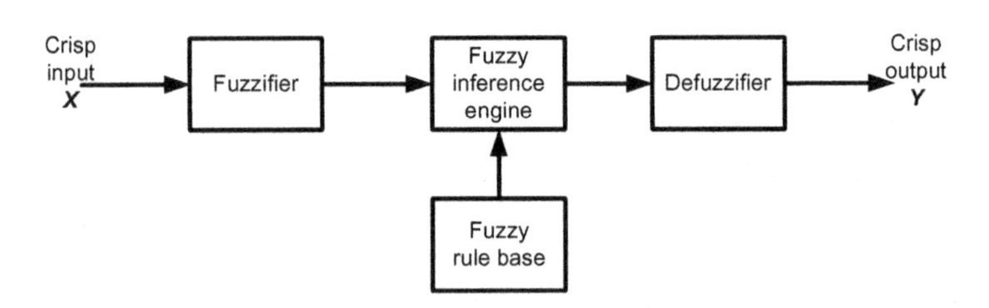

Figure 4.2. Block diagram of a fuzzy logic system

The block diagram of a fuzzy logic system is shown in Figure 4.2. From the figure it is seen that the fuzzy logic system takes the crisp input value (***X***) and this is then ***fuzzified*** (converted into corresponding membership grade in the input fuzzy sets), thereafter, it is fed to the fuzzy inference engine. Using the stored ***IF-THEN*** fuzzy rules from the rule base the inference engine produces a fuzzy output that undergoes further ***defuzzification*** to result in crisp output (***Y***).

In artificial intelligence, fuzzy logic systems were first styled as ***fuzzy rule systems*** and ***fuzzy expert systems***.

Fuzzy sets can be involved in a fuzzy logic system in a number of ways:

- in system description

- in specification of system parameters
- in representation of input, output and system states.

When involved in system description, fuzzy sets appear as linguistic terms or labels to represent the state of the linguistic variable in the fuzzy rule. An illustration of the first case can be presented considering once again Boyle's observations, which are described by three IF-THEN rules, as stated in the Section 4.2. In fact, any system can be described by a collection of such types of IF-THEN linguistic rules, also known as fuzzy rules. The fuzzy logic systems are actually a rule-based system and usually defined using the IF-THEN rules. The general form of such an IF-THEN rule is: IF *antecedent propositions* THEN *consequent propositions*. The example of fuzzy (antecedent) propositions can be "*Pressure* is *High*" or even "x is A". Here, the term "*High*" is a linguistic term or label, also called a fuzzy term, represented by a fuzzy set (membership function) on the universe of discourse (UD) of the linguistic variable "Pressure". Similarly, fuzzy set A is a representative of a linguistic label/term. Sometimes linguistic hedges (modifier) are used to modify the linguistic label/fuzzy set without redefining the fuzzy set completely. An example of the latter can be "very A" or "more or less A," *etc*.

When involved in specification of system parameters, fuzzy sets may appear as fuzzy numbers. Similarly, as an example of the second case, let us consider a system that can be described by algebraic or differential equations in which the parameters are approximate (fuzzy) numbers instead of exact real numbers. For instance, a linear system of the form $y = f(x)$, where x is the input to the system and y is the corresponding output from the system, can be represented by a linear equation, but one with fuzzy numbers such as $y = \tilde{2}x + \tilde{3}$, where the numbers $\tilde{2}$ and $\tilde{3}$ (with tilde symbol) represent the fuzzy numbers approximately 2 and approximately 3 respectively.

Finally, fuzzy sets may appear as the only means to express human perceptions or even noisy or uncertain data or information that have to be used as system input, output, and system state. As an illustration of the latter, consider the input of a system that can be noisy data (reading from unreliable sensors/transducers), or even human perceptions such as hot, warm, comfortable, uncomfortable, beautiful, and tasty, *etc*. Fuzzy-logic-based system can process such types of information by defining their suitable ranges and criteria with fuzzy sets.

A ***fuzzy inference system*** is the core part of a fuzzy logic system. In practice, the following fuzzy inference systems have most frequently been employed and have most frequently been the subject of theoretical study:

- Mamdani type fuzzy inference systems
- Takagi-Sugeno type fuzzy inference systems
- Relational (Pedrycz) fuzzy logic systems.

4.3.1 Mamdani Type of Fuzzy Logic Systems

Mamdani (1977) proposed the first fuzzy inference system with the objective to control a combination of a steam engine and a boiler, based on a set of linguistic control rules built as the extracted knowledge of a human expert.

When applied to Boyle's law, as described in Section 4.2, the following fuzzy linguistic rules can be written:

Rule-1: IF the pressure is HIGH and the temperature is CONSTANT, THEN the Volume is LOW.
Rule-2: IF pressure is MEDIUM and the temperature is CONSTANT, THEN the volume is MEDIUM.
Rule-3: IF the pressure is LOW and the temperature is CONSTANT, THEN the volume is HIGH.

These rules are known as ***Mamdani-type fuzzy rules*** (first introduced and used by Mamdani in 1977). The main features of such rules are that both the IF (antecedents) parts and the THEN (consequents) parts of the rules are fuzzy (imprecise) in nature. That is, fuzzy sets are used here in order to describe both the input and the output variables of the system.

As another example of the Mamdani-type fuzzy rules, consider a single input–single output system that describes the relationship between the heater current and the temperature trend as follows:

- IF the heater current is HIGH, THEN the temperature rise is FAST
- IF the heater current is MEDIUM, THEN the temperature rise is MODERATE
- IF the heater current is LOW, THEN the temperature rise is SLOW.

Note that in the above Mamdani-type fuzzy rules the *heater current* and *temperature rise* are the two ***linguistic variables*** (input and output of the system respectively), whereas HIGH, MEDIUM, and LOW are the three fuzzy sets, represented by suitable (triangular/Gaussian) ***membership functions*** and provide the means to express the states of the linguistic input variables. Similarly, FAST, MODERATE, and SLOW are the three output fuzzy sets – also represented by suitable membership functions – representative of the states of linguistic output variables of the systems.

4.3.2 Takagi-Sugeno Type of Fuzzy Logic Systems

With Takagi-Sugeno (TS) type fuzzy rules the IF (antecedent) part is fuzzy in nature, whereas the THEN (consequent) part is a ***crisp function*** of an antecedent variable (as a rule, a linear equation) rather than a fuzzy proposition. The example presented above for Boyle's law could be written correspondingly as:

Rule-1: IF P is LOW and T is CONSTANT, THEN $V = a_1P + b_1T + c_1$
Rule-2: IF P is HIGH and T is CONSTANT, THEN $V = a_2P + b_2T + c_2$
Rule-3: IF P is MEDIUM and T is CONSTANT, THEN $V = a_3P + b_3T + c_3$.

where a_l, b_l, and c_l parameters with $l = 1, 2, 3$ corresponding to Rule-1, Rule-2, and Rule-3 are constants.

As another example, we again take a single input–single output system and present it using Takagi-Sugeno rules:

- IF heaterCurrent is HIGH,
  THEN temperatureRise = a_H(heaterCurrent) + b_H
- IF heaterCurrent is MEDIUM,
  THEN temperatureRise = a_M(heaterCurrent) + b_M
- IF heaterCurrent is LOW,
  THEN temperatureRise = a_L(heaterCurrent) + b_L

Using similar kinds of rules, many real systems can be described and modeled very accurately, where each rule represents a local linear model of the system. Also, these types of rule enable the system output variables (real valued/crisp valued) to be very easily inferred, which is an advantage of the presentation.

Note that in the above rules if the first constant parameters are all set to zero (*i.e.* $a_H = 0$, $a_M = 0$, $a_L = 0$), then the rule's consequents are singleton fuzzy sets. Similarly, with Mamdani-type fuzzy rules if the consequent fuzzy sets are singleton type (a real value) then they are identical to the Takagi-Sugeno type fuzzy rules with singleton consequents (*i.e.* when $a_H = 0$, $a_M = 0$, $a_L = 0$).

4.3.3 Relational Fuzzy Logic System of Pedrycz

In relational fuzzy logic systems, similar to Mamdani-type fuzzy logic system, both the IF (antecedent) parts as well as the THEN (consequent) parts are fuzzy. However, there is a slight difference in the rule's representation: in this case, one particular antecedent proposition is allowed to be associated with several different consequent propositions via a fuzzy relation (Pedrycz, 1984). This can be explained, again, on the above single input–single output system, which is described now by the following rules:

- IF heater current is HIGH,
  THEN temperature rise is SLOW (0.0), MODERATE (0.1), FAST (0.9)
- IF heater current is MEDIUM,
  THEN temperature rise is SLOW (0.1), MODERATE (0.95), FAST (0.0)
- IF the heater current is LOW,
  THEN temperature rise is SLOW (1.0), MODERATE (0.1), FAST (0.0).

In the first relational fuzzy rule the consequent fuzzy set FAST (0.9) represents the output variable (temperature rise) belonging partially to the fuzzy set FAST with degree of affiliation (also called degree of membership) equal to 0.9. Similarly SLOW (0.0) and MODERATE (0.1) represent respectively that the same output variable does not belong to fuzzy set SLOW at all (as the degree of membership in SLOW is 0.0), whereas the same output belongs a little to fuzzy set MODERATE (partially with degree of membership 0.1). Following the same argument, one can see that in the third rule SLOW (1.0) indicates that the output variable (temperature rise) belongs fully to fuzzy set SLOW (as the degree of membership in SLOW is 1.0), whereas it (output) simultaneously belongs to the fuzzy set MODERATE partially with degree of membership equal to 0.1, but it does not belong to the fuzzy set FAST at all. It is important to note that in any of the above rules the summation of degree of membership of output variable in the consequent fuzzy sets need not always be 1.0. From the above three rules it is easy to understand that relational fuzzy rule can be regarded as a generalization of the Mamdani-type fuzzy rules.

4.4 Inferencing the Fuzzy Logic System

Inferencing refers to the process of generating the output fuzzy set when the fuzzy rules and the input set are given. Usually, inferencing of Mamdani-type linguistic fuzzy rules and relational (Pedrycz) fuzzy rules produces an output fuzzy set that is *not* directly compatible with a real-world signal (such as a control signal for an actuator within the range 4 to 20 mA) as it is fuzzy in nature. If a crisp (numerical) output value is required, which is directly compatible with a real-world signal, then the output fuzzy set must be defuzzified. Defuzzification is a transformation process that translates the output fuzzy set into a single numerical value representative of that fuzzy set. For this purpose, preferably the centre of gravity (COG) method is used.

Given a fuzzy set F represented in the point-wise form as

$$F = \left\{ \mu_F(x_1)/x_1, \mu_F(x_2)/x_2, \cdots, \mu_F(x_p)/x_p \right\},$$

the COG ***method*** helps in computing the x coordinate of the centre of gravity of the fuzzy set F as follows:

$$x' = \frac{\sum_{i=1}^{p} \mu_F(x_i) \cdot x_i}{\sum_{i=1}^{p} \mu_F(x_i)}$$

In contrast, fuzzification translates a crisp value into a corresponding fuzzy value (degree of membership). If the computed degree of membership of the crisp input in a fuzzy set F is exactly 1 or close to 1 or greater than some threshold value the input (crisp) is considered to be equivalent to that fuzzy set F.

4.4.1 Inferencing a Mamdani-type Fuzzy Model

Inferencing the Mamdani type of fuzzy model basically consists of four steps. For a single-input single-output model, however, if an input fuzzy set is given instead of a crisp input value, then the procedure is slightly altered, as shown below. Given the rule base, for instance with M fuzzy rules, as

R^1: IF x is G_1^1, THEN y is F_1^1
R^2: IF x is G_1^2, THEN y is F_1^2
: : : :
R^M: IF x is G_1^M, THEN y is F_1^M

Now, if x is $G_1'^1$, is given as the input fuzzy set and $G_1'^1 \neq G_1^1$, then the objective is to determine the corresponding output fuzzy set through the Mamdani rule inferencing mechanism. The procedure is as follows.

Each fuzzy rule above can be regarded as a fuzzy relation:

$$R^l : (X \times Y) \rightarrow [0,1]$$

computed as

$$\mu_{R^l}(X \times Y) = I\left(\mu_{G_1^l}(x), \mu_{F_1^l}(y)\right),$$

where the operator I can be either a ***fuzzy implication*** or a ***conjunction operator*** such as a ***t-norm***. It is to be noted that $I(.,.)$ is computed on the Cartesian product space $X \times Y$, *i.e.* for all possible pairs of x and y from the domain, using the Mamdani implication

$$I\left(\mu_{G_1^l}(x), \mu_{F_1^l}(y)\right) = \min\left(\mu_{G_1^l}(x), \mu_{F_1^l}(y)\right).$$

Once the fuzzy relation (R^l) is computed for each rule l = 1, 2, 3, ..., M, the fuzzy relation R for the entire rule base is computed taking the element-wise maximum of all (R^l) *i.e.* R is the union of all (R^l), for l = 1, 2, 3, ..., M. From this fuzzy relation the output fuzzy set is computed directly by applying a max-min composition and written as

$$F_{out}{}_1^l = G'^l_1 \ R.$$

Using the minimum *t-norm* operator, the max-min composition is obtained as

$$\mu_{F_{out}{}_1^l}(y) = \max_X \min_{X,Y}\left(\mu_{G'^l_1}(x), \mu_R(x,y)\right)$$

The final result of this max-min composition is nothing but the desired output fuzzy set. The COG of the output fuzzy set gives the equivalent crisp output (y coordinates).

The procedure described above can be circumvented by the following few steps:

- Step 1: compute the degree of fulfilment of each rule by

$$\beta^l = \max_X \left[\mu_{G'^1_1}(x) \wedge \mu_{G^l_1}(x) \right], 1 \le l \le M,$$

where $\wedge$ is the min operator. For a crisp input $x = x_0$, which is equivalent to a singleton fuzzy set, *i.e.*

$$\mu_{G'^1_1}(x) = 1, \text{ for } x = x_0;$$

and for all other points

$$x \ne x_0, \mu_{G'^1_1}(x) = 0.$$

So the degree of fulfilment of the *l*th rule is reduced to

$$\beta^l = \mu_{G^l_1}(x_0).$$

- Step 2: compute the each rule consequent set as given by

$$F'^l_1 = \beta^l \wedge F^l_1$$

- Step 3: aggregate all consequent fuzzy sets as shown by

$$F'_{aggr} = \bigcup_{l=1}^{M} F'^l_1 = \left(F'^1_1 \cup F'^2_1 \cup \cdots \cup F'^M_1 \right)$$

- Step 4: defuzzify the aggregated fuzzy set F'_{aggr} using the COG method.

The inferencing mechanism of the Mamdani type of fuzzy logic system can easily be explained on an *n*-input single-output system described by M numbers of Mamdani-type fuzzy rules

R^1:	IF x_1 is G^1_1 and… and x_n is G^1_n, THEN y is F^1_1
R^2:	IF x_1 is G^2_1 and… and x_n is G^2_n, THEN y is F^2_1
⋮	⋮ ⋮ ⋮ ⋮ ⋮
R^M:	IF x_1 is G^M_1 and… and x_n is G^M_n, THEN y is F^M_1.

For a given set of rules and inputs x_i, with $i = 1, 2, \cdots, n$; (also called the training sample), the objective is to determine the crisp output value which is the defuzzified value of the output fuzzy set. The inferencing of such a rule-based fuzzy system proceeds as follows.

- Step 1: compute the degree of fulfilment of each rule for any given input set (crisp) by

$$\beta^l = \mu_{G_1^l}(x_1) \wedge \mu_{G_2^l}(x_2) \wedge \cdots \wedge \mu_{G_n^l}(x_n),$$

where $\wedge$ is the min or product operator

- Step 2: compute the each rule consequent set as given by

$$F'^l_1 = \beta^l \wedge F^l_1$$

- Step 3: aggregate all consequent fuzzy sets as shown by

$$F'_{aggr} = \bigcup_{l=1}^{M} F'^l_1 = \left(F'^1_1 \cup F'^2_1 \cup \cdots \cup F'^M_1\right)$$

- Step 4: defuzzify the aggregated fuzzy set F'_{aggr} using the COG method.

The defuzzified value of the aggregated fuzzy set is the crisp output value from the Mamdani-type fuzzy model in response to the given input value. In Step 3 the aggregation is the union (standard/Zadeh's union) of the consequent fuzzy sets.

4.4.2 Inferencing a Takagi-Sugeno type Fuzzy Model

The inference formula of the Takagi-Sugeno model is only a two-step procedure, based on a weighted average defuzzifier. In the first step the degree of fulfilment, or firing strength (also called the degree of activation), of each rule is computed using the product operator. In the second step, the final output value of the system is calculated using the weighted average defuzzifier. This can, for the inference process of a Takagi-Sugeno type fuzzy logic system consisting of M rules, be presented as

R^1: IF x_1 is G_1^1 and … and x_n is G_n^1 THEN $y_{TS}^1 = \theta_0^1 + \theta_1^1 \cdot x_1 + \cdots + \theta_n^1 \cdot x_n$

R^2: IF x_1 is G_1^2 and … and x_n is G_n^2 THEN $y_{TS}^2 = \theta_0^2 + \theta_1^2 \cdot x_1 + \cdots + \theta_n^2 \cdot x_n$

: : : : :

R^M: IF x_1 is G_1^M and … and x_n is G_n^M THEN $y_{TS}^M = \theta_0^M + \theta_1^M \cdot x_1 + \cdots + \theta_n^M \cdot x_n$

The ***degree of fulfilment*** is now calculated using the ***product operator***, as was done when the set of Takagi-Sugeno rules with antecedent fuzzy sets and parameters are known for a given set of inputs

$$\beta^l = \prod_{i=1}^{n} \mu_{G_i^l}(x_i) = \mu_{G_1^l}(x_1) \times \mu_{G_2^l}(x_2) \times \cdots \times \mu_{G_n^l}(x_n); \quad l = 1, 2, \cdots, M.$$

The output value of the system is then given by

$$y_0 = \frac{\sum_{l=1}^{M} \beta^l \cdot y_{TS}^l}{\sum_{l=1}^{M} \beta^l} = \frac{\sum_{l=1}^{M} \beta^l \cdot \left(\theta_0^l + \theta_1^l \cdot x_1 + \cdots + \theta_n^l \cdot x_n\right)}{\sum_{l=1}^{M} \beta^l}$$

Alternatively, the final output of the system is represented by the normalized degree of fulfilment (normalized degree of activation) as

$$y_0 = \sum_{l=1}^{M} \gamma^l \cdot y_{TS}^l = \sum_{l=1}^{M} \gamma^l \cdot \left(\theta_0^l + \theta_1^l \cdot x_1 + \cdots + \theta_n^l \cdot x_n\right);$$

where $\gamma^l = \beta^l \Big/ \sum_{l=1}^{M} \beta^l$ is the normalized degree of fulfilment (activation).

4.4.3 Inferencing a (Pedrycz) Relational Fuzzy Model

The inference process of a relational (Pedrycz) fuzzy model

R^1: IF x_1 is G_1^1 … and x_n is G_n^1, THEN y is $F_1^1(\mu_1^1), F_2^1(\mu_2^1), \cdots, F_k^1(\mu_k^1)$

R^2: IF x_1 is G_1^2 … and x_n is G_n^2, THEN y is $F_1^2(\mu_1^2), F_2^2(\mu_2^2), \cdots, F_k^2(\mu_k^2)$

⋮ ⋮ ⋮ ⋮ ⋮ ⋮

R^M: IF x_1 is G_1^M … and x_n is G_n^M, THEN y is $F_1^M(\mu_1^M), F_2^M(\mu_2^M), \cdots, F_k^M(\mu_k^M)$.

consists of the following three steps:

- Step 1: compute the degree of fulfilment of each fuzzy rule by

$$\beta^l = \mu_{G_1^l}(x_1) \wedge \mu_{G_2^l}(x_2) \wedge \cdots \wedge \mu_{G_n^l}(x_n),$$

where $\wedge$ is the minimum or product operator.

- Step 2: apply the max-min relational composition operator to compute the relational composition $\omega = \beta \quad R$, where

$$\beta = \left[\beta^1, \beta^2, \cdots, \beta^M\right],$$

$$\omega^j = \max_{1 \le l \le M}\left(\min\left(\beta^l, R_{lj}\right)\right), \quad j = 1, 2, \cdots, K.$$

$$\omega = \left[\omega^1, \omega^2, \cdots, \omega^K\right]$$

with $R = \left[\mu_j^l\right]_{M \times k}$, a relational matrix of size $(M \times k)$, M is the number of given fuzzy rules and k is the number of output fuzzy sets/membership functions that make the partitioning of the output domain or output universe of discourse.

- Step 3: defuzzify the consequent fuzzy set by COG method to compute the crisp output value

$$y_0 = \left(\sum_{j=1}^{k} \omega^j \cdot y^j\right) \bigg/ \left(\sum_{j=1}^{k} \omega^j\right),$$

where $y^j = \mathrm{COG}\left(F_j^l\right)$, $F_j^l = j$th output fuzzy set for the lth rule, and $j = 1, 2, \cdots, k$.

To illustrate the above inference mechanism of a relational fuzzy model, let us again consider the n-input, single-output system described by the relational fuzzy rule- based model

R1: IF x_1 is G_1^1 and... and x_n is G_n^1,
THEN y is HIGH (0.9), y is MEDIUM (0.1), y is LOW (0.0)

R2: IF x_1 is G_1^2 and... and x_n is G_n^2,
THEN y is HIGH (0.1), y is MEDIUM (0.8), y is LOW (0.0)

R3: IF x_1 is G_1^3 and ... and x_n is G_n^3,
THEN y is HIGH (0.0), y is MEDIUM (0.7), y is LOW (0.2)

If the antecedent's fuzzy sets, *i.e.* G_i^l with $i = 1, 2, \cdots, n$; $l = 1, 2, \cdots, M$; and $M = 3$, are given, then for given values of input variables x_i; $i = 1, 2, \cdots, n$; we first determine the output fuzzy set through the inferencing mechanism as stated in the above three steps.

Now, the degree of fulfilment of the lth rule is computed using the product operator

$$\beta^l = \prod_{i=1}^{n} \mu_{G_i^l}(x_i) = \mu_{G_1^l}(x_1) \times \mu_{G_2^l}(x_2) \times \cdots \times \mu_{G_n^l}(x_n)$$

Therefore,

$$\beta^1 = \prod_{i=1}^{n} \mu_{G_i^1}(x_i) = \mu_{G_1^1}(x_1) \times \mu_{G_2^1}(x_2) \times \cdots \times \mu_{G_n^1}(x_n) = 0.5 \text{ (say)}.$$

Similarly, let the computed values of β^2 and β^3 for the second and the third rules, using a similar procedure, be 0.6 and 0.7 respectively. Therefore, the computed row vector will be $\beta = \left[\beta^1, \beta^2, \beta^3\right] = [0.5, 0.6, 0.7]$. Furthermore, for this example the relational matrix R is of size $(M \times k)$, where the number of rules $M = 3$ and the number of output fuzzy sets (*e.g.* HIGH, MEDIUM, and LOW) $k = 3$. The relational matrix is formulated using the degree of membership of each rule output in the output fuzzy set. Therefore,

$$R = \begin{bmatrix} 0.9 & 0.1 & 0.0 \\ 0.1 & 0.8 & 0.0 \\ 0.0 & 0.7 & 0.2 \end{bmatrix},$$

$$R_{lj} = \mu_{F_j^l}(y) \text{ and } l = 1, 2, \cdots, M; \quad j = 1, 2, \cdots, k.$$

Now applying the max-min relational composition, the output fuzzy set can be computed as follows:

$$\omega = \begin{bmatrix} 0.5 & 0.6 & 0.7 \end{bmatrix} \begin{bmatrix} 0.9 & 0.1 & 0.0 \\ 0.1 & 0.8 & 0.0 \\ 0.0 & 0.7 & 0.2 \end{bmatrix} = \begin{bmatrix} \max(\min(0.5, 0.9), \min(0.6, 0.1), \min(0.7, 0.0)) \\ \max(\min(0.5, 0.1), \min(0.6, 0.8), \min(0.7, 0.7)) \\ \max(\min(0.5, 0.0), \min(0.6, 0.0), \min(0.7, 0.2)) \end{bmatrix}^T.$$

This finally results in $\omega = \begin{bmatrix} 0.5 & 0.7 & 0.2 \end{bmatrix}$.

Supposing now that the COGs of the output fuzzy sets are known, *i.e.* if the $\mathrm{COG}(F_j^l)$; $j = 1, 2, \cdots, k$; and noting that $F_j^1 = F_j^2 = F_j^3$; are given respectively as $y^1 = 30$, $y^2 = 20$ and $y^3 = 10$, then the crisp output from the inference of the relational fuzzy-rule-based system will be

$$y_0 = \frac{(0.5 \times 30 + 0.7 \times 20 + 0.2 \times 10)}{(0.5 + 0.7 + 0.2)} = \frac{31}{1.4} = 22.142.$$

The various fuzzy inferencing mechanisms described in the Sections 4.4.1 to 4.4.3 can similarly be applied to time series forecasting applications when the corresponding fuzzy model (fuzzy rules) of a given time series is available.

4.5 Automated Generation of Fuzzy Rule Base

From the description of the various fuzzy logic systems it is well understood that the fuzzy inference system, *i.e.* the fuzzy inference engine requires a fuzzy rule base containing a complete set of well-consistent rules that model the system to be investigated. The automated generation of such a rule base, based on the time series data, and later its application to time series forecasting is our prime interest.

4.5.1 The Rules Generation Algorithm

The idea of data-driven automated rule generation, presented in this section, originates from Wang and Mendel (1992), who have proposed an adequate procedure for it's practical implementation. In addition, we have proposed a few modifications of those described by Wang and Mendel (1992), based on scaled and normalized time series data, partitioned into multi-input single-output data sets.

For example, for a two-input one-output fuzzy logic system using the Wang and Mendel's approach the input-output partitioning would be

$$\left(X_1^{(1)}, X_2^{(1)}, Y^{(1)}\right); \cdots ; \left(X_1^{(k)}, X_2^{(k)}, Y^{(k)}\right) \quad etc.$$

To generate the fuzzy rules automatically from these input-output partitioned data that represent the mapping of the input values to the respective output values, each X and Y domain will be divided into fuzzy regions and for each variable the universe of discourse (UD) obtained by considering the values [Min (X), Max (X)] or, [Min (Y), Max (Y)] of that variable. Thereafter, the UD is divided into a number of overlapping (fuzzy) regions and to each region a membership function, usually one of the triangular form, is assigned. This is followed by the fuzzification of crisp input-output values, in which a mapping of crisp input/output value from the domain into the unit interval is performed, and consequently for each membership function the corresponding label or the membership grade is obtained. Owing to overlapping of the fuzzy sets, more than one grade of membership may exist for each input or output value, out of which the fuzzy set with maximum grade is selected. The fuzzy input-output data pair, obtained for an individual input-output data set when connected through fuzzy logic operators, define the corresponding fuzzy rule. Here, however, conflict situations can arise when rules with the same antecedents, *i.e.* the same IF parts, but with different consequents (the THEN parts), are generated. To overcome this, to each conflicting rule a degree or a grade is assigned, for instance,

$$D(Rule) = \mu_A(X_1) \cdot \mu_B(X_2) \cdot \mu_C(Y),$$

for the given

Rule: IF ($X_1 = A$) AND ($X_2 = B$), THEN ($Y = C$).

Thereafter, the combined rule base and the rule grade table (RGT) are built, but the conflicting rules with the maximum value of *D(Rule)* are selected, whereas the conflicting rules with the lower value of *D(Rule)* are all rejected. It is to be noted that besides the conflicting rules the redundant rules, that have both identical IF parts as well as THEN parts, are also generated by this rule generation algorithm.

Since our final aim is to develop a fuzzy-logic-based predictor, or a fuzzy model that is capable of forecasting the future values of a given time series, in the following we will describe a fuzzy-rules generation algorithm based on a multi-input single-output partitioning of the time series data.

Given a time series $X = \{X_1, X_2, X_3, ..., X_q\}$, at time points $t = 1, 2, 3, \ldots, q$; our objective is to forecast the future values of this time series using a fuzzy-logic-based predictor. For this forecasting problem, usually a set of known values of the time series up to a point in time, say t, is used to predict the future value of the time series at some point, say $(t+L)$. The standard practice for this type of prediction is to create a mapping from D sample data points, sampled every d units in time, to a predicted future value of the time series at time point $(t+L)$. Therefore, for each t, the input data for the fuzzy logic predictor to be developed is a D-dimensional vector of the form:

$$XI(t)=[X\{t-(D-1)d\},\ X\{t-(D-2)d\},\ \ldots\ldots,\ X\{t\}]$$

Following the conventional settings (for predicting the Mackey-Glass time series), $D = 4$ and $d = L = 6$ have been selected and, therefore, the input data of the fuzzy predictor is a four-dimensional vector, *i.e.*

$$XI(t)=[X(t-18),\ X(t-12),\ X(t-6),\ X(t)].$$

The output data of the fuzzy predictor is a scalar and corresponds to the trajectory prediction:

$$XO(t)=[X(t+L)]$$

Therefore, for a four-input one-output fuzzy logic system the time series partition can be obtained as:

$$(X_{11}, X_{12}, X_{13}, X_{14}, Y_1);\ \ldots.;\ (X_{k1}, X_{k2}, X_{k3}, X_{k4}, Y_k);\ etc.$$

which can be represented in *XIO* matrix (multi-input single output) form as

$$XIO = \begin{bmatrix} X_{11} & X_{12} & X_{13} & X_{14} & Y_1 \\ \vdots & \vdots & \vdots & \vdots & \vdots \\ X_{k1} & X_{k2} & X_{k3} & X_{k4} & Y_k \end{bmatrix} \quad (4.1)$$

where, X_{k1}, X_{k2}, X_{k3}, X_{k4} are input values and Y_k as the corresponding output value for $k = 1, 2, 3, ..., m$. Note that *XIO* stands for the time series data

$X = \{X_1, X_2, \cdots, X_q\}$ are represented in input and output form. The objective is to generate, from the above data set, the IF-THEN rules that will construct the rule base of a fuzzy predictor system. This is carried out in the following steps.

Step 1: formation of fuzzy input and output regions

Suppose that the domain interval of X_{ki} is $[X_{i\_lo}, X_{i\_hi}]$ and that of (Y_k) is $[Y_{lo}, Y_{hi}]$, where k = 1, 2, 3, ... , m; and i = 1, 2, 3, 4; corresponding to the four inputs respectively.

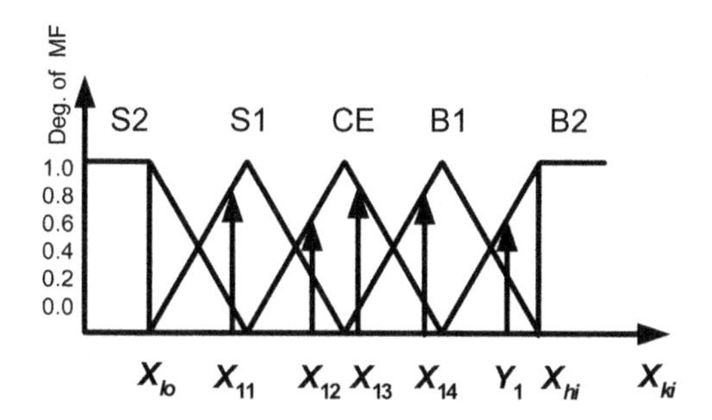

Figure 4.3(a). Division of input and output range in fuzzy regions

Taking into account that all input values and the output value belong to the same time series $X = \{X_1, X_2, X_3, ..., X_q\}$, for t = 1, 2, 3, ... , q, the domain intervals of all inputs and the output can be taken to be the same, say $[X_{lo}, X_{hi}]$ or $[Y_{lo}, Y_{hi}]$, for i = 1, 2, 3, 4. Each domain interval can be divided into $(2N + 1)$ fuzzy regions (see Figure 4.3(a) and 4.3(b)) like

> SN(Small N), ..., S2(Small 2), S1(Small 3), CE(Center), B1(Big 1), B2(Big 2), ..., BN(Big N).

Step 2: data fuzzification and rules generation

This includes the determination of the degrees of membership of X_{k1}, X_{k2}, X_{k3}, X_{k4}, Y_k in different fuzzy regions and assignment of a given X_{k1}, X_{k2}, X_{k3}, X_{k4}, Y_k for k = 1, 2, ... , m; to the region with the maximum degree. For example, for k = 1, X_{k1} in Figure 4.3(a) has degree of membership 0.8 in S1, 0.2 in S2 and 0 degrees in all other regions. Similarly, for k = 1, X_{k2} in Figure 4.3(a) has degree of membership 0.6 in CE, 0.4 in S1 and 0 degrees elsewhere. Again, for k = 1, X_{k1} in Figure 4.3(a) is considered to be S1 and X_{k2} in Figure 4.3(a) is considered to be CE.

Similarly, the fuzzy regions with maximum degree should be assigned to the X_{k3}, X_{k4}, Y_k for k = 1. Now, the corresponding rules can be obtained from the input-output data sets. According to Figure 4.3(a), for k = 1, X_{k1}, X_{k2}, X_{k3}, X_{k4}, and Y_k give [X_{11} (0.8 in S1, max), X_{12} (0.6 in CE, max), X_{13} (0.8 in CE, max), X_{14} (0.8 in B1, max); Y_1 (0.6 in B2, max)], and rule R1 is

R1: IF X_{11} is S1 AND X_{12} is CE AND X_{13} is CE AND X_{14} is B1 THEN Y_1 is B2.

Note that in the above rule X_{11}, X_{12}, X_{13}, and X_{14} actually represent the first, second, third and fourth inputs respectively of the system, whereas Y_1 represents the corresponding single output from the system.

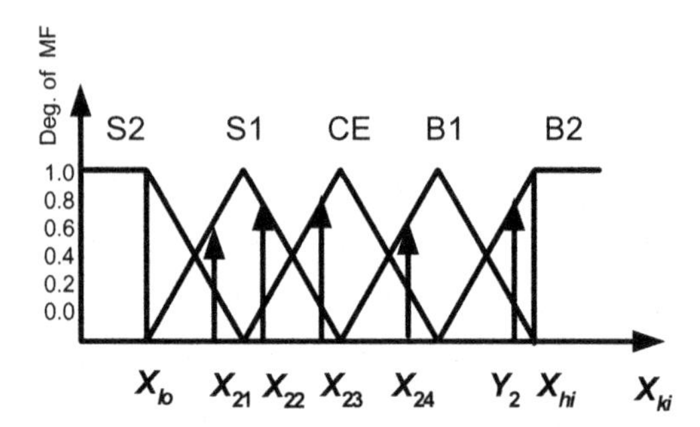

Figure 4.3(b). Division of input and output range in fuzzy regions

Furthermore, according to Figure 4.3(b), this gives [X_{21} (0.6 in S1, max), X_{22} (0.8 in S1, max), X_{23} (0.8 in CE, max), X_{24} (0.6 in B1, max); Y_2 (0.8 in B2, max)], *i.e.* rule R2 is

R2: IF X_{21} is S1 AND X_{22} is S1 AND X_{23} is CE AND X_{24} is B1 THEN Y_2 is B2.

Note that, as in the previous rule, here also X_{21}, X_{22}, X_{23}, and X_{24} actually represent the first, second, third and fourth inputs respectively to the system, and Y_2 represents the corresponding single output from the system.

Step 3: rules degree assignment

The large number of data pairs available generate a large number of rules, some of them being conflicting rules. To each rule the degree will be assigned and the conflicting rules with the highest degree retained. For example, the degree of the rule

Rule: IF x_1 *is* A *AND* x_2 *is* B *AND* x_3 *is* C *AND* x_4 *is* D, *THEN* y *is* E

is as follows:

$$D(Rule) = \mu_A(x_1).\ \mu_B(x_2).\ \mu_C(x_3).\ \mu_D(x_4).\ \mu_E(y),$$

so that the rules R1 and R2 above have the degrees:

$$D(Rule\ 1) = \mu_{S1}(X_{11}).\ \mu_{CE}(X_{12}).\ \mu_{CE}(X_{13}).\ \mu_{B1}(X_{14}).\ \mu_{B2}(Y_1)$$

$$= (0.8).(0.6).(0.8).(0.8).(0.6) = 0.18432,$$

$$D(Rule\ 2) = \mu_{S1}(X_{21}).\ \mu_{S1}(X_{22}).\ \mu_{CE}(X_{23}).\ \mu_{B1}(X_{24}).\ \mu_{B2}(Y_2)$$

$$= (0.6).(0.8).(0.8).(0.6).(0.8) = 0.18432$$

respectively, and both are found to be same in this example. However, they are usually different for realistic time series data.

Step 4: building of combined fuzzy rule base

A combined fuzzy rule base, built in the following way, is shown in Figure 4.3(c). It is a lookup table to be explained on the example of a two inputs $[X_{k1}, X_{k2}]$, one output Y_1 system for simplicity. Individual boxes are filled with fuzzy rules generated from input-output data, whereby the AND-rules fill only one box and the OR-rules fill all the boxes in the rows and/or columns corresponding to the regions of their IF parts. If there is more than one rule in one box, then the rule with the maximum degree is taken. For example, the rule

Rule: IF X_{k1} *is B1 OR* X_{k2} *is B2 THEN* Y_1 *is S2*

fills five boxes in the column of B1 and five boxes in the row of B2 with S2. The degrees of all the S2's in these boxes are equal to the degree of the OR-rule, whereas the same rule with AND, instead of OR, fills only the intersection of column B1 and row B2 with S2

X_{k2} \ X_{k1}	S1	S2	CE	B1	B2
S1		S1		S2	
S2			B1	S2	
CE		S1		S2	
B1	CE			S2	
B2	S2	S2	S2	S2	S2

Figure 4.3(c). Look-up table for fuzzy rule base

The combined rule base of Figure 4.3(c) describes the following fuzzy rules:

IF X_{k1} *is S2 AND* X_{k2} *is S1 THEN* Y_1 *is S1*
IF X_{k1} *is CE AND* X_{k2} *is S2 THEN* Y_1 *is B1*
IF X_{k1} *is S1 AND* X_{k2} *is B1 THEN* Y_1 *is CE*
IF X_{k1} *is S2 AND* X_{k2} *is CE THEN* Y_1 *is S1*
IF X_{k1} *is B1 OR* X_{k2} *is B2 THEN* Y_1 *is S2*

It is to be noted that for a system with more than two inputs the above table would require a multidimensional space for its presentation.

4.5.2 Modifications Proposed for Automated Rules Generation

The described rules generation algorithm requires much manual effort in handling the numerous data sets (say 500) and the large number of triangular membership functions (say 20 to 50). To reduce this, the modification of some of the operational steps and the use of Gaussian membership functions of the following form are proposed (Palit and Popovic, 1999):

$$f_j(x_i; c_j, \sigma_j) = \exp\left\{-(x_i - c_j)^2 / 2 \cdot \sigma_j^2\right\} \tag{4.2}$$

for simplification of computation of degree of membership for all values of X_i.

Dividing the domain interval $[X_{\text{lo}}, X_{\text{hi}}]$ into (n-1) overlapping fuzzy regions, with $n = 2N + 1$, N is some integer value, and assigning to each region a Gaussian membership function, the mean C and the variance σ will also have $n = 2N + 1$ values, such as C_1 to C_n and σ_1 to σ_n respectively. For ease of computer program implementation, the domain interval is divided into (n-1) equal regions such that $C_1 = X_{lo}$, $C_2 = C_1 + (X_{hi} - X_{lo})/2N$, $C_3 = C_2 + (X_{hi} - X_{lo})/2N$, ..., $C_n = X_{hi}$ and $\sigma_1 = \sigma_n = \sigma_a$, $\sigma_2 = \sigma_3 = \cdots = \sigma_{n-1} = \sigma_b$. For forecasting of Mackey-Glass chaotic time series, for example, $n = 17$, $\sigma_a = 0.08$, $\sigma_b = 0.04$, and the domain interval [0.4, 1.4] were selected. The fuzzy regions in this case are denoted by G_1, G_2, G_3, ..., G_n, *etc.* for convenience and G indicates the Gaussian membership functions (GMFs). For any input X_i within the domain interval the degree of membership $\mu_{G_j}(X_i) = f_j(X_i)$ will be within [0, 1], for j = 1, 2, 3, ..., n. If $X_i = C_j$ then $\mu_{G_j}(X_i) = 1$, whereas the degree of membership $\mu_{G_j}(X_i) = f_j(X_i)$ will be zero only if $X_i = \pm\infty$ and for other values of X_i in the domain interval the degree of membership can assume any value between 0 to 1. The fuzzy rules can now be generated in the usual way.

With the above modifications and after preprocessing the time series data, the automated fuzzy rules generation continues with fixing the domain interval as $[X_{lo}, X_{hi}] \equiv [\min(X), \max(X)]$ and with dividing the domain interval into (n-1) equal regions, where $n = 2N + 1$, and N is any suitable integer value such that each segment is of length $S = (X_{hi} - X_{lo})/(n-1)$, on which the accuracy of the forecast depends.

Now, the total number $n = 2N + 1$ of GMFs G_1 to G_n, over the entire domain with $C_1 = X_{lo}$, $C_2 = C_1 + S$, $C_3 = C_2 + S = C_1 + (3-1)S$, ..., $C_r = C_1 + (r-1)S$, ..., $C_n = C_1 + (n-1)S = X_{hi}$, and $\sigma_2 = \sigma_3 = \ldots = \sigma_{n-1} = \sigma_b$, whereas $\sigma_1 = \sigma_n = \sigma_a$, are assigned with suitably selected values of σ_a and σ_b. The integer n and, hence, the C_2 to $C_{n\text{-}1}$ and σ values are chosen such that two consecutive membership functions partially overlap. The forecasting accuracy also depends greatly on the extent of overlapping. It has been observed that the overlaps that are too large or too narrow may deteriorate the forecasting accuracy of the time series.

In the next step, the crisp input and output values are fuzzified. For any input value X_{ki}, or output value Y_k, the $f(X_{ki})$, or $f(Y_k)$, is calculated such that

$$\begin{aligned} \mu_{Gj}(X_{ki}) &\equiv f_j(X_{ki}) = \exp\left\{-(X_{ki}-c_j)^2/2\sigma_j^2\right\}, \\ \mu_{Gj}(Y_k) &\equiv f_j(Y_k) = \exp\left\{-(Y_k-c_j)^2/2\sigma_j^2\right\} \end{aligned} \tag{4.3}$$

where i = 1, 2, 3, 4 (corresponding to the first, second, third and fourth inputs in our case), j = 1, 2, 3, ..., n (corresponding to G_1 to G_n), k = 1, 2, 3, ..., m/2 (*i.e.* corresponding to the kth row of the *XIO* matrix and m being the total number of rows in *XIO* matrix in Equation (4.1)), and $\mu_{G_j}(X_{ki})$ is the degree of membership (μ) of X_{ki} in the jth Gaussian fuzzy set G_j. Hence, for any particular X_{ki} (or Y_k), *i.e.* for input X_{11} (or output Y_1), when $i = k = 1$, then $\mu_{G_j}(X_{11})$ will have n values (because j = 1, 2, ..., n) within the range [0, 1] and the same should be arranged in a column vector form of size $(n\times 1)$.

Similarly, the same procedure should be adopted for other inputs and output X_{12}, X_{13}, X_{14} and Y_1, *i.e.* $\mu_{G_j}(X_{ki})$ should be computed for all i = 1, 2, 3, 4, and, thereafter, should also be arranged in $(n\times 1)$ column vector form. When such column vectors, each of size $(n\times 1)$ for all the inputs (X_{11}, X_{12}, X_{13}, X_{14}) and output (Y_1), are arranged side by side sequentially, this results in a ***Mu-matrix*** of size $(n\times(i_{\max}+1))$, *i.e.* of $(n\times 5)$ size for our four-inputs one-output system. Now, the maximum value of degree of membership from each column of the ***Mu-matrix*** is selected and the corresponding row number is recorded.

For example, $\mu_{G_r}(X_{ki})$ is to be found out such that $\left(0<\mu_{G_r}(X_{ki})\le 1\right)$ and $\mu_{G_r}(X_{ki}) = \max\left(\mu_{G_j}(X_{ki})\right)$, for all j = 1, 2, 3, ..., n, the integer value of r $(1\le r\le n)$ should then be recorded, which is the key point of the automated rules generation algorithm. Once the r values are computed for all X_{ki} and Y_k, for i = 1, 2, 3, 4 and k = 1, 2, 3, ..., m/2, they should be recorded. For instance, it may be the case that

$$\begin{aligned} &\mu_{G_3}(X_{11}) = \max(\mu_{G_j}(X_{11})) = 0.95, \text{ i.e. } r = 3 \text{ implies } G_3 \\ &\mu_{G_5}(X_{12}) = \max(\mu_{G_j}(X_{12})) = 0.80, \text{ i.e. } r = 5 \text{ implies } G_5 \\ &\mu_{G_2}(X_{13}) = \max(\mu_{G_j}(X_{13})) = 0.98, \text{ i.e. } r = 2 \text{ implies } G_2 \\ &\mu_{G_6}(X_{14}) = \max(\mu_{G_j}(X_{14})) = 0.90, \text{ i.e., } r = 6 \text{ implies } G_6 \\ &\mu_{G_2}(Y_1) = \max(\mu_{G_j}(Y_1)) = 0.97, \text{ i.e. } r = 2 \text{ implies } G_2 \end{aligned}$$

In the next step, the fuzzy rules are built based on the values of r and degrees of membership. For instance, in the above example the degree of membership (μ) of X_{11} assumes a maximum value of 0.95 in G_3 (because r = 3). Similarly, the degrees of membership (μ) of X_{12}, X_{13}, X_{14} and Y_1 assume maximum values of 0.8 in G_5

(because $r = 5$), 0.98 in G_2 (because $r = 2$), 0.9 in G_6 (because $r = 6$), and 0.97 in G_2 (because $r = 2$) respectively. Hence, the corresponding fuzzy rule R3 will be:

R3: IF X_{11} is G_3 AND X_{12} is G_5 AND X_{13} is G_2 AND X_{14} is G_6 THEN Y_1 is G_2

The same fuzzy rule can also be written as [3 5 2 6 2 *Drule FOP*], where the numbers correspond to G_3, G_5, G_2, G_6, G_2 of the rule respectively whereas, (*Drule*) and *FOP* stand for the degree of the rule and fuzzy operator (AND) respectively. *Drule* = 1 if no degree of rule is specified or all rules have the same degree. If only the AND operator is used, then *FOP* has the value 1. Otherwise, for the OR operator the value 0 is used. Thus, for no degree of rule and for the AND operator the same rule is rewritten as [3 5 2 6 2 1 1]. The rules built in this way are used to build the rule list of the fuzzy system. If any two rules in the rule list create a conflict situation, the rule with the higher *Drule* value is taken and the other one is rejected from the rule list. For example, for the conflicting rules

[3 5 2 6 2] and [3 5 2 6 4]

$$\begin{aligned}\text{Drule3} &= \mu_{G_3}(X_{11}).\mu_{G_5}(X_{12}).\mu_{G_2}(X_{13}).\mu_{G_6}(X_{14}).\mu_{G_2}(Y_1)\\ &= (0.95).(0.80).(0.98).(0.90).(0.97) = 0.65 \text{ (say)},\end{aligned}$$

$$\begin{aligned}\text{Drule4} &= \mu_{G_3}(X_{21}).\mu_{G_5}(X_{22}).\mu_{G_2}(X_{23}).\mu_{G_6}(X_{24}).\mu_{G_4}(Y_2)\\ &= (0.90).(0.50).(0.80).(0.60).(0.70) = 0.15 \text{ (say)},\end{aligned}$$

so that because Drule3 > Drule4 the rule3 is selected and the second one rejected. However, for redundant (duplicate rules) rules from a list of several such rules any one is selected. Rules generated in this way are actually Mamdani-type fuzzy rules and the complete procedure of such automated rule generation is summarized in Algorithm 4.1.

However, a small modification in the final stage will also generate fuzzy relational rules (fuzzy relational model/Pedrycz model), *i.e.* to generate the fuzzy relational rule, the r values from the ***Mu-matrix*** are recorded for all four inputs as mentioned earlier and these generate as usual the antecedent part of the fuzzy relational rule. Now, the consequent part of the fuzzy relational rule is generated from the complete last column (fifth column) of the ***Mu-matrix*** that contains the degree of membership of output Y_k in the G_1 to G_n fuzzy sets. Note that for the last column of the *Mu-matrix* we do not record the r value for the output Y, whereas the entire column is recorded for rule generation. Therefore, the corresponding fuzzy relational rule can be written as

R5: IF X_{k1} is G_3 AND X_{k2} is G_5 AND X_{k3} is G_2 AND X_{k4} is G_6
THEN Y_k is G_1 ($\mu_{G_1}(Y_k)$), Y_k is G_2 ($\mu_{G_2}(Y_k)$), ..., Y_k is G_n ($\mu_{G_n}(Y_k)$).

Similarly, the antecedent part of the Takagi-Sugeno fuzzy rule is also generated in the same way, whereas the linear consequent parameters of the TS rules are generated by least squares error (LSE) estimation as described in Section 4.5.3.

Algorithm 4.1. Automated rules generation algorithm for time series prediction

Given the time series $\mathbf{X} = \{X_1, X_2, X_3, ..., X_q\}$ *for* $t = \{1, 2, 3, ..., q\}$, *the Mamdani-type fuzzy rules are generated as follows:*

- ***Step 1.*** *Partition the time series data into MISO form*
  $XI(t) = [X\{t-(D-1)d\}, X\{t-(D-2)d\}, ..., X\{t-d\}, X\{t\}]$
  $XO(t) = [X(t+L)]$,
  For four-inputs system $D = 4$, *and select sampling interval* $d = 6$, *and lead time of forecast* $L = 6$.
- ***Step 2.*** *Divide the domain interval* $[X_{lo}, X_{hi}]$
  into $(n-1) = 2N$ *overlapping fuzzy regions.*
  $X_{lo} = min(X)$, $X_{hi} = max(X)$.
  Assign to each region a GMF and denote them as $G_1, G_2, ..., G_n$.
- ***Step 3.*** *Compute* $S = (X_{hi} - X_{lo})/2N$, *so that the mean parameters of GMFs are:*
  $C_1 = X_{lo}$, $C_2 = C_1 + S$, ..., $C_r = C_1 + (r-1)S$, ...,
  $C_n = C_1 + (n-1)S = X_{hi}$.
  and variance parameters of GMFs are:
  $Sigma\_{G1} = Sigma\_{Gn} = Sigma_1$,
  $Sigma\_{G2} = Sigma\_{G3} =, ..., = Sigma\_{G(n-1)} = Sigma_2$.
  Select two suitable values for $Sigma_1$, $Sigma_2$, *so that two adjacent fuzzy regions partially overlap.*
- ***Step 4.*** *Fuzzify all the crisp inputs and output.*
  For any input X_{ki} *or output* Y_k *compute the degree of membership in all Gaussian regions.*
  $0 < \mu_{G_j}(X_{ki}) = f_j(X_{ki}) = exp(-0.5.(X_{ki}-C_j)^2/(Sigma\_{Gj})^2) \leq 1$.
  Say, for $i = 1$, $k = 1$, *and for* $j = 1, 2, 3, ..., n$.
- ***Step 5.*** *Arrange all degrees of membership in an* $(n\times 1)$ *column vector.*
  Similarly, compute the degree of membership for $i= 2, 3, 4$ *and for* Y_k, *etc., when* $k = 1$, *and* $j =1,2, 3, ...,n$; *etc.*
  Arrange them all in a column vector form of size $(n\times 1)$.
  When all such column vectors each of size $(n\times 1)$ *are arranged side by side sequentially they result in a Mu-matrix of size* $\{n\times(i_{max}+1)\}$. *For four-inputs and one-output system the Mu-matrix is of size* $(n\times 5)$.
- ***Step 6.*** *Select the maximum value of degree of membership from each column and record the corresponding row number i.e. integer value of r, such that* $0 < \mu_{G_r}(X_{ki}) = max(\mu_{G_j}(X_{ki})) \leq 1$, $j = 1, 2, 3, ..., n$ *and* $1 \leq r \leq n$.
  Note the value of r from each column of the Mu-matrix, such that $1 \leq r$ = *integer* $\leq n$. *This is the key step of the automated rule generation algorithm. Now create the fuzzy rule based on the r values and the corresponding degree of membership.*
- ***Step 7.*** *Create the rule list and solve the rule conflict problems (if any) using the degree of rule. Also remove the redundant rule from the rule list (if any).*

Once the fuzzy rule base with well-consistent and non-redundant rules is determined, the final step is to check the quality of the rule base generated. For this purpose, the first 50% data from the remaining data sets (*XIO* matrix) are used as validation data and, thereafter, by applying the Mamdani rules inferencing mechanism described above, the corresponding crisp output values for the given input data sets are determined. The crisp values generated are then compared with the desired output data and, consequently, the SSE or RMSE values are computed for these validation data sets. If the computed values of SSE or RMSE are less than the acceptable limit, then the rule base generated is considered as final and is stored for the forecasting test. Otherwise, with finer or coarser partitioning of universes of discourse of inputs and outputs and adopting a similar procedure, a new rule base is built.

Alternatively, after the rule generation, in the final step the defuzzification strategy recommended by (Wang and Mendel, 1992), usually the center of area strategy, can be selected and, consequently, the output control Y for given inputs $(X_{k1}, X_{k2}, X_{k3}, X_{k4})$ is determined by computing the degree of fulfilment of rule or, degree μ_O of the output control corresponding to $(X_{k1}, X_{k2}, X_{k3}, X_{k4})$ as:

$$\mu^1{}_{O_1} = \mu I^1{}_1 (X_{k1}).\ \mu I^1{}_2 (X_{k2}).\ \mu I^1{}_3 (X_{k3}).\ \mu I^1{}_4 (X_{k4}). \tag{4.4}$$

where O_l denotes the output region of rule l, and $I^l{}_i$ represents the input region for ith component of the rule l. For example,

$$\mu_{B2} = \mu_{S1}(X_{11}).\ \mu_{CE}(X_{12}).\ \mu_{CE}(X_{13}).\ \mu_{B1}(X_{14}). \tag{4.5}$$

represents the degree of fulfilment of the Rule-1. The crisp output value y is then determined using the center average defuzzification formula

$$y = \left(\sum_{l=1}^{M} y^l \cdot \mu^l O_l\right) \Big/ \left(\sum_{l=1}^{M} \mu^l O_l\right) \tag{4.6}$$

where y^l is the center value of region O_l and M represents the number of fuzzy rules in the combined fuzzy rule base.

4.5.3 Estimation of Takagi-Sugeno Rule's Consequent Parameters

Using Wang and Mendel's approach, or it's proposed modifications, the antecedent's fuzzy sets of the Takagi-Sugeno rules similar to Mamdani rules can be determined easily. Once the IF parts (antecedents) of the Takagi-Sugeno type of fuzzy rules are determined, the linear rule's consequent parameters of the Takagi-Sugeno rule can be estimated by applying the least squares error (LSE) technique.

In order to describe the LSE method for rule's consequent parameter estimation, Takagi-Sugeno type of fuzzy rules of a multi-input single-output system are once again considered:

R_l: *IF* x_1 *is* $G^l{}_1$ *and* ..... *and* x_n *is* $G^l{}_n$

THEN $y_{TS}^{1} = \theta_0^1 + \theta_1^1 \cdot x_1 + \cdots + \theta_n^1 \cdot x_n$

R_2: *IF* x_1 *is* G^2_1 *and ..... and* x_n *is* G^2_n

THEN $y_{TS}^{2} = \theta_0^2 + \theta_1^2 \cdot x_1 + \cdots + \theta_n^2 \cdot x_n$

: : : : :

R_M: *IF* x_1 *is* G^M_1 *and ..... and* x_n *is* G^M_n

THEN $y_{TS}^{M} = \theta_0^M + \theta_1^M \cdot x_1 + \cdots + \theta_n^M \cdot x_n$

Therefore, for a given set of inputs the corresponding Takagi-Sugeno inference will be

$$y_0 = \frac{\sum_{l=1}^{M} \beta^l \cdot y_{Ts}^l}{\sum_{l=1}^{M} \beta^l} = \frac{\left(\beta^1 \cdot y_{Ts}^1 + \beta^2 \cdot y_{Ts}^2 + \cdots + \beta^M \cdot y_{Ts}^M\right)}{\left(\beta^1 + \beta^2 + \cdots + \beta^M\right)} \tag{4.7}$$

where β^l is the degree of fulfilment or firing strength of the lth rule, which is computed for the n-(multiple) input system using the product operator as

$$\beta^l = \prod_{i=1}^{n} \mu_{G_i^l}(x_i) = \mu_{G_1^l}(x_1) \times \mu_{G_2^l}(x_2) \times \cdots \times \mu_{G_n^l}(x_n). \tag{4.8}$$

Therefore,

$$y_0 = \left(\gamma^1 \cdot y_{TS}^1 + \gamma^2 \cdot y_{TS}^2 + \cdots + \gamma^M \cdot y_{TS}^M\right), \tag{4.9}$$

where the normalized degree of fulfilment for lth rule is

$$\gamma^l = \frac{\beta^l}{\sum_{l=1}^{M} \beta^l} = \frac{\beta^l}{\left(\beta^1 + \beta^2 + \cdots + \beta^M\right)}, \tag{4.10}$$

or the corresponding Takagi-Sugeno inference for sth training sample will be

$$\begin{aligned} y_{0s} &= \gamma_s^1 \cdot \left(\theta_0^1 + \theta_1^1 \cdot x_{1\_s} + \cdots + \theta_n^1 \cdot x_{n\_s}\right) \\ &+ \gamma_s^2 \cdot \left(\theta_0^2 + \theta_1^2 \cdot x_{1\_s} + \cdots + \theta_n^2 \cdot x_{n\_s}\right) + \ \ldots\ldots \\ &+ \gamma_s^M \cdot \left(\theta_0^M + \theta_1^M \cdot x_{1\_s} + \cdots + \theta_n^M \cdot x_{n\_s}\right). \end{aligned} \tag{4.11}$$

Now, by appending 1 along with n inputs in XIe_s, which takes care of θ_0^l from the rule consequent, the sth extended training sample is given as (4.12)

$$XIe_s = \left[1, x_{1\_s}, x_{2\_s}, \cdots, x_{n\_s}\right] \tag{4.12}$$

and the consequent's parameters as

$$\theta^l = \left[\theta_0^l, \theta_1^l, \cdots, \theta_n^l\right]^T = \begin{bmatrix} \theta_0^l \\ \vdots \\ \theta_n^l \end{bmatrix}, \tag{4.13}$$

then in the matrix form the corresponding Takagi-Sugeno inference, for 1 through N extended training samples, can be written as

$$\begin{bmatrix} y_{01} \\ y_{02} \\ \vdots \\ y_{0N} \end{bmatrix} = \begin{bmatrix} \gamma_1^1 \cdot XIe_1 & \gamma_1^2 \cdot XIe_1 & \cdots \gamma_1^M \cdot XIe_1 \\ \gamma_2^1 \cdot XIe_2 & \gamma_2^2 \cdot XIe_2 & \cdots \gamma_2^M \cdot XIe_2 \\ \vdots & \vdots & \vdots \\ \gamma_N^1 \cdot XIe_N & \gamma_N^2 \cdot XIe_N & \cdots \gamma_N^M \cdot XIe_N \end{bmatrix} \cdot \begin{bmatrix} \theta^1 \\ \theta^2 \\ \vdots \\ \theta^M \end{bmatrix} \tag{4.14}$$

where the corresponding vectors and matrices are of the dimensions

$$[Y] = N \times 1, \quad [\gamma XIe] = N \times ((n+1) \cdot M), \quad [\theta] = ((n+1) \cdot M) \times 1.$$

It is, therefore

$$[Y] = [\gamma XIe] \cdot [\theta], \tag{4.15}$$

or,

$$[\gamma XIe]^T \cdot [\gamma XIe] \cdot [\theta] = [\gamma XIe]^T \cdot [Y],$$

or finally

$$[\theta] = \left([\gamma XIe]^T \cdot [\gamma XIe]\right)^{-1} \cdot [\gamma XIe]^T \cdot [Y]. \tag{4.16}$$

Note that in this case the dimension of

$$[\theta] = \left(M \cdot (n+1) \times N\right) \cdot (N \times 1) = \left(M \cdot (n+1) \times 1\right).$$

For instance, when the rule base has $M = 7$ rules and the system has $n = 4$ inputs, the resulting dimension is

$$[\theta] = (7 \cdot (4+1) \times N) \cdot (N \times 1) = (7 \cdot (4+1) \times 1) = (35 \times 1).$$

and the resulting column vector

$$[\theta] = \left[\theta_0^1, \theta_1^1, \ldots, \theta_n^1; \cdots; \theta_0^M, \theta_1^M, \ldots, \theta_n^M\right]^T.$$

So, it is to be noted that once the parameters of antecedent fuzzy sets (using the Wang-Mendel's approach, or its modification, or by fuzzy clustering) are determined, which are required for computation of degree of fulfilment of each rule for a given set of (N training samples) inputs, the linear TS rule's consequent parameters can be determined easily by LSE technique as described above.

4.6 Forecasting Time Series Using the Fuzzy Logic Approach

In the forecasting examples described below, the shifted values $X(t + L)$ are the predicted values based on sampled past values of a time series up to the point t, *i.e.* $[X\{t-(D-1)d\}, X\{t-(D-2)d\}, ..., X(t-d), X(t)]$. Therefore, the predictor to be implemented should map from D sample data points, sampled at every d time units, as its input values to the predicted value as its output. Depending on the availability of the time series data and on its complexity, the D, d, and L values are selected. In our case $D = 4$ and $d = L = 6$ have been selected, corresponding to a four-inputs system, and to a sampling interval of six time units.

Hence, for each $t > 18$, the input data represent a four dimensional vector and the output data a scalar value

$$XI(t) = [X(t-18), X(t-12), X(t-6), X(t)]$$
$$XO(t) = [X(t+6)].$$

Supposing that there are m input-output data sets, we generally use the first $m/2$ input-output data values (training samples) for fuzzy rule generation and the remaining $m/2$ input data sets for verification of forecasting accuracy with the fuzzy logic approach.

4.6.1 Forecasting Chaotic Time Series: An Example

As an example, forecasting of a chaotic time series is considered in this section. The chaotic series are generated from deterministic nonlinear systems that are sufficiently complicated as they appear to be random, however, because of the underlying nonlinear deterministic maps that generate the series, chaotic time series are not random time series (Wang and Mendel, 1992). The chaotic series, for our experiment, is obtained by solving the Mackey-Glass differential equation (Junhong, 1997; MATLAB, 1998). Lapedes and Farber (1987) also used feedforward neural networks for the prediction of the same chaotic time series and reported that the neural network gave the best predictions in comparison with

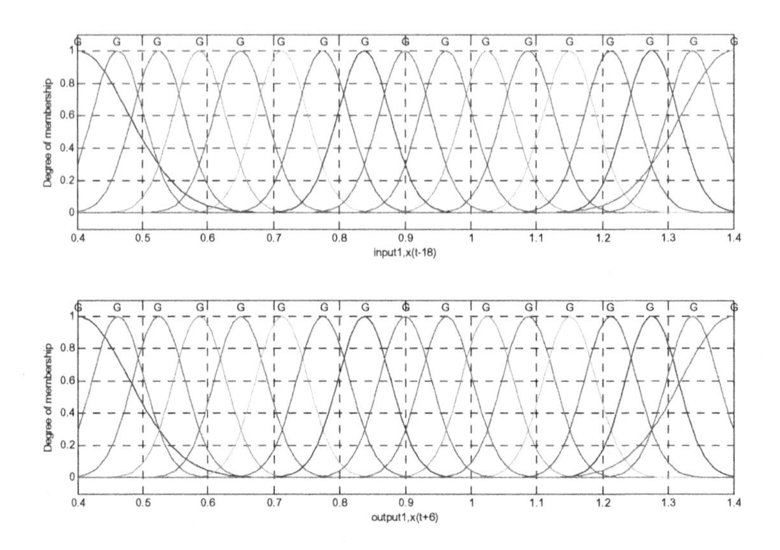

Figure 4.4(a). Input and output domains partitioning by 17 GMFs for rule generation

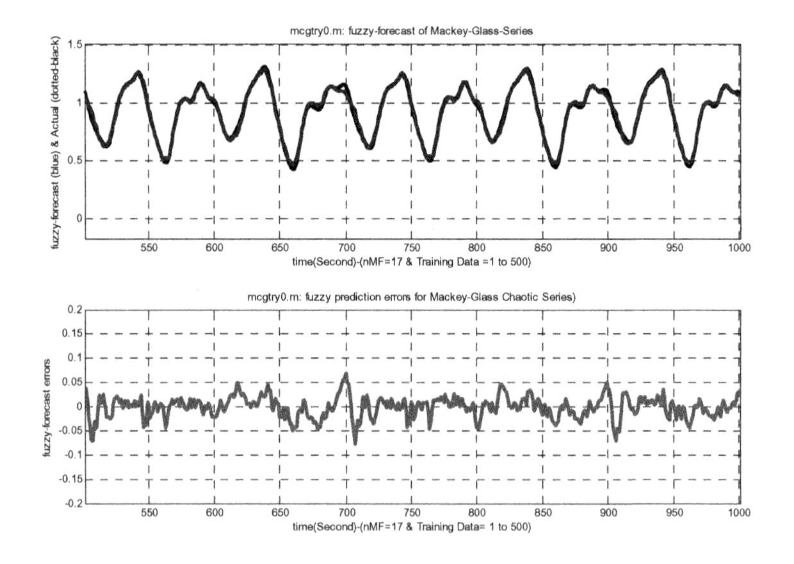

Figure 4.4(b). Forecasting chaotic series with fuzzy predictor with $n = 27$ GMFs

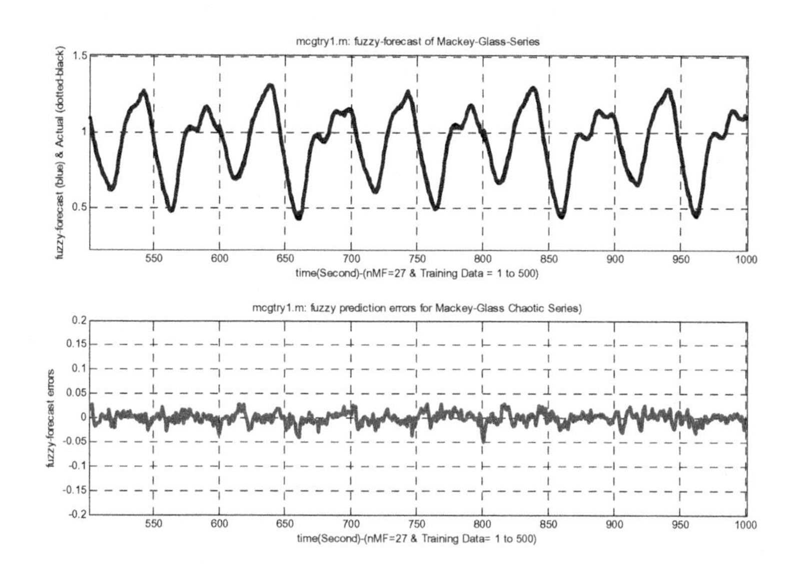

Figure 4.4(c). Forecasting chaotic series with fuzzy predictor with $n = 27$ GMFs

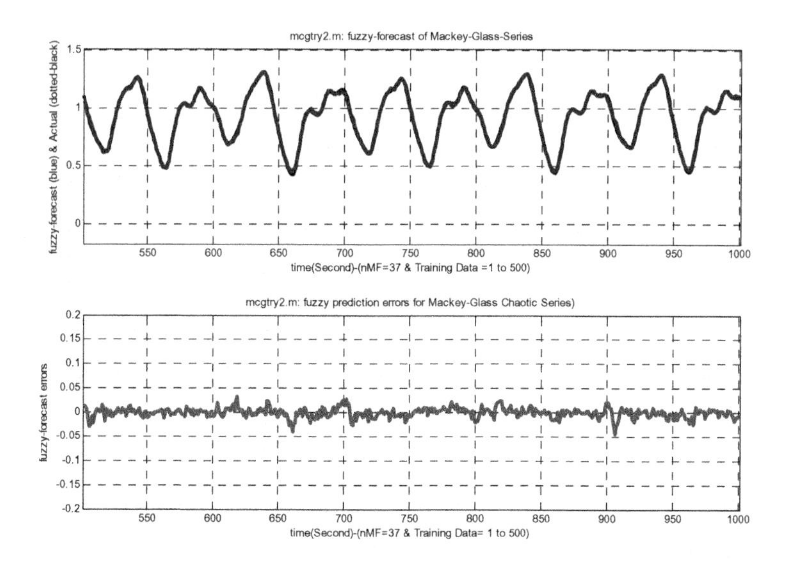

Figure 4.4(d). Forecasting chaotic series with fuzzy predictor with $n = 37$ GMFs

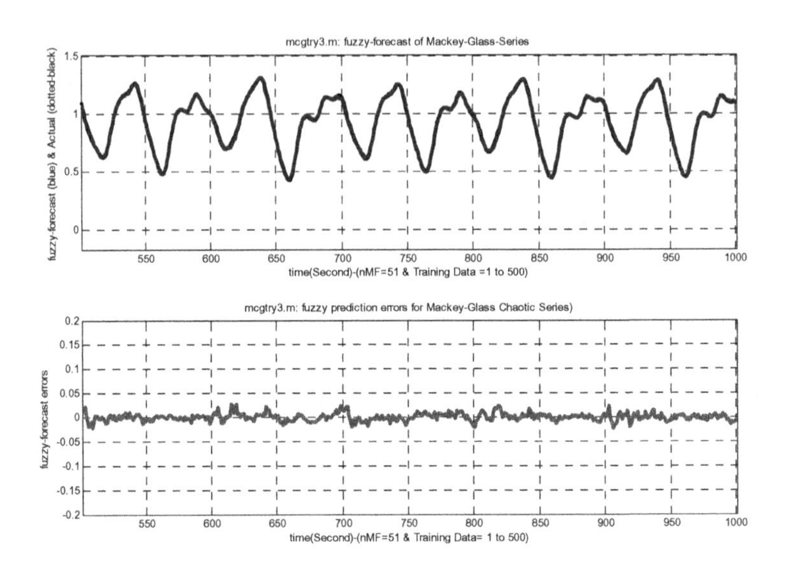

Figure 4.4(e). Forecasting chaotic series with fuzzy predictor with $n = 51$ GMFs

Table 4.1. Forecasting performance comparisons for various fuzzy predictors with 500 to 1000 input data sets (evaluation data) for Mackey-Glass chaotic series

Sl. No.	No. of GMFs	SSE	RMSE
1.	17	0.1252	0.0224,
2.	27	0.0389	0.0125
3.	37	0.0255	0.0101
4.	51	0.0164	0.0081

conventional approaches, like the linear predictive method and Gabor polynomial method, *etc*. Here, we apply the fuzzy predictor to forecast the future values of Mackey-Glass chaotic time series data that were directly obtained from MATLAB version 5.2 as "mgdata.dat" (MATLAB, 1998).

From the numerical data of the chaotic series, a fuzzy logic system capable of forecasting the future values of the above time series was developed. For this purpose, neglecting the first 100 transient data points of the chaotic series, with the remaining data 1000 rows of the *XIO* matrix have been built, out of which the first 500 rows (training data) are used for rules generation and the remaining 500 rows (validation data) are used for verification of forecasting accuracy, and for all four inputs and for the output the domain interval $[X_{lo}, X_{hi}] \equiv [0.4, 1.4]$ have been selected. Four different fuzzy logic systems with 17, 27, 37, and 51 Gaussian membership functions (GMFs) have been investigated.

For the fuzzy predictor, the Mamdani-type fuzzy rules were initially generated using the first 500 rows (training data) of the *XIO* matrix; thereafter, redundant and conflicting rules were removed from the rule list. For that purpose, only $c_1 = 0.4$ and $c_n = 1.4$ were selected and the values $c_2, c_3, \ldots, c_{n-1}$ were calculated for equal divisions of all (n-1) intervals. For the first system, *i.e.* with $n = 17$ GMFs, $\sigma_a = 0.08$ and $\sigma_b = 0.04$ were selected. Similarly, $\sigma_a = 0.08$ and $\sigma_b = 0.02$ were selected for the second and third systems (with $n = 27$ and 37 GMFs), whereas $\sigma_a = \sigma_b = 0.02$ were selected for the fourth (with $n = 51$ GMFs) fuzzy predictor. Figure 4.4(a) shows the partitioning of universes of discourse for the first input and output of the predictor with the $n = 17$ GMFs, and Figure 4.4(b) through Figure 4.4(e) show the results of forecasting, along with the forecasting errors, for the investigated systems. Note that, because of good prediction accuracy, forecasted series can hardly be distinguished from the original chaotic series except for Figure 4.4(b).

The performance functions like SSE ($0.5E^T.E$), with E as a column vector of errors and T indicating transposition of the E vector, and RMSE indicating the efficiency of the individual fuzzy system investigated, are also computed and listed in Table 4.1 for mutual comparison. The results from the Table 4.1 confirm the high suitability of the proposed approach, based on automatically generated fuzzy rules for forecasting of Mackey-Glass chaotic time series. From Table 4.1 it also follows that, when the number of GMFs is increased from 17 to 51, the forecasting accuracy is significantly increased.

4.7 Rules Generation by Clustering

Automated data driven rule generation, as described above, works considerably well for nonlinear time series modelling and forecasting. However, the fuzzy rule base generated in this way is generally very large, because each set of input-output pair generates a fuzzy rule. This is true even after the removal of redundant and conflicting rules from the rule base generated. For instance, using the first 500 input-output data sets of Mackey-Glass chaotic time series and using 27 number of GMFs, which are used for partitioning of input and output universes of discourse, the generated fuzzy rules, after the removal of conflicting and redundant rules, are still of the order of 350. This definitely imposes a large amount of computational load for fuzzy inferencing. To avoid this, an alternative approach, based on a ***fuzzy clustering algorithm*** was proposed that, for instance, uses only a few fuzzy rules for nonlinear time series modelling and forecasting.

4.7.1 Fuzzy Clustering Algorithms for Rules Generation

Clustering algorithms are mathematical tools useful in identifying the natural groupings of data, based on common similarities, from a large data set to produce a concise representation of a system's behaviour. Most clustering algorithms are unsupervised and do not rely on assumptions common to statistical classification methods, such as the statistical data distribution. They are, therefore, very appropriate for situations where little *a priori* knowledge exists. The data classification capability of clustering algorithms has been widely exploited in pattern recognition, image processing, and nonlinear system modelling. In what follows, we will introduce the reader to the clustering theory and present some fuzzy clustering algorithms, based on the ***c-means functional***. For an in-depth treatment of fuzzy clustering, readers may refer to the classical monograph by Jain and Dube (1988); for an overview of different clustering algorithms, refer to Bezdek and Pal (1992), Babuška (1996), and Setnes (2000).

4.7.1.1 Elements of Clustering Theory

Clustering techniques essentially try to group data samples in feature space and they form the basis of many classification and system modelling algorithms. They are applied to data that could be numerical (quantitative), qualitative (categorical), or a mixture of both. Our attention here will be focused on clustering of quantitative data, which might be observations of some physical process, such as time series data. It will be supposed that each observation consists of n variables, grouped into an n-dimensional column vector

$$\mathbf{Z}_s = [Z_{1s}, Z_{2s}, \cdots, Z_{ns}]^T, \mathbf{Z}_s \in \mathbb{R}^n.$$

A set of N observations is described by

$$\mathbf{Z} = \{Z_s | s = 1, 2, \cdots, N\},$$

and is represented by $n \times N$ ***pattern matrix*** Z:

$$Z = \begin{bmatrix} z_{11} & z_{12} & \cdots & z_{1N} \\ z_{21} & z_{22} & \cdots & z_{2N} \\ \vdots & \vdots & & \vdots \\ z_{n1} & z_{n2} & \cdots & z_{nN} \end{bmatrix}.$$

The rows and columns of the ***pattern matrix***, in pattern recognition terminology, are respectively called ***features*** (or ***attributes***) and ***patterns*** (or ***objects***). The pattern matrix Z is also called the ***data matrix***, and in control engineering, for example, each row of a data matrix may represent one of the ***process variables*** like pressure, temperature, flow, *etc.*, whereas the columns may indicate the time point of sampling.

Clusters are usually defined as groups of objects mutually more similar within the same groups than with the members of other clusters (Bezdek,1981; Jain and Dube, 1988), whereby the term "similarity" should be understood as mathematical similarity, measured in some well-defined sense. In metric spaces, similarity is often defined by means of a distance norm that is measured among the data vectors themselves, or as a distance from a data vector to some prototypical object or center of the cluster. The cluster centers are usually not known beforehand and are, therefore, determined simultaneously by the clustering algorithm while partitioning the data. The prototypes may be a vector of the same dimension as the data objects, and they can also be defined as geometrical objects, such as linear or nonlinear subspaces or functions. Data can reveal clusters of different geometrical shapes, sizes, and densities, such as spherical, ellipsoid, or as linear and nonlinear subspaces of data space.

Various clustering algorithms have been proposed in the literature, and these can be classified according to whether the clusters – seen as subsets of the entire data set - are fuzzy or crisp. Clustering algorithms, based on classical set theory, classify the individual objects according to their belonging or not belonging to a cluster, which is known as ***hard clustering***. Here, the partitioning of data is such that any particular object can be a member of only one particular subset of data or of a particular cluster.

Fuzzy clustering algorithms, however, allow the objects to belong to several clusters simultaneously, but with different degrees of membership, which in many situations is more natural than hard clustering. For instance, in this case the objects on the boundaries between several clusters are not forced to belong fully to one of the classes, but rather are assigned membership degrees between 0 and 1, indicating their partial membership.

On the other hand, the discrete nature of hard partitioning also causes difficulties with algorithms based on analytic functionals, since these functionals are not differentiable. Clustering algorithms may use an objective function to measure the desirability of partitions. Nonlinear optimization algorithms are used to search for local optima of the objective function. The concept of fuzzy partition is essential for cluster analysis, and consequently also for the identification techniques based on fuzzy clustering.

4.7.1.2 Hard Partition

A hard partition can be considered as a group of subsets formulated in terms of classical sets. The objective of hard clustering is to partition the given data set $Z = \{z_1, z_2, \cdots, z_N\}$ into c clusters, also called groups or classes. We initially assume that the number of clusters, *i.e.* c is known a priori, based on some prior knowledge about the dynamics of the system that generated the data set Z. Using classical sets, a hard partition of Z can be defined as a family of subsets $\{A_g \mid 1 \le g \le c\}$ with the following properties (Bezdek, 1981):

$$
\begin{aligned}
&\bigcup_{g=1}^{c} A_g = Z, \\
&A_g \cap A_h = 0, \quad 1 \le g \ne h \le c, \\
&0 \subset A_g \subset Z, \quad 1 \le g \le c.
\end{aligned}
\tag{4.17}
$$

The first of the above equations implies that the union of all subsets A_g contains all the data. The second and third equations respectively suggest that the intersection of the subsets must be a void set, *i.e.* subsets are disjoint, and none of the subsets is empty or contains all the data contained in *Z*. In terms of membership functions, a partition can be conveniently represented by the partition matrix:

$$U = \left[\mu_{gs}\right]_{c \times N}.$$

That is, the gth row of this partition matrix contains the values of the membership function μ_g of the gth subset A_g of *Z*. Therefore, it can be represented as

$$U = \begin{bmatrix} \mu_{11}\ \mu_{12} \cdots \mu_{1N} \\ \mu_{21}\ \mu_{22} \cdots \mu_{2N} \\ \vdots \quad \vdots \quad \vdots \\ \mu_{c1}\ \mu_{c2} \cdots \mu_{cN} \end{bmatrix}_{c \times N} \tag{4.18}$$

It follows from the above equation that the elements of the *U* partition matrix must satisfy the following conditions:

$$\mu_{gs} \in \{0,1\},\ 1 \le g \le c; 1 \le s \le N; \tag{4.19a}$$

$$\sum_{g=1}^{c} \mu_{gs} = 1,\ 1 \le s \le N; \tag{4.19b}$$

$$0 < \sum_{s=1}^{N} \mu_{gs} < N,\ 1 \le g \le c. \tag{4.19c}$$

The space of all possible hard partition matrices for *Z*, called the hard partitioning space (Bezdek, 1981), is thus defined by

$$M_{hc} = \left\{ U \in \mathbb{R}^{c \times N} \middle| \mu_{gs} \in \{0,1\}, \forall g,s; \sum_{g=1}^{c} \mu_{gs} = 1, \forall s; 0 < \sum_{s=1}^{N} \mu_{gs} < N, \forall g \right\}.$$

In the following, let us illustrate the hard partitioning concept by an example with the given data set $Z = \{z_1, z_2, \cdots, z_N\}$, where $N = 10$. Suppose that the given data set is hard partitioned into three clusters A_1, A_2 and A_3. The partition matrix *U* in this case may look like:

$$U = \begin{bmatrix} 1,1,1,0,0,0,0,0,0,0 \\ 0,0,0,1,1,1,0,0,0,0 \\ 0,0,0,0,0,0,1,1,1,1 \end{bmatrix}$$

From the U matrix it is seen that the elements z_1, z_2, and z_3 possibly belong to cluster A_1 (as the first three entries in the first row are 1), and z_4, z_5, and z_6 belong to cluster A_2, whereas the remaining data elements z_7 to z_{10} belong to cluster A_3. Here, note that the sum of each column of the partition matrix U is always 1.

4.7.1.3 Fuzzy Partition

A fuzzy partition can be considered as a generalization of the hard partition and this follows directly by allowing μ_{gs} to attain any real values within [0,1] (Babuška, 1996). Similar to hard partitioning the conditions for a fuzzy partition matrix are described by Ruspini (1970):

$$\mu_{gs} \in [0,1], \ 1 \le g \le c; 1 \le s \le N; \tag{4.20a}$$

$$\sum_{g=1}^{c} \mu_{gs} = 1, \ 1 \le s \le N; \tag{4.20b}$$

$$0 < \sum_{s=1}^{N} \mu_{gs} < N, \ 1 \le g \le c. \tag{4.20c}$$

Similar to hard partitioning, the gth row of the partition matrix U contains the values of the membership function μ_g of the gth subset A_g of Z. The fuzzy partitioning space for Z is the set

$$M_{fc} = \left\{ U \in \mathbb{R}^{c \times N} \,\middle|\, \mu_{gs} \in [0,1], \forall g, s; \sum_{g=1}^{c} \mu_{gs} = 1, \forall s; 0 < \sum_{s=1}^{N} \mu_{gs} < N, \forall g \right\}.$$

Let us now illustrate the fuzzy partitioning concept by an example with the same data set $Z = \{z_1, z_2, \cdots, z_N\}$, where $N = 10$, as used in the hard partitioning example. Suppose that the given data set is fuzzy partitioned into three clusters A_1, A_2 and A_3. The partition matrix U in this case may look like

$$U = \begin{bmatrix} 0.82, & 0.90, & 0.96, 0.20, 0.10, 0.02, 0.03, 0.05, 0.1, 0.02 \\ 0.05, & 0.06, & 0.02, 0.75, 0.85, 0.90, 0.17, 0.25, 0.3, 0.08 \\ 0.13, & 0.04, & 0.02, 0.05, 0.05, 0.08, 0.80, 0.70, 0.6, \ 0.90 \end{bmatrix}.$$

Here, the elements in the first row of the matrix correspond to the degrees of membership of the elements z_1, z_2, ..., z_{10} respectively in the cluster or subset A_1. Similarly, entries in the second row and third row of the U matrix represent the degrees of membership of the data elements z_1, z_2, ..., z_{10} in the clusters A_2 and A_3 respectively. In addition, the entries in the U matrix are not restricted to 0 and 1 but can take any real value within 0 and 1. Moreover, the sum of each column of the U matrix is also equal to 1 in this case. If this restriction is relaxed, *i.e.* the sum of degrees of membership of any particular data element in the various clusters need not be 1, then we have ***possibilistic partition***, a special case of fuzzy partition and very useful in identifying ***outliers***. Outliers are data points that are neither a member of any cluster nor are they in the boundary of any cluster, but they are far apart from any cluster. These can be easily detected from the partition matrix of a possibilistic partition, as the member element of a particular cluster will have degree of membership 1.0 and the boundary points of two clusters may have a degree of membership close to 0.5 for both the two clusters, whereas outliers may have degree of membership as low as 0.01 in all clusters, indicating that the said data point (probably noise) is far off from all clusters.

4.7.2 Fuzzy c-means Clustering

The fuzzy *c*-means clustering algorithm is one of the most popular clustering algorithms used for data-driven automated fuzzy rules generation. The minimization of the *c*-means functional (4.21) represents a nonlinear optimization problem that can efficiently be solved using genetic algorithms; here, however the method chosen is a simple Picard iteration through the first-order conditions for stationary points of (4.21), known as the fuzzy *c*-means (FCM) algorithm.

The stationary points of the objective function (4.21) can be found by adjoining the constraint (4.20b) to J by means of Lagrange multipliers

$$\bar{J}(Z;U,V,\lambda)=\sum_{g=1}^{c}\sum_{s=1}^{N}\left(\mu_{gs}\right)^{m}D_{gsA}^{2}+\sum_{s=1}^{N}\lambda_{s}\left[\sum_{g=1}^{c}\mu_{gs}-1\right], \tag{4.21}$$

and by setting the gradients of J with respect to U, V and λ to zero. It can be proven that if $D_{gsA}^{2}>0, \forall g,s$ and $m>1$, then $(U,V)\in M_{fc}\times\mathbb{R}^{n\times c}$ may minimisz (4.21) only if

$$\mu_{gs}=\frac{1}{\sum_{h=1}^{c}\left(D_{gsA}/D_{hsA}\right)^{2/(m-1)}},\ 1\le g\le c; 1\le s\le N; \tag{4.22a}$$

and

$$v_{g}=\frac{\sum_{s=1}^{N}\left(\mu_{gs}\right)^{m}\cdot Z_{s}}{\sum_{s=1}^{N}\left(\mu_{gs}\right)^{m}};\quad 1\le g\le c. \tag{4.22b}$$

It is to be noted that this solution also satisfies the remaining constraints (4.20a) and (4.20c). Equations (4.22a) and (4.22b) are first-order necessary conditions for stationary points of the functional (4.21). The FCM algorithm iterates through Equations (4.22a) and (4.22b). Sufficiency of (4.21) and the convergence of the FCM algorithm is reported by Bezdek (1980). Also, note that Equation (4.22b) gives V_g as the weighted mean of the data items that belong to a cluster, where the weights are the membership degrees in the clusters. This being the reason why the algorithm is called "fuzzy *c*-means." The FCM algorithm is described next.

4.7.2.1 Fuzzy c-means Algorithm

Given the data set $Z=\{Z_1, Z_2, \cdots, Z_N\}$, select the number of clusters $1<c<N$, the weighting exponent (also called fuzziness exponent) $m > 1$, the termination tolerance $\varepsilon > 0$ and the norm-inducing matrix A. Initialize the partition matrix randomly, such that

$$U^{(l=0)} \in M_{fc}.$$

Repeat for iterations $l = 1, 2, 3, \ldots$.

- **Step 1:** compute the cluster prototypes or cluster centres (means)

$$v_g^{(l)} = \frac{\sum_{s=1}^{N} \left(\mu_{gs}^{(l-1)}\right)^m \cdot Z_s}{\sum_{s=1}^{N} \left(\mu_{gs}^{(l-1)}\right)^m}; \quad 1 \le g \le c.$$

- **Step 2:** compute the distances:

$$D_{gsA}^2 = \left(Z_S - v_g^{(l)}\right)^T A\left(Z_s - v_g^{(l)}\right), \quad 1 \le g \le c; 1 \le s \le N;$$

- **Step 3:** update the partition matrix

  if $D_{gsA} > 0$, for all $g = 1, 2, 3, \ldots, c$.

$$\mu_{gs}^{(l)} = \frac{1}{\sum_{h=1}^{c} \left(D_{gsA}/D_{hsA}\right)^{2/(m-1)}}, \quad 1 \le g \le c; 1 \le s \le N,$$

  else,

$$\mu_{gs}^{(l)} = 0 \text{ and } \mu_{gs}^{(l)} \in [0,1], \text{ with } \sum_{g=1}^{c} \mu_{gs}^{(l)} = 1$$

  until

$$\left\| U^{(l)} - U^{(l-1)} \right\| < \varepsilon.$$

Listed below are a few general remarks on the fuzzy c-means algorithm.

1. The "if and else" branch at step 3 takes care of singularity that occurs in fuzzy c-means when the distance term $D_{gsA} = 0$ for some Z_s and certain

cluster prototypes v_g. In such case, the membership degree cannot be computed at all and, therefore, zero is assigned to that μ_{gs} and the memberships are distributed arbitrarily among other clusters subject to the constraint that the sum of the degree of membership in each column of the U partition matrix must be one.

2. The fuzzy c-means algorithm converges to a local minimum of the c-means functional. Therefore, different initialization may lead to different results.

3. While steps 1 and 2 are straightforward, step 3 is a bit more complicated, as a singularity in the fuzzy c-means occurs when distance $D_{gsA} = 0$ for some Z_s and one or more v_g, though it is very rare in practice.

4. In the above iterative optimization scheme used by fuzzy c-means loops through the estimates,

$$U^{(l-1)} \rightarrow v^{(l)} \rightarrow U^{(l)}$$

and terminates as soon as

$$U^{(l)} - U^{(l-1)} < \varepsilon .$$

Alternatively, the algorithm can be initialized with $v^{(0)}$, loop through

$$v^{(l-1)} \rightarrow \mathbf{U}^{(l)} \rightarrow v^{(l)},$$

and terminate when

$$\left(v^{(l)} - v^{(l-1)}\right) < \varepsilon .$$

The error norm (termination tolerance) in the termination criterion is usually chosen as

$$\underset{gs}{Max}\left(abs\left(\mu_{gs}^{(l)} - \mu_{gs}^{(l)}\right)\right).$$

Different results may be obtained with the same values of termination tolerance, since the termination criterion used in the algorithm requires that more parameters become close to one another.

4.7.2.1.1 Parameters of Fuzzy c-means Algorithm

The following parameters must be specified before the fuzzy c-means algorithm is executed: the number of clusters c, the fuzziness exponent m, the termination tolerance and norm-inducing matrix A. Moreover, the fuzzy partition matrix U must be initialized. The choices for these parameters are now described next.

Number of clusters
The total number of clusters c is the most important parameter, as the remaining parameters have little influence on the resulting partition: when clustering real data without any prior information about the structures in the data, one usually has to make assumptions about the number of underlying clusters. The clustering algorithm chosen then searches for c clusters regardless of whether they are really present in the data or not. Two main approaches to determining the appropriate number of clusters in the data can be distinguished:

A. Validity measures
Validity measures are scalar indices that assess the goodness of the partition obtained. Clustering algorithms generally aim at locating well-separated and compact clusters. When the number of clusters is chosen equal to the number of groups that are actually present in the data, it is expected that the clustering algorithm will identify them correctly. When this is not the case, misclassifications appear, and the clusters are not likely to be well-separated and compact. Hence, most cluster validity measures are open to interpretation and can be formulated in different ways. Consequently, many validity measures have been introduced in the literature (Bezdek, 1981; Gath and Geva, 1989; Pal and Bezdek, 1995). For the FCM algorithm, the ***Xie-Beni index*** (Xie and Beni, 1991)

$$\chi(Z;U,V)=\frac{\sum_{g=1}^{c}\sum_{s=1}^{N}\mu_{gs}^{m}\cdot\left\|Z_s-v_g\right\|^2}{c\cdot\min_{g\neq h}\left(\left\|Z_s-v_g\right\|^2\right)} \tag{4.23}$$

has been found to perform well in practice. This index can be interpreted as the ratio of the total within-group variance and the separation of the cluster centers. The best partition minimizes the value of $\chi(Z;U,V)$.

B. Iterative merging
In the iterative cluster merging, one starts with a sufficiently large number of clusters and successively by merging clusters, that are similar (compatible) with respect to some well-defined criteria (Krishnapuram and Freg, 1992; Kaymak and Babuška, 1995), the number of clusters is reduced. One can also adopt the opposite approach, *i.e.* start with a small number of clusters and iteratively insert clusters in the region where the data points have a low degree of membership in the existing clusters (Gath and Geva, 1989).

Fuzziness parameter
The fuzziness exponent or weighting exponent m is a rather important parameter that is to be selected properly as well. This is because it significantly influences the fuzziness of the resulting partition. As m approaches to one, the partition becomes hard partition ($\mu_{gs}\in\{0,1\}$) and v_g are ordinary means of the clusters. On the other hand, as $m\to\infty$, the partition becomes completely fuzzy (μ_{gs} = $1/c$) and the

cluster means are all equal to the mean of Z. These limit properties of fuzzy c-means functionals are independent of optimization method used (Pal and Bezdek, 1995). Usually, m is selected as 2.

Termination criterion

The FCM algorithm stops iterating when the norm of the difference between U in two successive iterations is smaller than the termination tolerance parameter. The usual choice of a termination tolerance is 0.001. The termination tolerance of 0.01 also works well in most cases, while it drastically reduces the computing times.

Norm-inducing matrix

The shape of the clusters is dependent on the choice of the norm-inducing matrix A in the distance measure. A common choice of the norm-inducing matrix A is the identity matrix I, which gives the standard Euclidean norm:

$$D_{gs}^2 = \left(Z_s - v_g\right)^T \left(Z_s - v_g\right).$$

Another choice of the norm-inducing matrix A is a diagonal matrix that accounts for different variances in the directions of the coordinate axes of Z:

$$A = \begin{bmatrix} \sigma_1^{-2} & 0 & \cdots & 0 \\ 0 & \sigma_2^{-2} & \cdots & 0 \\ \vdots & \vdots & & \vdots \\ 0 & 0 & \cdots & \sigma_n^{-2} \end{bmatrix}$$

This matrix induces a diagonal norm on $\mathbb{R}^n$. Finally, A can be defined as the inverse of the covariance matrix of Z: $A = R^{-1}$, with $R = \frac{1}{N}\sum_{s=1}^{N}\left(Z_s - \bar{Z}\right)\left(Z_s - \bar{Z}\right)^T$.

Here, $\bar{Z}$ denotes the mean of the data. In this case A induces the ***Mahalanobis norm*** on $\mathbb{R}^n$. The norm influences the clustering criterion by changing the measure of dissimilarity. The Euclidean norm induces hyperspherical clusters (hyperspheres are surfaces of constant memberships). Both the diagonal and the Mahalanobis norm generate hyperellipsoidal clusters. With the diagonal norm, the axes of the hyperellipsoids are parallel to the coordinate axes, while with the Mahalanobis norm the orientation of the hyperellipsoids is arbitrary. A common limitation of clustering algorithms based on a fixed distance norm is that it forces the objective function to prefer clusters of a certain shape even if they are not present in the data.

Initial partition matrix

The partition matrix is usually initialized at random, such that $U \in M_{fc}$. A simple approach to obtain such U is to initialize the cluster centers v_g at random and compute the corresponding U by computing the distance and, thereafter, using the last step of FCM algorithm.

4.7.3 Gustafson - Kessel Algorithm

In order to detect clusters of different geometrical shapes in one data set, the standard FCM clustering algorithm is extended by employing an adaptive distance norm (Gustafson and Kessel, 1979). In this case, each cluster has it's own norm-inducing matrix A_g, which yields the following inner-product norm:

$$D_{gsA_g}^2 = \left(Z_s - v_g\right)^T A_g \left(Z_s - v_g\right), \; 1 \le g \le c; 1 \le s \le N; \tag{4.24a}$$

The matrices A_g are used as optimization variables in the c-means functional, thus allowing each cluster to adapt the distance norm to the local topological structure of the data. The objective functional of the Gustafson-Kessel algorithm is defined by:

$$J\left(Z;U,V,\{A_g\}\right) = \sum_{g=1}^{c} \sum_{s=1}^{N} \left(\mu_{gs}\right)^m D_{gsA_g}^2 \tag{4.24b}$$

This objective function cannot be directly minimized with respect to A_g, since it is linear in A_g. To obtain a feasible solution, A_g must be constrained in some way. The usual way of accomplishing this is to constrain the determinant of A_g:

$$\det\left(A_g\right) = \rho_g, \; \rho_g > 0, \forall g. \tag{4.24c}$$

Allowing the matrix A_g to vary, with it's determinant fixed, corresponds to optimizing the cluster's shape while it's volume remains constant. By using the Lagrange-multiplier method, the following expression for A_g is obtained (Gustafson and Kessel, 1979):

$$A_g = \left[\rho_g \det\left(F_g\right)\right]^{1/n} \cdot F_g^{-1} \tag{4.24d}$$

where F_g is the fuzzy covariance matrix of the gth cluster given by

$$F_g = \frac{\sum_{s=1}^{N} \left(\mu_{gs}\right)^m \cdot \left(Z_s - v_g\right)\left(Z_s - v_g\right)^T}{\sum_{s=1}^{N} \left(\mu_{gs}\right)^m}; \; 1 \le g \le c. \tag{4.24e}$$

Note that the substitution of Equations (4.24d) and (4.24e) into (4.24a) gives a generalized squared Mahalanobis distance norm, where the covariance is weighted by the membership degrees in U. The Gustafson-Kessel algorithm is given in next section. The Gustafson-Kessel algorithm is computationally more expensive than

FCM, since the inverse and the determinant of the cluster covariance matrix must be calculated in each iteration.

4.7.3.1 Gustafson-Kessel Clustering Algorithm

Given the data set $Z=\{Z_1,Z_2,\cdots,Z_N\}$, select the number of clusters $1<c<N$, the weighting exponent or fuzziness exponent parameter $m > 1$, the termination tolerance $\varepsilon>0$ and the cluster volumes ρ_g. Initialize the partition matrix randomly, such that

$$U^{(l=0)} \in M_{fc}.$$

Repeat for iterations $l = 1, 2, 3, \ldots$

- **Step 1** compute the cluster prototypes or cluster centres (means)

$$v_g^{(l)} = \frac{\sum_{s=1}^{N}\left(\mu_{gs}^{(l-1)}\right)^m \cdot Z_s}{\sum_{s=1}^{N}\left(\mu_{gs}^{(l-1)}\right)^m}; \quad 1 \le g \le c.$$

- **Step 2** compute the cluster covariance matrices

$$F_g = \frac{\sum_{s=1}^{N}\left(\mu_{gs}^{(l-1)}\right)^m \cdot \left(Z_s - v_g^{(l)}\right)\left(Z_s - v_g^{(l)}\right)^T}{\sum_{s=1}^{N}\left(\mu_{gs}^{(l-1)}\right)^m}; \quad 1 \le g \le c.$$

- **Step 3** compute the distances

$$D_{gsA_g}^2 = \left(Z_s - v_g^{(l)}\right)^T \left[\rho_g \det\left(F_g\right)^{1/n} \cdot F_g^{-1}\right]\left(Z_s - v_g^{(l)}\right),$$

$$1 \le g \le c; 1 \le s \le N;$$

- **Step 4** update the partition matrix:

for $1 \le s \le N$

if $D_{gsA} > 0$ for all $g = 1, 2, \cdots, c$;

$$\mu_{gs}^{(l)} = \frac{1}{\sum_{h=1}^{c}\left(D_{gs\mathbf{A}_g}/D_{hs\mathbf{A}_g}\right)^{2/(m-1)}}, 1 \le g \le c; 1 \le s \le N;$$

else

$$\mu_{gs}^{(l)} = 0 \text{ and } \mu_{gs}^{(l)} \in [0,1], \text{ with } \sum_{g=1}^{c} \mu_{gs}^{(l)} = 1 .$$

until

$$\left\| U^{(l)} - U^{(l-1)} \right\| < \varepsilon.$$

4.7.3.1.1 Parameters of Gustafson-Kessel Algorithm

The same parameters must be specified beforehand in the Gustafson-Kessel (GK) clustering algorithm as for the fuzzy c-means algorithm (except for the norm-inducing matrix A, which is automatically adapted): the number of clusters c, the fuzziness exponent m, and the termination tolerance parameters. Additional parameters are the cluster volumes ρ_g. Without any prior knowledge, the cluster volumes is simply fixed at 1 for each cluster. Due to this constraint, the Gustafson-Kessel algorithm can only find clusters of approximately equal volumes. This is a drawback of this setting.

4.7.3.1.2 Interpretation of Cluster Covariance Matrix

The cluster covariance matrix provides important information about the shape and orientation of the cluster. The ratio of the lengths of the cluster's hyperellipsoids axes is given by the ratio of the square roots of the eigenvalues of the covariance matrix. The directions of the axes are given by the eigenvectors of covariance matrix. The Gustafson-Kessel algorithm can be used to detect clusters along linear subspaces of the data space. These clusters are represented by flat hyperellipsoids, which can be regarded as hyperplanes. The eigenvector corresponding to the smallest eigenvalue determines the normal to the hyperplane, and can be used to compute optimal local linear models from the covariance matrix.

4.7.4 Identification of Antecedent Parameters by Fuzzy Clustering

Using the given data, the identification of antecedent parameters of the Takagi-Sugeno model is usually done in two steps. In the first step, the antecedent fuzzy sets of the rules are determined. This can be done manually, from knowledge of the process, by interviewing the human experts, or by some data-driven technique, such as a neuro-fuzzy technique, or by the fuzzy clustering method described earlier, which produces a partitioning of the antecedent (input) space. Once the fuzzy antecedent parameters are determined, the LSE estimate described earlier is then applied in order to determine the consequent parameters of the Takagi-Sugeno rules. After obtaining both the antecedents fuzzy sets and rule's consequent parameters, the corresponding fuzzy rule base can be built easily. When observations have been obtained from a system or a process, an input matrix X and an output vector y can be constructed as follows:

$$X = \left[x_1, x_2, \cdots, x_{N_s}\right]^T, \quad y = \left[y_1, y_2, \cdots, y_{N_s}\right]^T$$

where N_s is the number of training data samples available for fuzzy identification. Now, for correct selection of input and output variables, the unknown nonlinear function $y = f(X)$ can be learnt from the data samples by means of regression techniques. The variables $x = (x_1, x_2, \cdots, x_n)^T \in \mathbb{R}^n$ and $y \in \mathbb{R}$ are called the regressor and regressand respectively. In order to determine the antecedent fuzzy sets of the Takagi-Sugeno rules, Babuška and Verbruggen (1995) proposed to apply either of the fuzzy clustering methods mentioned above in the Cartesian product space of $X \times y$ in order to partition the training data into characteristic regions, where the system's behaviours are approximated by a local linear model (rules). The pattern matrix Z to be clustered is formed by X and y as follows:

$$Z^T = [X, y]$$

Given the data Z and the number of clusters c, the fuzzy clustering algorithm can be applied to obtain the partitions of Z into c fuzzy clusters. A fuzzy partition can be represented as a $c \times N_s$ matrix U, whose entries are $\mu_{gs} \in [0,1]$ as described earlier. For the computation of the fuzzy partition matrix and the corresponding cluster prototypes (centers) GK clustering algorithm is usually applied, as it applies adaptive distance norms in order to detect clusters of different geometrical shapes, unlike the popular fuzzy c-means algorithm, which always identifies spherical-shape clusters in the data because of it's fixed distance norm. Because each cluster has it's own distance norm, induced by the fuzzy covariance matrix, that allows to adapt the local structures of the data. This evidently makes Gustafson-Kessel clustering superior for identifying subspaces of data (hyperplanes) that can be effectively modeled by the rules in the Takagi-Sugeno model.

Each cluster represents a certain operating region of the system, and the number of cluster centers or clusters c sought in the data equals the number of fuzzy rules implemented. Often, this number is not known *a priori*; thus, the optimum number of clusters is determined using suitable cluster validity measures.

The membership functions of the fuzzy sets in the premise of rules are obtained from the fuzzy partition matrix U, whose (g,s)th element $\mu_{gs} \in [0,1]$ is the membership degree of the input-output combination in the sth column of Z in cluster or data group g. To obtain the one-dimensional fuzzy set G_{gj}, the multidimensional fuzzy sets defined point-wise in the gth row of the partition matrix U are projected onto the space of input variables x_j:

$$\mu_{G_{gj}}\left(x_{js}\right) = \mathrm{proj}_j^{N_{n+1}}\left(\mu_{gs}\right),$$

where "proj" is the point-wise projection operator (Kruse *et al.*, 1994). The point-wise fuzzy sets G_{gj} are typically non-convex. However, the core and the corresponding left and right parts of the set can be recognized.

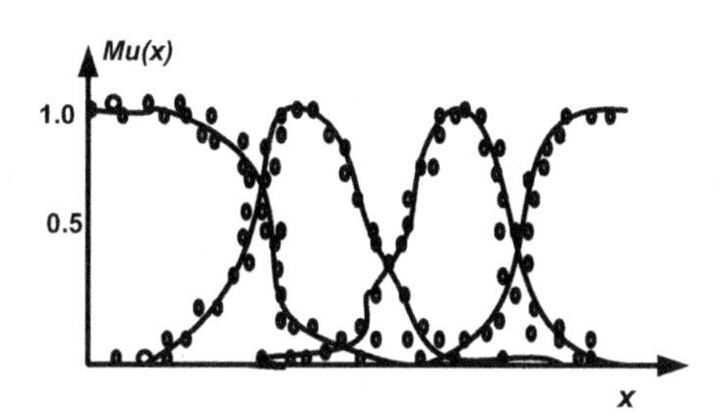

Figure 4.5. Parametric function fitting (solid line) to obtain one-dimensional antecedent fuzzy sets from point-wise projection (dots) of rows of fuzzy partition matrix

To obtain convex (unimodal) fuzzy sets, for the computation of $\mu_{G_{gj}}(x_j)$ for any value of x_j, the fuzzy sets are approximated by fitting suitable parametric membership functions (say, Gaussian type) to the point-wise projection (Babuška, 1996) as illustrated in Figure 4.5. After determination of the antecedent fuzzy sets, the LSE estimate is applied, as usual, to determine the rule consequent parameters.

4.7.5 Modelling of a Nonlinear Plant

In order to demonstrate the efficiency of the clustering-based fuzzy model, the second-order nonlinear plant (4.25) that was studied by Wang and Yen (1999) and Roubos and Setnes (2001) is considered here.

$$y(k) = g\left(y(k-1), y(k-2)\right) + u(k),$$

with,

$$g\left(y(k-1), y(k-2)\right) = \frac{y(k-1)\,y(k-2)\left(y(k-1)-0.5\right)}{1+y^2(k-1)\,y^2(k-2)} \tag{4.25}$$

The goal is to approximate the nonlinear component $g(y(k\text{-}1), y(k\text{-}2))$ of the plant with the fuzzy model. For this experiment, 400 data points were available, of which 200 samples of identification data were obtained with a random input signal $u(k)$ uniformly distributed in [-1.5, 1.5], followed by 200 samples of evaluation data obtained by using a sinusoidal input signal.

Table 4.2(a). Cluster centers (V) generated by Gustafson-Kessel algorithm

v_1 for input u, or X_1	v_2 for input y, or X_2	v_3 for output g
-0.4099	1.0024	0.2146
-0.8820	-0.8681	-0.3636
1.1066	-0.1326	-0.0055

Table 4.2(b). Variance parameters of GMFs determined from fitting the projected data

Serial number of antecedent GMFs	For input u, or X_1	For input y, or X_2
First GMF	2.1398	1.5698
Second GMF	1.0178	1.6112
Third GMF	0.9319	2.4221

Table 4.2(c). Consequents' parameters of Takagi-Sugeno rules

Theta0	Theta1	Theta2
-0.4706	0.0750	-0.0765
0.5056	0.1282	0.1685
-0.1839	0.4057	0.3783

Here, we apply the Gustafson-Kessel clustering algorithm to construct the desired fuzzy model using the first two columns of the $XIO = [u, y, g]$ matrix as the input data and the third column as desired output data, *i.e.* the data (pattern) matrix here is constructed as $Z = [XIO]^T$. The first 200 (training) samples (rows of XIO matrix) were used for fuzzy rules generation by applying the Gustafson-Kessel clustering algorithm using the following parameter settings: number of clusters $c = 3$, fuzziness exponent $m = 2$ and termination tolerance = 0.001. Accordingly, three clusters with cluster centers $V = [v_1, v_2, v_3]$ and partition matrix U of size 3×200 were obtained. Projecting the first two rows of the U matrix on to the input dimension and, thereafter, by fitting the Gaussian function of the form $y = \exp\{4 \cdot \log(2) \cdot (-(x - v_i)^2 / \sigma^2)\}$, three antecedent fuzzy membership functions for each input were obtained (Figure 4.6(a)).

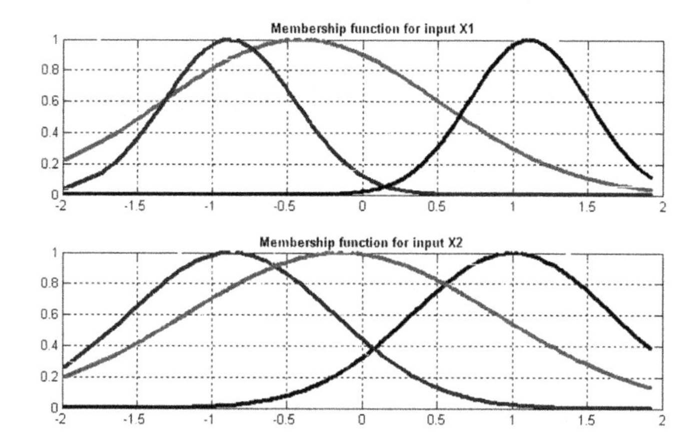

Figure 4.6(a). GMFs (three antecedent fuzzy sets) for input u (top) and input y (bottom).

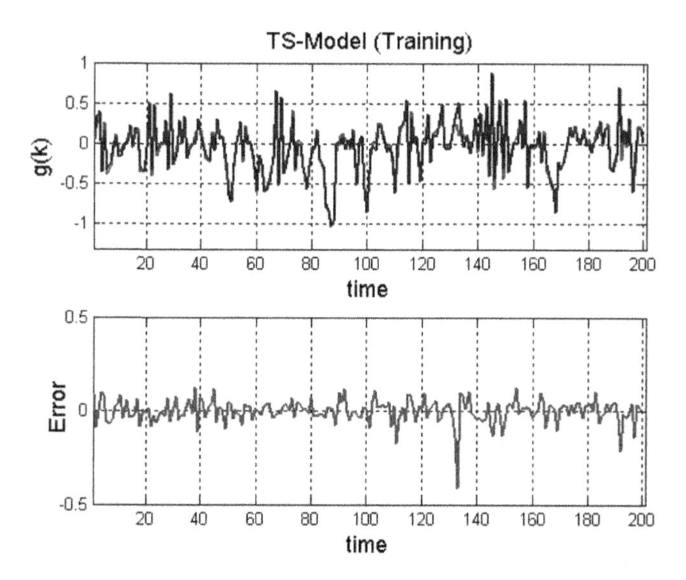

Figure 4.6(b). Actual output and fuzzy model predicted output with training data (top),

Thereafter, using the antecedent fuzzy sets and LSE estimation on the training data Takagi-Sugeno-type fuzzy rules' consequents were determined. Finally, the efficiency of the model was tested by applying the generated fuzzy rules on the evaluation data. The simulation results achieved are illustrated in Table 4.2(a) to Table 4.2(e) and in Figure 4.6(a) to Figure 4.6(c).

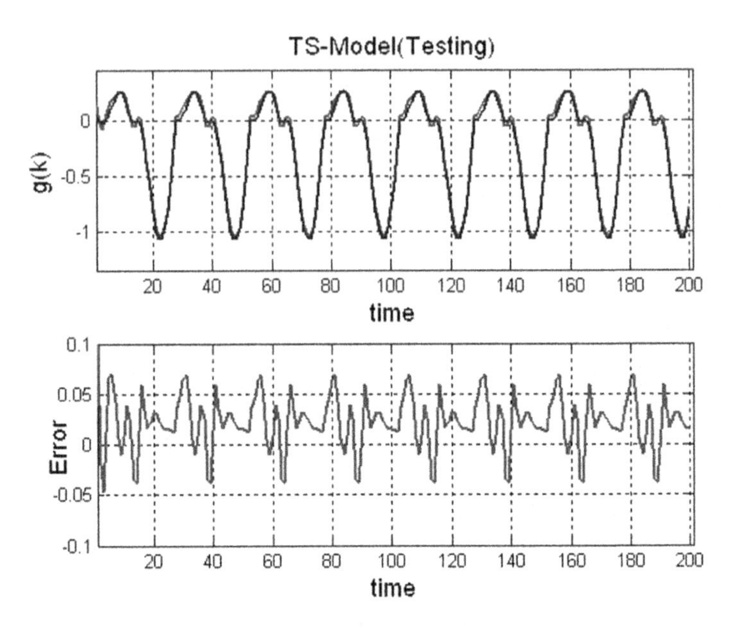

Figure 4.6(c). Performance of the Gustafson-Kessel clustering-based fuzzy model with evaluation data (top) and prediction error (bottom)

Table 4.2(d). Simulation results for nonlinear plant modelling

With training data	With evaluation data
SSE(train) = 0.3973 MSE(train) = 0.0040	SSE(eval.) = 0.1215 MSE(eval.) = 0.0012

It is to be noted that the fuzzy model generated used only three Takagi-Sugeno fuzzy rules and six antecedent fuzzy sets (for two inputs), which are much less than that generated by the Wang-Mendel method or its modified approach.

4.8 Fuzzy Model as Nonlinear Forecasts Combiner

The need to combine forecasts of a time series has been well understood for a long time. It has already been mentioned in Chapter 3 that not just any arbitrary combination of forecasts is decisive in providing an improved forecast, but it is essential that it is a nonlinear combination of various forecasts of a given time series. The latter has been reconfirmed by many studies, which have revealed that only the nonlinearity provides the combination with the guarantee to produce better forecasts than either of the combination components separately. This is mainly because, here, we have a kind of synergic effect.

In this section we describe an application example where a fuzzy model has been used as a nonlinear forecasts combiner. For this purpose, we consider once again the temperature series discussed in Chapter 3, along with its two forecasted series. The temperature series selected is a non-stationary, non-seasonal time series. Moreover, the original temperature series with 226 observations was obtained from a chemical process by temporarily disconnecting the controllers from the pilot plant involved and recording the subsequent temperature fluctuation every minute (Box and Jenkins, 1976).

The two separate forecasts of the selected temperature time series were made, one by applying the Box-Jenkins ARMA/ARIMA method (Box and Jenkins, 1976) and the other by applying Holt's exponential smoothing technique (Chatfield, 1980). In order to utilize the fuzzy model as a nonlinear forecasts combiner, here, we used both the forecasted series as two inputs to the fuzzy model to be developed, and the original temperature series as the desired output from the fuzzy model. The two forecasted series and the original time series have been rearranged as the first, second and the third columns respectively of a HBXIO matrix. Thereafter, the first 150 rows from the HBXIO matrix were used as training data and the remaining rows, *i.e.* 151 to 224 rows of HBXIO matrix were used as test samples to evaluate the efficiency of the combination approach described (Palit and Popovic, 2000). It is to be noted that by applying conventional forecasting methods on the original temperature series we obtained only 224 data points in both cases.

Using the modified and automated rule-generation algorithm, Mamdani-type fuzzy rules were generated from the training data based on the implemented $n = 21$ GMFs, and fixing $X_{\text{lo}} = 18$, $X_{\text{hi}} = 28$, $\sigma_a = 0.4$, and $\sigma_b = 0.2$. Care has been taken to make the rule base somewhat compact by eliminating the conflicting rules and unnecessary redundant rules. Thereafter, a nonlinear combination of forecasts with the fuzzy model was generated, based on the above rule base and utilizing only the input data from the validation data sets (see Figure 4.7(b)). Finally, the performance of the approach was measured by computing performance indices, such as SSE, RMSE *etc.*, for the validation data set as illustrated in Table 4.3. From Table 4.3 it can be seen that the SSE and RMSE achieved with the proposed fuzzy model is much better than the individual forecast generated either by the Box-Jenkins method or by Holt's exponential smoothing technique. The reported result obviously confirms the high suitability of the fuzzy logic approach as a nonlinear forecasts combiner.

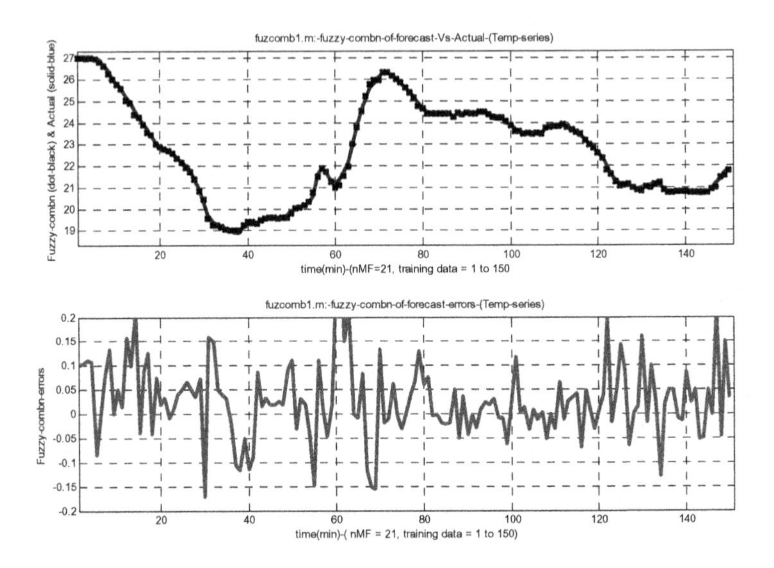

Figure 4.7(a). Nonlinear combination of forecasts using fuzzy model (with training data). Dots: fuzzy model output; solid line: desired output (upper part), prediction error (bottom)

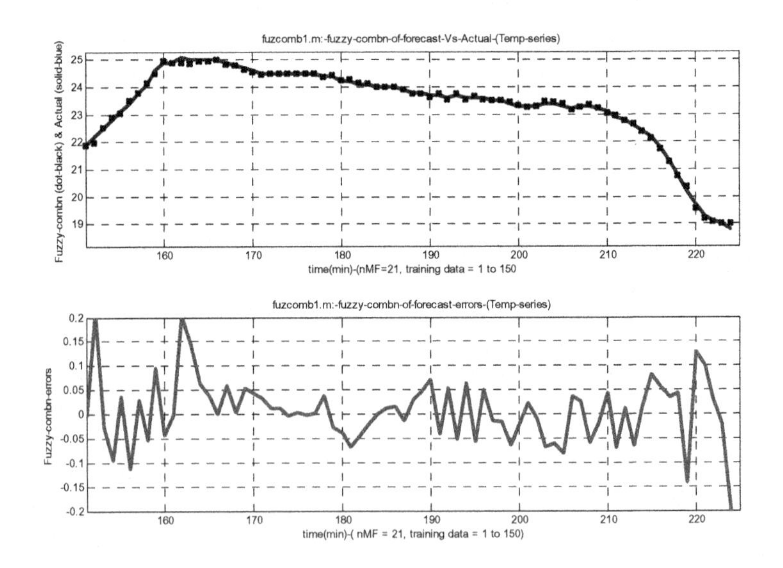

Figure 4.7(b). Nonlinear combination of two forecasts using fuzzy model (with test data). Dots: fuzzy model output; solid line: desired output (upper part), prediction error (bottom)

Table 4.3. Performance of the fuzzy model generated as nonlinear forecasts combiner

Sl. No.	Forecasts	Data from HBXIO matrix	SSE / RMSE
1.	Box and Jenkins	151–224 (column-1)	SSE = 0.4516 RMSE = 0.112
2.	Holt's method	151–224 (column-2)	SSE = 0.3174 RMSE = 0.0933
3.	Fuzzy model (Mamdani), 21 GMFs	1– 150 (training data)	SSE = 0.5155 RMSE = 0.0832
4.	Fuzzy model (Mamdani), 21 GMFs	151–224 (validation data)	SSE = 0.1680 RMSE = 0.0678

4.9 Concluding Remarks

In this chapter, various fuzzy models, such as the Mamdani model, the Takagi-Sugeno model and the relational fuzzy model, along with their corresponding inferencing mechanisms have been described. As the fuzzy inferencing mechanism relies on a well-consistent set of fuzzy rules, in order to generate the proper output in response to an unknown input set from the universe of discourse, various rule-generation algorithms based on Wang and Mendel's approach, or it's modification, and fuzzy clustering have also been presented in the chapter. The effectiveness of the fuzzy models generated has been tested on two application examples, namely the forecasting of chaotic time series and nonlinear plant modelling. In addition, a fuzzy model has also been applied as a nonlinear forecasts combiner. It is important to note that the primary objective of using a fuzzy model is to achieve an inspectable or interpretable model, unlike the black-box model of neural networks. However, it should be emphasized here that the fuzzy modelling approach described in this chapter rather primarily focuses on the function approximation accuracy than the inspectability of the model and, in fact, none of the methods presented in the chapter can guarantee model transparency issues. Therefore, the fuzzy model generated is eventually nothing but a replica of a neural-networks-like model, and needs to be treated further as discussed in Chapter 7, where the primary attention is paid to the improvement of model transparency.

References

[1] Babuška R (1996) Fuzzy Modelling for control, Ph.D thesis, Delft University of Technology, Netherlands.

[2] Babuška R and Verbruggen HB (1995) Identification of composite linear models via fuzzy clustering, Proc. of European Control Conference, Rome, Italy, pp. 1207-1212.

[3] Bezdek JC (1980) A convergence theorem for the fuzzy isodata clustering algorithms, IEEE Trans. Pattern Analysis Machine Intelligence, PAMI-2(1): 1-8.

[4] Bezdek JC (1981) Pattern recognition with fuzzy objective functions, Plenum Press, New York.

[5] Bezdek JC and Pal SK (Eds.) (1992) Fuzzy models for pattern recognition, IEEE Press, New York.

[6] Box GEP and Jenkins GM (1976) Time series analysis, Forecasting and Control, Holden Day.

[7] Chatfield C, (1980) The analysis of time series: An Introduction, Chapman and Hall, London, Second edition.

[8] Gath I and Geva AB (1989) Unsupervised optimal fuzzy clustering, IEEE Trans. on Pattern Analysis and Machine Intelligence, 11(7): 773-781.

[9] Gustafson and Kessel (1979) Fuzzy clustering with a fuzzy covariance matrix. In proceedings IEEE CDC, San Diego, CA, USA, pp. 761-766.

[10] Jain AK and Dubes RC (1988) Algorithm for clustering data, Prentice Hall, Englewood Cliffs.

[11] Junhong N (1997) Nonlinear time series forecasting: A fuzzy-neural approach, Neurocomputing, Elsevier, vol. 16, 63-76.

[12] Kaymak U and Babuška R (1995) Compatible cluster merging for fuzzy modelling, In Proc. of FUZZ-IEEE/IFES'95, Yokohama, Japan, pp. 897-904.

[13] Krishnapuram R, Freg CP (1992) Fitting an unknown number of lines and planes to image data through compatible cluster merging, Pattern Recognition, 25(4): 385-400.

[14] Kruse R, Gebhardt J and Klawonn F (1994) Foundations of fuzzy systems, John Wiley and Sons, Chichester.

[15] Lapedes A and Farber R (1987), Nonlinear signal processing using neural network: prediction and system modelling, LA-UR-87-2662.

[16] Mamdani EH (1977) Application of fuzzy logic to approximate reasoning using linguistic systems, Fuzzy Sets and Systems 26: 1182-1191.

[17] MATLAB (1998), Fuzzy Logic Toolbox, User's guide, version 2, revised for Matlab 5.2, The Math Works Inc., Natick, MA.

[18] Pal NR and Bezdek JC (1995) On cluster validity for the fuzzy *c*-means model, IEEE trans. Fuzzy Systems, 3(3): 370-379.

[19] Palit AK and Popovic D (1999) Fuzzy logic based automatic rule generation and forecasting of time series, Proc. of FUZZ-IEEE, 1: 360-365.

[20] Palit AK and Popovic D (2000) Nonlinear combination of forecasts using artificial neural networks, fuzzy logic and neuro-fuzzy approaches, Proc. of FUZZ-IEEE, San Antonio, Texas, 2: 566-571.

[21] Pedrycz W (1984) An identification algorithm in fuzzy relational systems, Fuzzy Sets and Systems, 13: 153-167.

[22] Roubos JA and Setnes M (2001) Compact and transparent fuzzy models and classifiers through iterative complexity reduction, IEEE Tran. Fuzzy Syst. 9: 516-524.

[23] Ruspini E (1970) Numerical methods for fuzzy clustering, Inform. Scien., 2: 319-350.

[24] Setnes M (2000) Supervised fuzzy clustering for rule extraction, IEEE Trans. on Fuzzy Systems, 8(5): 509-522.

[25] Wang L and Yen J, (1999) Extracting fuzzy rules for system modelling using a hybrid of genetic algorithms and Kalman filters. Fuzzy Sets and Systems 101:353-362

[26] Wang LX and Mendel JM (1992) Generating fuzzy rules by learning from examples, IEEE Trans. on Systems, Man and Cybernetics, 22(6): 1414-1427.

[27] Xie XL and Beni GA, (1991) Validity measure for fuzzy clustering. IEEE Trans. on Pattern Analysis and Machine Intelligence. 3(8): 841-846.

[28] Zadeh LA (1965) Fuzzy Sets, Information and Control 8: 338-353.

5

Evolutionary Computation

5.1 Introduction

The study of evolutionary behaviour of biological processes has produced a constructive impact on development of a new intelligent computational approach for solving complex optimization problems in mathematics, natural sciences, engineering and in real-life in general. The algorithms developed under the common term of ***evolutionary computation*** are mainly based on selection of a population as a possible initial solution of a given problem. Through stepwise processing of initial population using ***evolutionary operators***, such as ***crossover***, ***recombination***, ***selection***, and ***mutation***, the fitness of the initial population steadily improves. Following this evolutionary concept, various computational algorithms have been elaborated, such as ***genetic algorithms***, ***genetic programming***, ***evolutionary strategies***, and ***evolutionary programming***, that are capable of solving complex problems, where the traditional mathematical methods cannot be applied easily. Depending on the nature of the problem in hand, the most adequate algorithm is to be selected.

The primary application area of evolutionary algorithms that we are interested in concerns the forecasting of time series data, but also evolving neural networks and fuzzy logic systems. It will be shown that the synergetic effects of combinations of different computational technologies – the neural networks, fuzzy logic, and evolutionary computation – help in designing the improved intelligent systems and also help in improving the accuracy and the convergence speed of evolutionary algorithms themselves. This will be discussed in Part 3 of the book, which is dedicated to the hybrid computational technologies.

It is well known that evolutionary computation is a category of algorithms, based on Darwin's idea of evolution of living creatures. According to this idea, every living creature has a single predecessor that has to adapt steadily to the changing environment in the attempt to survive. According to Darwin, the idea of adaptation is strongly connected with the principle of natural selection, because the creatures that adapt best to the changed environment will be selected by nature to survive.

5.1.1 The Mechanisms of Evolution

The subject of evolutionary theory is actually the genetic evolution of individuals within a ***population***. The population stores multiple solutions of the given problem, with each solution being a **member** of this population.

Associated with each member is its ***fitness***, which is simply a measure of how well this solution solves the problem. Throughout the search for the optimal solution, a ***survival of the fittest*** procedure is used, which means that a solution with a high fitness is chosen over one with a lower fitness. The main difference between individual evolutionary algorithms is the way in which new solutions (or ***offspring***) are generated from the existing members. There are two possible ways: two solutions are ***mated*** to form two new solutions or each member of the population generates an offspring by ***mutation***.

Genetic operators are operators (or mechanisms) that produce a change in the genetic code of genes. The most common of them is mutation in its various forms causing various effects like:

- ***Deletion***, *i.e.* a part of the code is deleted. Deletions in genes usually cause genetic disaster.
- ***Duplication***, *i.e.* a part of the code is actually duplicated. Again, this also causes some major problems.
- ***Cross-over***, *i.e.* the physical exchange of parts of a gene for parts of another gene. This is more commonly known as the ***exchange of genetic material*** and, much like mutation, promotes variation in an individual. In fact, this is the principle way in which children often get a combination of genes from both parents.
- ***Reproduction***, *i.e.* the most important genetic operation described below.

In nature, there are two types of reproduction: ***asexual reproduction*** and ***sexual reproduction***. Asexual reproduction is actually the splitting of a single individual into two new individuals, *e.g.* as with bacteria. In sexual reproduction, two individuals of the same species, the ***male*** and the ***female***, produce an offspring. The key difference between these two types of reproduction is that sexual reproduction includes the exchange of genetic material of both parents, whereas asexual does not, because the daughter cells, produced by splitting into two new cells, are genetically identical with the original cells of the mother. Hence, in sexual reproduction the offspring is a combination of its parents, having some traits from its father and some from its mother, and even some traits that are the combination of traits from both mother and father.

5.1.2 Evolutionary Algorithms

The study of evolutionary behaviour of biological processes has produced a qualitatively new background knowledge and a constructive impact on development of new intelligent computational approaches to solving complex optimization problems, valuable in mathematics, natural sciences, and engineering. The earliest attempts to map Darwin's ideas on to real-life problems was made by John Holland and David Goldberg, who modeled many such problems. They developed ***classifier systems***, which are the predecessors of ***evolutionary systems***. Thereafter, accelerated work on evolutionary methods across the world was started.

The algorithms developed under the common term of evolutionary computation generally start with the selection of an initial population as a possible initial set of the problem solution. This is followed by stepwise iterative changing of the selected population by random selection and use – in each iteration step – of evolutionary operators like crossover, recombination, selection, and mutation in order to improve the fitness of initial individual population members. Although simple in principle, the evolutionary concept of computation has proven to be very efficient in solving complex application problems that are not easily solvable using traditional mathematical approaches.

In the meantime, depending on the nature of the problem to be solved, adequate evolutionary algorithms have been developed, such as

- ***genetic algorithms*** (Holland, 1975), related to direct modelling of genetic evolutionary processes
- ***genetic programming*** (Koza, 1992 and 1994), an extension of genetic algorithms in which the population individuals are replaced by programs
- ***evolutionary strategies*** (1973), which model the ***evolution of evolution*** by tuning the strategic parameters that control the changes in the evolutionary process
- ***evolutionary programming*** (Fogel *et al.*, 1966), based on simulation of adaptive behaviour of the evolution process
- ***differential evolution*** (Storn and Price 1995, 1996), a population-based search strategy for optimizing real-valued, multi-modal objective functions.

As shown in this chapter, evolutionary algorithms are a special category of ***random search algorithms***. In contrast to traditional search algorithms like gradient methods, which become impractical with the growing size of the search space, evolutionary algorithms, because they are based on the ***population concept*** and are operating with the genetic terms and operators, retain more or less the same size of population over the generations and remain mathematically well manageable.

5.2 Genetic Algorithms

Genetic algorithms (GAs) are gradient free, parallel, robust search and optimization techniques based on the laws of natural selection and genetics. The GAs have confirmed their application power in solving practical problems which are generally ill-defined, complex, and with multimodal objective functions. This optimization technique is similar to its associated algorithms, such as ***simulated annealing*** and other guided random techniques. GAs employ random search algorithms aimed at directed location of the global optimum of the solution. The algorithms are superior to the "gradient descent" methods that are not immune against being trapped in local minima. On the other hand, GAs differ from pure random search algorithms in that they, from the very beginning, search for the relatively "prospective" regions in the search space.

Typically, GAs are characterized by the following features:

- genetic representation, *i.e.* encoding of the feasible solutions of given optimization problems
- a population of encoded solutions
- a fitness function that evaluates the optimality or quality of each solution
- genetic operators that generate a new population from the existing population
- control parameters.

A typical execution of a GA involves the following steps:

- Random generation of an initial population $X(t)$: $= (x_1, x_2, \ldots, x_N)$ with N individuals at $t = 0$.
- Computation of fitness $F(x_j)$ of each individual x_j in the current population $X(t)$.
- Checking whether the termination condition is met.

  1. If YES, then pick up the best individual, *i.e.* the one with the highest fitness value and stop the search process.

  2. If NO, then create new population $X(t+1)$ with N new individuals, applying the reproduction, mutation and crossover genetic operators, from the current population $X(t)$ and start the new iteration step with a fitness computation.

In the recent past, GAs have been used, along with other evolutionary algorithms, to train neural networks (Harrald and Kamastra, 1997) and neuro-fuzzy networks (Palit and Popovic, 2000), as well as for the design of fuzzy-rule-based systems through fuzzy clustering (Klawonn, 1998), for identification, modeling and classification (Roubos and Setnes, 2001), *etc*. In the following, the application of binary-coded GA in training neuro-fuzzy networks is presented. The simple two-step approach that combines ***fuzzy clustering*** for initial modeling and a real-coded GA for fine-tuning and optimization of the fuzzy rule base can be found in detail in Panchariya *et al*., (2004).

The structure of the GA implemented for the neuro-fuzzy network training is shown in Figure 5.1, in which $P(C)$, $P(M)$, and $P(R)$ stand for operators of the ***adaptive genetic algorithm*** (AGA) as described in Chapter 9.

5.2.1 Genetic Operators

In what follows, a short description of individual GA operators is given.

5.2.1.1 Selection

Individuals or chromosomes are selected from the ***mating pool***, based on a ***roulette wheel (RW) selection procedure***. This selection emulates the survival-of-the-fittest mechanism in nature. It is expected that a fitter chromosome will give rise to a higher number of offspring and thus will have a higher chance of surviving in the subsequent generation. There are many ways to achieve effective selection, including ranking, tournament, and proportionate schemes (Tang *et al.*, 1996), but the key assumption is to give preference to fitter individuals. The ***RW selection*** procedure commonly used to implement the proportionate scheme can be described as follows.

- Sum the fitness of all population members, termed total fitness F_s.
- Generate a random number r between 0 and 1 and multiply this by the total fitness, *i.e.*

$$0 < r < 1 \quad \text{and} \quad 0 < rF_s < F_s \tag{5.1a}$$

- Pick up the ith population member whose fitness added to the sum of the fitness of the preceding population members is greater than or equal to rF_s, as expressed by

$$r \cdot F_s \leq f_i + \sum_{j=1}^{i-1} f_j, \quad i \leq N_{\text{pop}} \tag{5.1b}$$

5.2.1.2 Reproduction

In the reproduction process, once an individual is selected, this is simply reproduced (copied) into the next generation's population if a certain test condition is satisfied. For example, the individual j selected from the mating pool is simply copied into the next generation if a random number generated is greater than the probability of reproduction (a small number less than 1). If a new individual is generated through reproduction then the population counter is incremented by 1 starting with a 0 value. Using the reproduction operator, only 20% of the total population is created for the next generation.

5.2.1.3 Mutation

Mutation is an operator that introduces variations into the chromosomes. The variation can be global or local. The operation occurs occasionally (usually with small probability $P(M)$) but randomly alters the value of the string position. In the mutation process, any particular bit location of an individual is changed to 1 if it was 0, or *vice versa*. Once an individual is selected, then the particular bit of the same chromosome is simply mutated if a certain condition is satisfied, *i.e.* if it passes the probability test condition. For example, the bit location 1 of an individual j will undergo mutation if a random number generated is greater than probability of mutation (a very small number less than 1). Otherwise, that particular bit remains unaffected. The same process is continued from bit location

1 to the last bit of the same individual. Since the probability of mutation is generally very low, only a very few bits may undergo mutation of an individual.

If a new individual is generated through mutation then the population counter is incremented by 1, starting with $0.2 \times N_{pop}$, where N_{pop} is the total number of the population in a particular generation. Otherwise the population counter remains the same. Using the mutation operator, only 30% of the population, in our case, is created for next generation (see Figure 5.1).

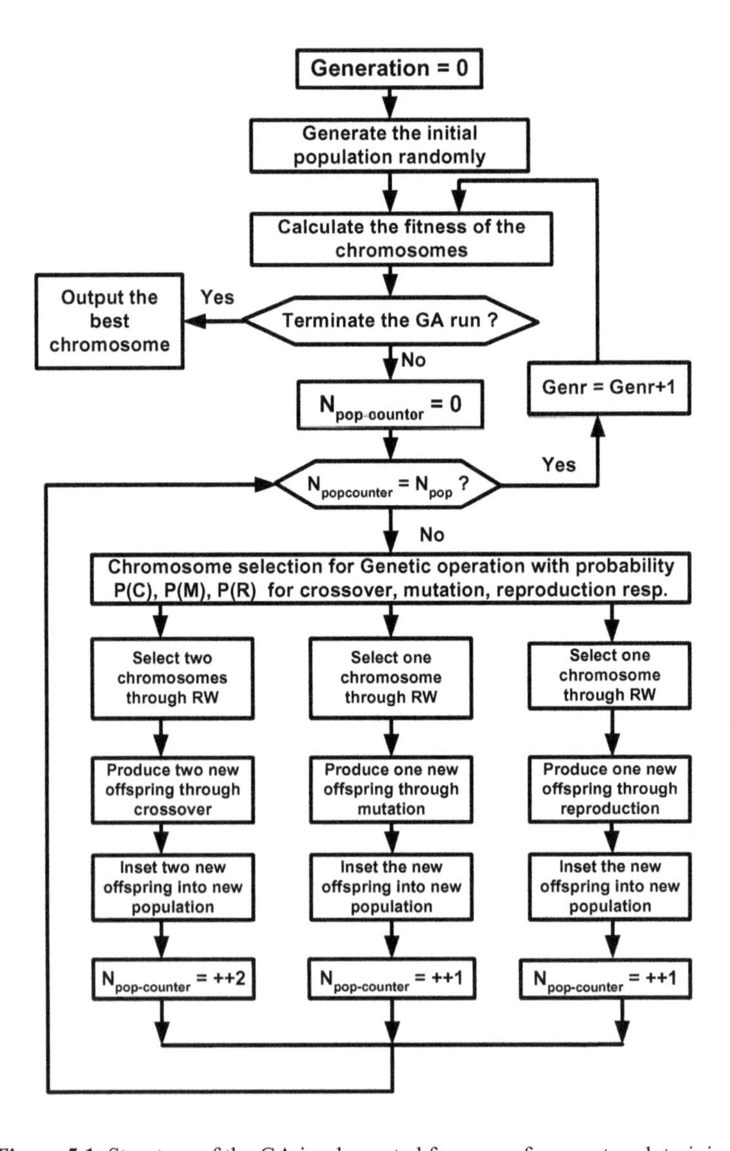

Figure 5.1. Structure of the GA implemented for neuro-fuzzy network training

5.2.1.4 Crossover

Crossover is a recombination operator that combines subparts of two parent chromosomes to produce offspring that contain some parts of both parents' genetic material. A crossover probability term $P(C)$ is set to determine the operation rate. Many GA practitioners consider the crossover operator to be the determining factor that distinguishes the GA from all other optimization algorithms.

The power of a GA arises from crossover, which causes a structured, yet randomized exchange of genetic materials between solutions, with the possibility that "good" solutions can generate the "better" ones. In the crossover process, two individuals called parent 1 and parent 2 are required. A crossover operation between parent 1 and parent 2 takes place with respect to a particular bit location (called the ***crossover point***) selected randomly and the portions of the chromosomes beyond this point are exchanged to form offspring. Hence, a crossover generates two new individuals of the next generation.

In our experiment, the best individual or chromosome from all generations is always selected as parent 1. Parent 2 is selected through the RW selection procedure from the mating pool. The crossover operation between two individuals takes place if a certain condition is satisfied. For example, two individuals undergo a crossover operation if a random number generated is greater than the probability of crossover (a small number). Otherwise, both individuals remain unaffected. If two new individuals are generated through crossover, then the population counter is incremented by 2, starting with $0.5 \times N_{\text{pop}}$. Otherwise the population counter remains the same. In our case, using the crossover operator, only 50% of the population is created for the next generation (see Figure 5.1).

5.2.2 Auxiliary Genetic Operators

In addition to the above ***standard genetic operators***, the following operators are also used in the GA experiment.

5.2.2.1 Fitness Windowing or Scaling

Regulation of the number of copies of superfit or extraordinary individuals is especially important in small-population GAs. At the beginning of the GA runs it is common to have a few extraordinary individuals in a population of mediocre colleagues. However, if left to the normal proportionate selection rule, say to the RW, the extraordinary individuals would take over a significant proportion of the finite population in a single generation, and this is undesirable, as it leads to premature convergence. This is because without the fitness scaling during the matured run of the GA most of the individuals may converge and maintain a small diversity, giving rise to a small difference between their fitness value even though the GA run may not have located the desired global optimum. Therefore, the crossover operation in this case produces new offspring practically without much improvement in their fitness value during the matured run of the GA. Only the mutation operator tries to maintain a small diversity and explores the new region randomly. As a remedy for this premature convergence, fitness scaling or fitness windowing can generally be applied. This prevents any super-fit individual from always taking over and suppressing the lower fitness individual during the RW selection. Hence, scaling involves a readjustment of fitness values to sustain a steady selective pressure in the population and to prevent the premature convergence of the populations.

Various techniques are available for fitness scaling or fitness windowing. Let us assume that the objective value of the worst chromosome in the population is f_w, and that each chromosome can be assigned a fitness value proportional to the cost difference between the chromosome i and the worst chromosome w, *i.e.*

$$V_i = k \pm k_f \cdot (f_i - f_w) \tag{5.2}$$

where f_i is the objective value or raw fitness of ith chromosome, f_w is the raw fitness of the worst chromosome, k and k_f are two constants. If a maximization problem is encountered, then a positive sign is adopted in Equation (5.2), whereas for the minimization problem negative sign is adopted. In our experiment we set the value of $k = 10$ and $k_f = 2$.

Alternatively, the fitness scaling can be implemented using linear scaling, *i.e.* the linear relationship between f and V

$$V_i = a \cdot (f_i) + b \tag{5.3}$$

where f is the raw fitness and V the corresponding scaled fitness. The coefficients a and b may be chosen in a number of ways; however, in all cases, the average scaled fitness is equal to the average raw fitness, *i.e.* $V_{avg} = f_{avg}$.

In the following example we use $V_{max} = C_{mult} f_{max}$, select $C_{mult} = 2$, and $V_{min} = f_{min}$. Towards the end of a GA run, this choice of C_{mult} stretches the raw fitness significantly. In turn, this may cause difficulty in applying the above linear relationship, when we cannot scale to the desired multiple C_{mult}; in this case, scaling is performed still keeping $V_{avg} = f_{avg}$ and then stretching the fitness until the minimum value maps to zero, *i.e.* $f_{min} = 0$. The entire scaling procedure is performed in three routines, namely pre-scale, scale, and scale-pop. This includes calculation of $f_{max}, f_{min}, f_{avg}$, *etc.*

Now, we check the following relation (Goldberg,1989):

$$f_{min} \geq (C_{mult} \cdot f_{avg} - f_{max}) / (C_{mult} - 1). \tag{5.4}$$

If the last relationship holds, then the calculation of a and b will be

$$a = (C_{mult} - 1) \cdot f_{avg} / (f_{max} - f_{avg}) \tag{5.5}$$

$$b = (1 - a) \cdot f_{avg}. \tag{5.6}$$

Otherwise, if relationship (5.4) does not hold then calculation of a and b will be as follows:

$$a = f_{\text{avg}} \big/ \left(f_{\text{avg}} - f_{\min} \right) \tag{5.7}$$

$$b = (1 - a) \cdot f_{\text{avg}} \tag{5.8}$$

Once the values of a and b are calculated, the scaling is done as per Equation (5.3). In order to avoid the division by zero situation, when the denominator of (5.5) and (5.7) are close to zero during the matured run of the GA, a very small constant (k_1) of the order of $k_1 = 0.0001$ can be added to their denominators.

5.2.3 Real-coded Genetic Algorithms

Genetic algorithms, being gradient-free and parallel optimization algorithms, have immense advantages over the conventional search methods.

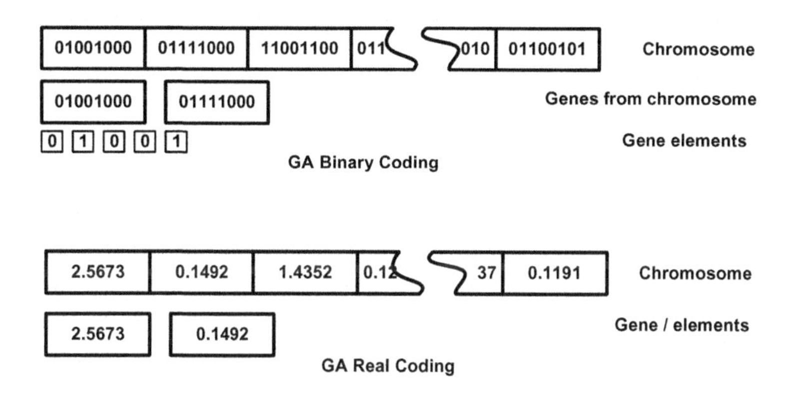

Figure 5.2. Coding in genetic algorithms

Genetic algorithms, like other parallel optimization algorithms, use a performance criterion for evaluation and a population of possible solutions to search for a global optimum (Michalewicz, 1994). In each search step, the algorithms select the prosperous solutions and manipulate them using appropriate genetic operators to achieve new, and possibly better solutions. The manipulations are carried out on chromosomes in which the parameters of possible solutions are encoded. In each generation of the GA, the new population replaces the solutions in the population that are selected for deletion.

The chromosomes can be represented or encoded either by binary values (Goldberg, 1989) or by real numbers (Michalewicz, 1994). The genetic algorithms with binary coded chromosomes, when applied to multidimensional, high-precision or continuous complex problems, are less efficient because, in such situations, the bit-strings can become very long. Furthermore, CPU time is lost for the conversion between the binary and real representation. Here, other alphabets, like real-coding, can favourably be applied to parameters' or variables' presentation in the continuous domain of values.

For instance, in real-coded GAs the parameters (such as mean and variance parameters of Gaussian membership functions, and singleton rule consequents in the training of a neuro-fuzzy network) or the variables appear directly in the chromosomes (see Figure 5.2) and are modified using special genetic operators. Various real-coded GAs were recently reviewed by Herrera *et al.* (1998). The main aspects of the proposed GA are discussed below, and implementation for compact, transparent and accurate fuzzy models is also summarized.

5.2.3.1 Real Genetic Operators

Two classical operators, simple arithmetic crossover and uniform mutation, and four special real-coded operators are used in this GA application. These operators have been successfully applied by Michalewicz (1998), Setnes and Roubos (1999), and Roubos and Setnes (2001).

In the following, $r \in [0,1]$ is a random number (uniform distribution), $g = 0, 1, 2, \ldots, G$ is the generation number, $l = 1, 2, 3, \ldots, N_{\text{pop}}$ is the chromosome number in a generation, S_a and S_b are two chromosomes selected for operation, $k \in \{1, 2, \cdots, L_{\text{chrom}}\}$ is the position of an element in the chromosome, and $\left(a_k^{\min}, a_k^{\max}\right)$ and $\left(b_k^{\min}, b_k^{\max}\right)$ are the lower and upper bounds of the parameter encoded by the kth element of chromosomes S_a and S_b, respectively.

5.2.3.1.1 Selection Function

The purpose of the selection function is to create a steady evolutionary pressure; this, to some extent, favours the well-performing chromosome to have a higher chance of survival. The RW selection method is used to select n_c chromosomes for various genetic operations (Michalewicz, 1994). The chance of winning on a spin of the RW is given by $f_l \Big/ \sum_{l=1}^{N_{\text{pop}}} f_l$, implying that the higher the ratio of fitness f_l of the chromosome S_l is with respect to total fitness of all chromosomes in the population, then the larger is the chance that chromosome S_l will be selected through the RW. The fitness f_l of the chromosome S_l is defined as

$$f_l = (1/J_l)^2, \quad l \in \{1, 2, \ldots, N_{pop}\},$$

where J_l is the performance of the model encoded in chromosome S_l measured in terms of the mean-squared error (MSE):

$$J = \frac{1}{N_s} \sum_{i=1}^{N_s} (y_i - \hat{y}_i)^2,$$

where y is the desired output, $\hat{y}$ is the model output, and N_s is the number of training samples. Notice that because of the reciprocal form and square term in right-hand side of the fitness function, a small difference in MSE values will be greatly amplified, *i.e.* if the MSE difference between two chromosomes is 0.1 then the corresponding fitness difference will be 100. The inverse of the selection function is used to select chromosomes for deletion, *i.e.* n_c old chromosomes are deleted and the population is refilled by n_c new chromosomes that are formed by selection. The best chromosome is always preserved in the population (elitist selection).

The probability that a selected chromosome will undergo a crossover operation is 95%, whereas the probability of mutation is selected as 5%. When a chromosome is selected for crossover (or mutation), one of the crossover (or mutation) operators described below is applied with equal probability.

5.2.3.1.2 Crossover Operators for Real-coded Genetic Algorithms

For crossover operations, the chromosomes are selected in pairs (s_a, s_b):

- Simple arithmetic crossover, in which s_a^g and s_b^g are crossed over at the *k*th position such that the resulting two offspring are:

  $$s_a^{g+1}=\left(a_1,\cdots,a_k,b_{k+1},\ldots,b_{L_{\text{chrom}}}\right) \text{ and } s_b^{g+1}=\left(b_1,\cdots,b_k,a_{k+1},\ldots,a_{L_{\text{chrom}}}\right),$$

  where *k* is selected randomly from {2, 3, ..., (L_{chrom} -1)}.

- Whole arithmetic crossover, in which a linear combination of s_a^g and s_b^g results in

  $$s_a^{g+1}=r\left(s_a^g\right)+(1-r)s_b^g \text{ and } s_b^{g+1}=r\left(s_b^g\right)+(1-r)s_a^g.$$

- Heuristic crossover, in which s_a^g and s_b^g are combined such that

  $$s_a^{g+1}=s_a^g+r\left(s_b^g-s_a^g\right) \text{ and } s_b^{g+1}=s_b^g+r\left(s_a^g-s_b^g\right).$$

It is to be noted that the heuristic crossover described above is very similar to the trial vector of differential evolution of type one (DE1; see Section 5.5), except for *r*, which is a random number within 0 to 1 here, whereas in DE1 it is a constant within the same 0 to 1 range.

5.2.3.1.3 Mutation Operators

Similar to crossover, there are various mutation operators. However, for the mutation operation only one chromosome is selected through the RW.

- Uniform mutation, in which a randomly selected element a_k, $k\in\{1,2,\cdots,L_{\text{chrom}}\}$, is replaced by a'_k, which is a random number in the range $\left[a_k^{\min},a_k^{\max}\right]$. The resulting chromosome is $s_a^{g+1}=\left(a_1,\cdots,a'_k,\cdots,a_{L_{\text{chrom}}}\right)$.
- Multiple uniform mutation is a uniform mutation of *n* randomly selected elements, where *n* is selected at random and $n\in\{1,2,\cdots,L_{\text{chrom}}\}$.

- Gaussian mutation, in which all elements of a chromosome are mutated such that $s_a^{g+1} = \left(a'_1, \cdots, a'_k, \cdots, a'_{L_{\text{chrom}}}\right)$, where $a'_k = a_k + f_k$ and f_k is a random number drawn from a Gaussian distribution with zero mean and an adaptive variance $\sigma_k = \left(\frac{G-g}{G}\right)\left(\frac{a_k^{\max} - a_k^{\min}}{3}\right)$. It can be seen that σ_k decreases as the generation counter g increases. Therefore, parameter tuning performed by a Gaussian mutation operator becomes finer as the generation counter g increases.

5.2.4 Forecasting Example

In this section we briefly describe a binary-coded GA that can be used to train a neuro-fuzzy system that will be considered once again in Chapter 6. For convenience we restrict our discussion to a Takagi-Sugeno-type neuro-fuzzy network, but with singleton rules consequent only, which has been used extensively by Wang (1994) for a variety of identification and modeling applications. Furthermore, the fuzzy logic system selected is based on GMFs, the product inference rule and a weighted-average defuzzifier. Mathematically, the Takagi-Sugeno-type fuzzy logic system selected can be written as

$$y = \sum_{l=1}^{M} y^l \beta^l \Big/ \sum_{l=1}^{M} \beta^l \text{ , where } \beta^l = \prod_{i=1}^{n} \exp\left\{-\left(x_i - c_i^l\right)^2 \Big/ \left(\sigma_i^l\right)^2\right\}$$

with

$$i = 1, 2, ..., n; \text{ and } l = 1, 2, ..., M.$$

Here, we assume that $c_i^l \in U_i$, $\sigma_i^l > 0$ and $y^l \in V$, where U_i and V are the input and output universes of discourse respectively.

The corresponding lth rule of the fuzzy logic system can be written as follows:

R^l: If x_1 is G_1^l and x_2 is G_2^l and ... and x_n is G_n^l Then y is y^l

where x_i with $i = 1, 2, ..., n$ represent the n number of inputs to the system, $l = 1, 2, ..., M$ are the M number of fuzzy rules that construct the fuzzy system, G_i^l with $i = 1, 2, ..., n$ and $l = 1, 2, ..., M$ are the GMFs with corresponding mean and variance parameters c_i^l and σ_i^l respectively that partition the ith input domain, and y^l represents the (singleton) output from the lth rule. It will be shown in Chapter 6 that a similar fuzzy system can be represented as a three-layer multi-input single-output feedforward network form. Because of neuro implementation of fuzzy logic systems, the same feedforward network actually represents a Takagi-Sugeno-type neuro-fuzzy network.

Given a set of N input-output training samples of the form $\left(X^p, d^p\right)$, where the input pattern $X^p = \left[x_1^p, x_2^p, ..., x_n^p\right] \in U \subset \mathbb{R}^n$ and the corresponding desired output

$d^p \in V \subset \mathbb{R}$ and with $p = 1, 2, ..., N$, the objective is to determine the fuzzy logic system described above such that the performance function of the network, *i.e.* sum square error (SSE) is minimized by optimal settings of the network's free parameters c_i^l, σ_i^l and y^l. The SSE of the network is defined as

$$S = 0.5 \cdot \sum_{p=1}^{N} \left(e^p\right)^2 = 0.5 \cdot E^{\mathrm{T}} E.$$

where, $e^p = \left(y^p - d^p\right)$, represents the approximation error of the network and y^p represents the output y of the network due to the presentation of pth input pattern X^p. We further assume that M, which corresponds to the number of implemented GMFs for the partitioning of the input domain and also the number of implemented fuzzy rules, is already given. Therefore, in order to train the network, *i.e.* for the optimal settings of the network's free parameters, the binary-coded GA can be applied.

For this purpose all the free parameters of the network are encoded in a binary bit string or chromosome. For M fuzzy rules and n inputs to the system the total number of mean parameters plus variance parameters of the GMFs along with singleton rules' consequents will be of size $2(M \times n) + (M \times 1)$. Therefore, for a network with $n = 2$ inputs and with $M = 5$ rules, each chromosome must encode $2(5 \times 2) + (5 \times 1) = 25$ parameter values.

Now if each parameter (say mean parameter of the GMF) of the network is represented by N_{p} bits, which include the first one bit as a sign bit, followed by N_{c} characteristic bits and N_{m} mantissa bits, then $N_p = (1 + N_c + N_m)$ and in this case $N_p = 12$ bits is selected.

Therefore, the entire bit length of each chromosome will be $L = \{2(M \times n) + (M \times 1)\} N_{\mathrm{p}} = 300$ bits. For example, if a parameter c_i^l assumes a decimal value -2.4256, then the first digit (2) before the decimal point is known as the characteristic part and the remaining four digits (4256) after the decimal point represent the mantissa part. Therefore, in order to represent any decimal number within +3.99 to -3.99 we can use a 12-bit binary number, where the first bit, say 0, will represent the "+ve" sign and 1 will represent the "-ve" sign, followed by the next two bits, which can represent only four decimal numbers 0, 1, 2 or 3, and the remaining nine bits represent the mantissa part.

For instance, the 12-bit number (1111 1111 1111) can represent the parameter value $(1{*}2^1 + 1{*}2^0 + 1{*}2^{-1} + 1{*}2^{-2} + \ldots + 1{*}2^{-9}) = -3.9980$, whereas the 12-bit number (0111 1111 1111) represents the decimal number +3.9980. Similarly, any other combination of such 12 bits will represent any number between -3.9980 and +3.9980.

Alternatively, the parameters within the above range can be encoded as equivalent binary numbers as follows. Suppose the number -2.55 or +2.55 has to be encoded into the equivalent 12-bit binary number, then just represent the -255 as (1000 1111 1111) and similarly +255 as (0000 1111 1111), neglecting the position of the decimal point during the encoding. However, during decoding multiply all the decoded numbers by 0.01, which will once again set all the parameters with signs and decimal points. In the following example, however, the first kind of encoding was found to give better results.

Therefore, for a chromosome (binary string) of total bit length L, the first $M*n*N_p$ bits represent the mean parameters (of the GMFs) in the order of each rule, and the next $M*n*N_p$ bits similarly represent the variance parameters of the GMFs; lastly, the remaining $M*1*N_p$ bits represent the M centres of fuzzy regions (singleton consequents). It is assumed that the total number of populations N_{pop} (*i.e.* total number of chromosomes in each generation) is fixed and is selected as 20 for our experiment.

Hence, at the beginning of the GA run, N_{pop} = 20 chromosomes or binary strings, each of length L bits, are generated randomly, which all represent the potential solutions of the network optimum parameter settings. Then the fitness of each chromosome is computed as the reciprocal of the SSE of the network for a given set of network training samples. Therefore, mathematically, the fitness is computed as

$$\text{Fitness} = (1/\ \text{SSE}).$$

In order to compute the fitness function of each chromosome, the binary data are decoded and rearranged into the corresponding parameter matrix of mean, variance and centres of fuzzy regions; the SSE, and hence the fitness, is computed for the above parameter values.

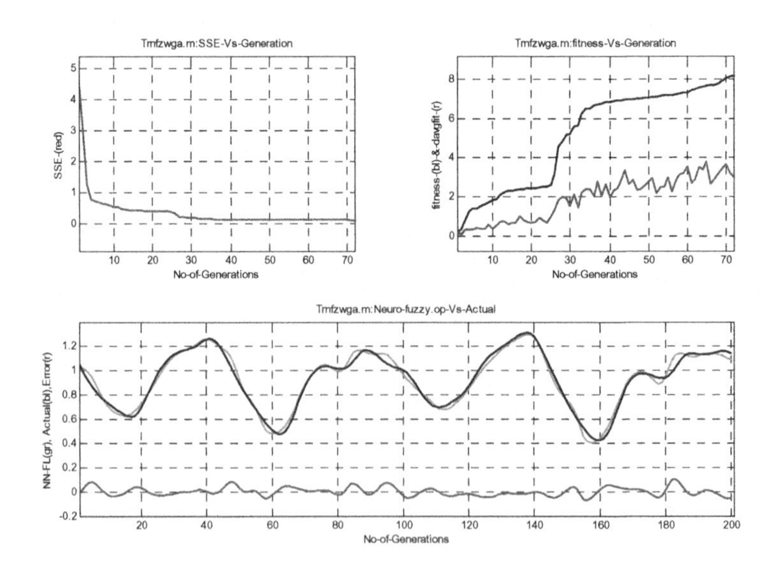

Figure 5.3. Training of neuro-fuzzy network with binary-coded GA

Once the fitness values are computed for all the chromosomes they are arranged in descending order. If the best fitness is greater than or equal to the desired fitness, then the corresponding chromosome is picked and, thereafter, is decoded into network parameter values. The network parameters generated are the final output of the GA run and a further GA run is not required (Figure 5.1). However, if the best fitness is less than the desired fitness, then the GA run is further continued as per Figure 5.1. Once the fitness values are all arranged in descending orders, the best 70% population are collected to form the mating pool in this example. Now, from the mating pool, the next-generation populations are created by applying the various genetic operations as described earlier in this chapter.

In order to test the efficiency of the GA-based neuro-fuzzy network training the Mackey-Glass chaotic time series was considered. The network in this case, as usual, had four inputs and only five rules were implemented. As described above, only 20 populations were selected in each generation. It can be seen from the Figure 5.3 that in only a few generations the GA could improve the fitness function to 8.4149, which corresponds to SSE = 0.1188 or MSE = 0.0012. However, because of the very slow progress of the generation run, the GA run was terminated after only a few generations. If a higher fitness value (say, a few hundred) is required, then the GA run may have to be continued for several thousands of generations so that the network can correctly approximate the nonlinear chaotic time series model.

5.3 Genetic Programming

Koza (1992) proposed an evolutionary algorithm for solving intelligent computational problems by automated generation of computer programs required for problem solution. He viewed the new algorithm as a model for ***machine learning*** in the space of programs and, therefore, named it ***genetic programming***.

Figure 5.4. Example of a LISP program "$a + b*c$"

Instead of operating with individuals, genetic programming operates with the computer programs and uses computer languages, preferably functional programming languages, for its implementation. Functional programming languages are based on syntax suitable for presenting ***parse trees*** used in genetic programming algorithms. This is due to the tree forms of LISP S-expressions which are equivalent to parse trees. For example, the LISP program chunk "*a*+*b***c*" is presented as a parse tree in Figure 5.4.

Koza gives a program example that presents the LISP expression

(+ 1 2 (IF (> TIME 10) 3 4))

as the corresponding tree structure (Figure 5.5).

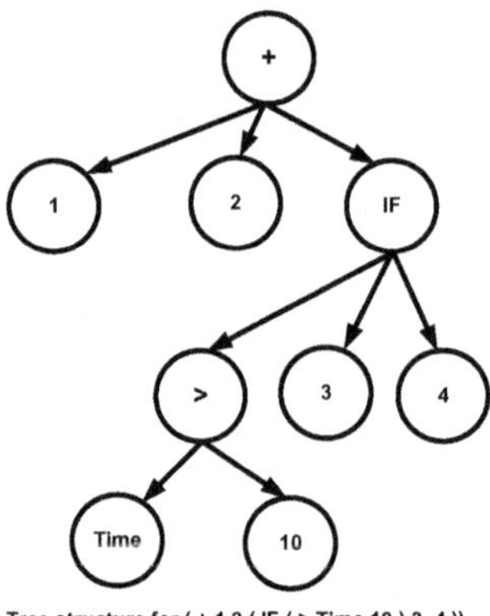

Figure 5.5. Tree structure of a LISP expression (+ 1 2 (IF (> TIME 10) 3 4))

Genetic programs run by executing ***program induction***, *i.e.* they automatically learn within the search space what programs are required in order to improve the problem solution and in this way finally find the best one. The search space here is the space of all possible programs, including the user-defined function set (programming or arithmetic operations, mathematical, logic, and other domain-specific functions) and the terminals set (containing variables and constants appropriate to the problem domain). While searching for the best solution, the genetic programming algorithm makes use of the ***statistical closure property*** of functions to accept as arguments the function return values of any other functions and the data from the terminal set.

5.3.1 Initialization

The first operational step of genetic programming is its initialization, which mainly includes generation of the initial population, *i.e.* of the random composition of the function and terminal sets. In fact, at this point a collection of random trees is generated representing the initial program configurations. Later, the trees will be the subject of specific successive handling by genetic operators (reproduction, crossover, *etc.*) They are generated by firm allocation of the function root node. Thereafter, the children are created and a recourse through the tree carried out during which the functions and terminals are randomly picked from their sets, until all branches end in terminals.

5.3.2 Execution of Algorithm

Once the initial population is obtained, the execution of the kernel algorithm procedure starts with the execution of all programs of the initial population generated, assigning the fitness values according to the fitness measures. Thereafter, a new population of programs is created through

- reproduction of existing programs and by their copying into a new population
- crossover of new programs generated from existing programs by genetic recombination of their randomly chosen parts, and by executing the crossover operation on two recombined programs
- mutation of a randomly chosen part of the program created from an existing program.

After the run of genetic programming the best computer program in the population is, for the time being, considered as the best, or nearly best one for the problem solution. The program run can be finished or continued in order to check whether a still better program can be found.

However, it should also be mentioned here that, like in genetic algorithms, the mutation operation is very sparingly used.

5.3.3 Fitness Measure

So far, we have not considered one of the most principal issues in genetic programming applications, *i.e.* the ***fitness measure***. It is a tool that helps calculate how well the individual programs of the population contribute to the evolutionary progress of finding the problem solution. In practice, the fitness measure is determined subjectively, so that it is viewed as a more obscure action than as an exact definition. Also, formulation of the fitness measure is strongly problem dependent. For the majority of problems it is understood as the error delivered by the programs after their execution. This is true for every program run, so that it is expected that the initial programs will most probably produce the lowest fitness value, but some among them could have higher values than the rest of the population. This triggers the evolutionary process. The offspring population, after undergoing treatment through genetic operational steps, could replace the parent population and undergoes a fitness check that is the basis for the next evolutionary step. This continues until the best solution of the problem is found.

5.3.4 Improved Genetic Versions

Koza (1994) reported about a second, amended version of genetic programming capable of evolving multipart programs by integrating the reusable, parameterized subprograms into the main program. The subprograms are termed ***automatically defined functions***. Each such program can contain ***function defining branches***, which are capable of communicating with the automatically defined branches, and ***result-producing branches***, and are also capable of calling the automatically defined functions. Koza (1994) has shown that genetic programming with automatically defined functions is scalable, enabling genetic programming to determine the size and the shape of the problem solution automatically (*i.e.* of the program tree).

However, when multipart programs and automatically defined functions are integrated, the problem arises as to how to tailor the architecture of the evolved programs. This problem has been solved through dynamic evolutionary selection of the architecture of the overall program while running the genetic programming.

5.3.5 Applications

Application examples of genetic programming are numerous. Apart from abundant mathematical applications, such as applications in symbolic regression, many practical applications have been reported in engineering, particularly in pattern classification, vehicle control, robotics, *etc*. For the reader, of direct interest is genetic programming application in time series prediction (Santini and Tettamanzi, 2001), where two problem solution strategies have mostly been applied:

- a neural network model has been optimally tuned by genetic programming (Zang *et al.*, 1997)
- appropriate programs have been evolved using genetic programming for computing the future values of a given time series, given its last values (Yoshichra *et al., 2000*).

The first strategy belongs to the category of evolving neural networks using evolutionary computation in general, which will be treated in detail in Part 3 of the book. In the following, our attention will be focused on the strategy used by Santini and Tettamanzi (2001), mainly achieved by

- evolving the individuals made up of some different expressions, one for each prediction step
- developing of special crossover and mutation operators adapted to the generated individuals of population
- calculating the fitness based on given time series data.

Mulloy *et al.* (1996) used the genetic programming approach in the prediction of chaotic time series.

5.4 Evolutionary Strategies

Evolutionary approaches that are very similar to genetic algorithms are the ***evolutionary strategies*** developed by Rechenberg and Schwefel (Rechenberg, 1973) while working on the design of an optimal jet nozzle that produces the most powerful propulsion at the lowest fuel consumption. They came to the idea of developing a new solution concept that starts with commercially available jet nozzles and, using a genetic evolution process, ends with finding the optimal nozzle shape. The solution concept used by them was termed ***evolutionary strategy***.

Evolutionary strategy also relies on the mechanism of evolutionary computation, but it uses it in an original way. In contrast to genetic algorithms, which aim at solving discrete and integer optimization problems, the objectives of evolutionary strategies are more focused on solving the problems of continuous parameter optimization. The evolutionary strategy achieves this through the search from one population of solutions to another, rather than like genetic algorithms searching from individual to individual. Also, the evolutionary strategy uses selection, recombination, and mutation as separate genetic activities for generating a new solution (*i.e.* the new generation), which is actually the major difference with the genetic algorithms.

The basic idea of evolution strategies relies on the hypothesis that, during biological evolution, the laws of heredity have been developed for rapid ***phylogenetic adaptation***. This is actually a considerable improvement of the genetic algorithm concept, which traditionally does not consider the effects of genetic procedures on the phenotype. The presumption for coding the variables in the evolution strategy is the realization of a sufficiently strong causality effect (*i.e.* that small changes in the cause must create small changes in the effect).

The climax of the theory of evolution strategy is the discovery of an ***evolution window***, stating that evolutionary progress takes place only within a very narrow band of the mutation step size. This fact indicates the need for a rule of self-adaptation of the mutation step size. These genetic operators were taken straight from biological evolution and rely strongly on the principle of mutation. In the problem at hand, a mutation was simply a small change in the overall make-up of a jet nozzle.

In their experiments, Rechenberg and Schwefel tested the performance of the evolved jet nozzles after every mutation. After many repeated trial runs of this kind, they succeeded in producing a jet nozzle that was better than any of the jet nozzles at that time available on the market. It is remarkable that, for jet nozzle optimization, no mathematics dealing with fluid dynamics and propulsion was taken into account. For the experiments, some nozzles available on the market were taken and evolved further in order to produce, with every evolutionary step, a better problem solution.

5.4.1 Applications to Real-world Problems

Evolutionary strategies, instead of a step-by-step search for a single problem solution, from the very beginning deal with a set of potential problem solutions. The strategies start with a set of initial solutions and improve them through repeated evolutionary steps until the best solution has been found. After every step, the degree of improvement is evaluated using some fitness criteria. Before initiating the next evolutionary step, a decision is made as to what genetic operators should be selected. Two such operators are dominant here, *i.e.* mutation and crossover, whereby mutation is the most frequently used because it offers prospective changes in the problem solution. The crossover operator, however, promotes the process of reproduction by mating two given solutions and producing a new one. It is expected that in this way good offspring are generated.

To estimate how far the generated offspring are good, the selected fitness criteria are used. This process is repeated, producing better and better offspring by mating and mutation operations. Although the evolutionary strategies are valuable search concepts, they still have their limitations and drawbacks: in practical applications, many decisions have to be made in the selection of an initial solution set, the application of appropriate genetic operators at each evolutionary step, the definition of an adequate fitness function, *etc*.

5.5 Evolutionary Programming

L.J. Fogel (Fogel *et al*., 1966), in his search for a new evolutionary method for developing artificial intelligence, elaborated a stochastic optimization methodology relying on genetic principles that was later formulated by D. Fogel (1994) as ***evolutionary programming***. The new methodology differs substantially both from genetic algorithms and genetic programming in that it evolves ***behavioural models*** rather than ***genetic models***. Hence, the objective of evolutionary programming is to find a set of best behavioural models from a space of possible behavioural models.

Like other evolutionary methods, evolutionary programming also relies on some repeated operational steps that are interrupted (before the next step commences) by the evaluation of the results achieved using a fitness function. But still, evolutionary programming is different from other genetic methods in that it uses a ***population of parents***, each of them producing a single offspring through mutation, because in evolutionary programming no crossover operator is implemented.

The algorithm of evolutionary programming can be outlined as follows:

- Generate randomly the initial generation as a set of initial problem solutions and calculate the fitness value of each individual of the population.
- For each individual (problem solution)
    - generate a new solution set by copying the set and changing it genetically
    - calculate the fitness of each individual/new solution
    - Store the new solution and fitness.
- From the new generation select the solution with the largest fitness and delete the rest.
- If the best or nearly the best solution is found, stop the evolutionary process; otherwise continue.

In practice, before the above search for the optimal solution runs, the population size and the number of iterations (*i.e.* number of generations) have to be fixed. Also, the mutation operator to be used for generation of the next solution is to be determined. This can be extended by integrating a randomly selected ***maturation***

operator procedure and eventually by applying the ***elitist strategy***, which could amend the selection of parents for the next generation.

5.5.1 Evolutionary Programming Mechanism

In evolutionary programming, each offspring is generated from its parent by changing one or more alleles in the chromosome. In biological terms, this represents a mutation. Now, because the selection of a new parent is based on the fitness of the organism, the Darwinian procedure of "***survival of the fittest***" is applied. Therefore, the procedure listed above can be described as a living organism that produces one or more offspring through mutation. A survival-of-the-fittest procedure helps in selecting the best parents for the next generation, so that the organism evolves by trying to maximize its fitness (*i.e.* trying to solve the given problem as best as possible).

5.6 Differential Evolution

Differential evolution is a population-based search strategy and an evolutionary algorithm that has recently proven to be a valuable method for optimizing real-valued multi-modal objective functions (Storn and Price 1995, 1996). It is a parallel direct search method having good convergence properties and simplicity in implementation. The method utilizes N_{pop} parameter vectors $\underline{X}_{i,G}$ as a population for each generation G, where $i = 0, 1, 2, \cdots, N_{\text{pop}-1}$. The number of parameter vectors, *i.e.* N_{pop}, does not change during the optimization (minimization) process and the initial population is chosen randomly, unless a preliminary solution is available. Where a preliminary solution is available, then the remaining population of the starting generation is often generated by adding normally distributed random deviations to the nominal solution.

The crucial idea behind the differential evolution is a new scheme for generating trial parameter vectors by adding the weighted difference vector between two population members to a third member. If the newly generated vector results in a lower objective function value (higher fitness) than the predetermined population member, then the resulting vector replaces the vector with which it was compared. The comparison vector can, but need not essentially, be part of the above generation process. In addition, the best parameter vector is evaluated for every generation G in order to keep track of the progress that is made during the minimization process. Extracting both distance and direction information from the population to generate random deviations results in an adaptive scheme that has excellent convergence properties (Storn and Price, 1995).

There are several variants of differential evolution, with the two most promising variants being

- DE1, the first variant of differential evolution
- DE2, the second variant of differential evolution.

5.6.1 First Variant of Differential Evolution (DE1)

The first variant of differential evolution works as follows: for each vector in generation *G*, *i.e.* $\underline{X}_{i,G}$, $(i = 0, 1, 2, \cdots, N_{\text{pop}-1})$, a trial vector $\underline{X}_{v,G}$ is generated as

$$\underline{X}_{v,G} = \underline{X}_{a_1,G} + K \cdot \left(\underline{X}_{a_3,G} - \underline{X}_{a_2,G}\right)$$

with $a_1, a_2, a_3 \in \left[0, N_{\text{pop}-1}\right]$. The integers a_1, a_2 and a_3 are mutually different from each other, and $K > 0$.

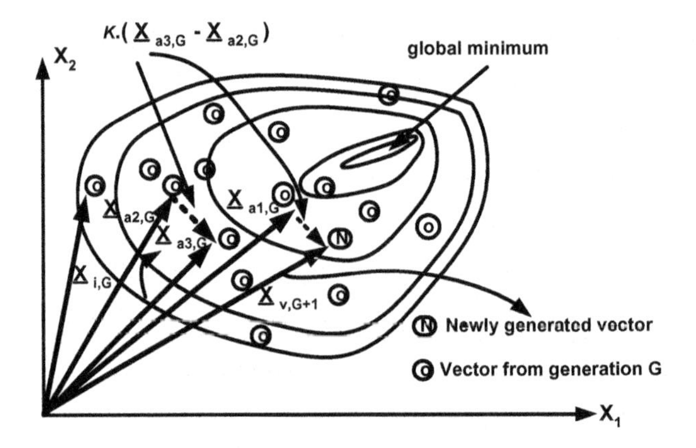

Figure 5.6. First variant of differential evolution (DE1)

Furthermore, the integers a_1, a_2 and a_3 are chosen randomly from the interval $[0, N_{\text{pop-1}}]$ such that they are different from the running index *i*. The real constant factor *K* controls the amplification of the differential variation $\left(\underline{X}_{a_3,G} - \underline{X}_{a_2,G}\right)$. Figure 5.6 shows a two-dimensional example that illustrates the different vectors and generation of a trial vector which play an important role in DE1.

In order to increase the potential diversity of the perturbed parameter vectors, crossover is introduced. The crossover operation generates the perturbed vector as follows:

$$\underline{X}_{u,G+1} = \left(X_{u_0,G+1}, X_{u_1,G+1}, \cdots, X_{u_{(D-1)},G+1}\right)$$

with,

$$X_{u_j,G+1} = \begin{cases} X_{v_j,G+1} & \text{for} \quad j = \langle n \rangle_D, \langle n+1 \rangle_D, \cdots, \langle n+L-1 \rangle_D. \\ X_{ji,G} & \text{for all other} \quad j \in [0, D-1]. \end{cases}$$

is generated. The angle brackets $\langle\ \rangle_D$ denote the modulo function with modulus D. The starting index n in the above equation is a randomly chosen integer from the interval [0, D-1]. The integer L, which denotes the number of parameters that are going to be exchanged, is drawn from the interval [1, D]. The algorithm that determines L works according to the following lines of pseudo code, where *rand*() is supposed to generate a random number within the interval [0, 1]:

```
L = 0;
        do {
                L = L+1;
                } while (( rand() < CR) and ( L < D));
```

Hence, the probability $\Pr(L \geq v) = (CR)^{v-1}$, $v > 0$. CR is taken from the interval [0, 1] and constitutes a control variable in the design process. The random decisions for both n and L are always made afresh for each newly generated vector $\underline{X}_{u,G+1}$.

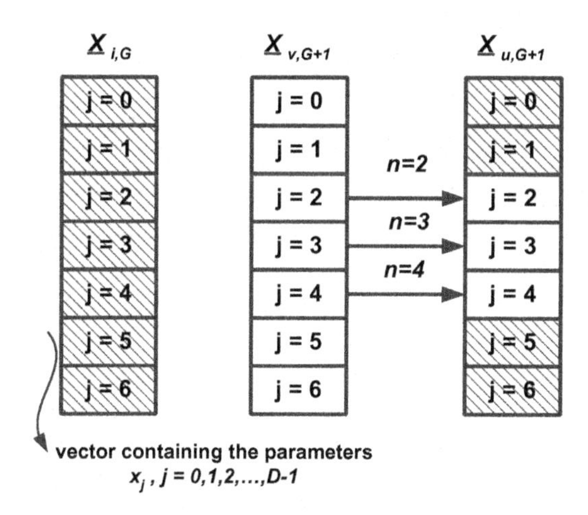

Figure 5.7. Crossover process in DE1 for $D = 7$, $n = 2$, $L = 3$ for new vector generation

Note that, in Figure 5.7, since L = 3, three parameters are exchanged; they are numbered as (n = 2), (n+1 = 3), (n+L-1 = 4), because the modulo function (n and D) = 2, modulo function (n+1 and D) = 3 and modulo function (n+L-1 and D) = 4, for D = 7.

To decide whether or not the newly generated vector should become a member of generation G+1, the new vector $\underline{X}_{u,G+1}$ is compared with $\underline{X}_{i,G}$. If the newly generated vector yields a smaller objective value than $\underline{X}_{i,G}$, then $\underline{X}_{i,G+1}$ is set to $\underline{X}_{u,G+1}$, otherwise the old vector $\underline{X}_{i,G}$ is retained.

5.6.2 Second Variant of Differential Evolution (DE2)

Basically the second variant of differential evolution also works in the same way as the first variant DE1, but it generates a new trial vector $\underline{X}_{v,G+1}$ according to

$$\underline{X}_{v,G+1} = \underline{X}_{i,G} + \eta \cdot \left(\underline{X}_{best,G} - \underline{X}_{i,G} \right) + K \cdot \left(\underline{X}_{a_3,G} - \underline{X}_{a_2,G} \right),$$

because $\underline{X}_{v1,G+1} = \underline{X}_{i,G} + \eta \cdot \left(\underline{X}_{best,G} - \underline{X}_{i,G} \right)$, introducing an additional control variable η.

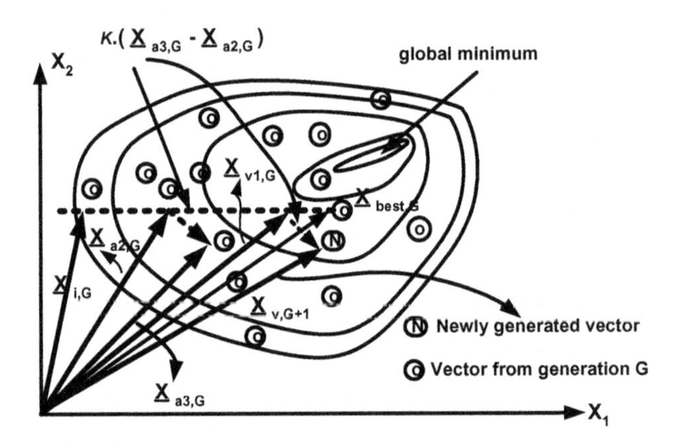

Figure 5.8. Second variant of differential evolution (DE2)

The idea behind η is to provide a means to enhance the greediness of the scheme by incorporating the best vector from the current generation. In order to reduce the differential evolution control parameters, $\eta = K$ is usually set. This feature can be useful for non-critical objective functions. Figure 5.8 illustrates the trial vector generation process for the generation G+1, defined by the above equation. The construction of perturbed vector $\underline{X}_{u,G+1}$ through a crossover operation from trial vector $\underline{X}_{v,G+1}$ and randomly selected $\underline{X}_{i,G}$ vector, as well as the decision process, are exactly same as the first variant of differential evolution.

References

[1] Fogel DB (1994) Evolutionary Programming: An introduction and some current directions, Statistics and Computing, vol. 4: 113-129.

[2] Fogel LJ, Owens AJ, and Walsh MJ (1966) Artificial intelligence through Simulated Evolution. Wiley, New York.

[3] Goldberg DE (1989) Genetic algorithms in search, optimization and machine learning, Addison-Weseley publishing co. Inc., Reading, MA.

[4] Harrald PG and Kamastra M (1997) Evolving artificial neural networks to combine financial forecasts, IEEE Trans. on Evolutionary Computation, 1(1): 40-51.
[5] Herrera F and Locano M (1998) Fuzzy genetic algorithms: Issues and models. Tech. Report DECSAI-98116, Univ. of Granada, dept. of computer science and AI.
[6] Holland JH (1975) Adaptation in Natural and Artificial Systems. Ann Arbor, University of Michigan Press.
[7] Klawonn F and Keller A (1998) Fuzzy clustering with evolutionary Algorithms, Internat. J. of Intell. Systems 13: 975-991. In: Goldberg GA Book, Reading, MA.
[8] Koza JR (1992) Genetic Programming. The MIT Press, Cambridge, MA.
[9] Koza JR (1994) Genetic Programming II: Automatic Discovery of Reusable Programs. Cambridge, MA: MIT Press, MA.
[10] Michalewicz Z (1994) Genetic Algorithms + Data Structures = Evolution Programs, second edition, Springer-Verlag, New York.
[11] Michalewicz Z (1998): Real Coded-GA Book, Springer-Verlag. Berlin.
[12] Mulloy BS, Riolo RL, and Savit RS (1996) Dynamics of genetic programming and chaotic time series prediction. In: Koza JR, Goldberg DE, Fogel DB, and Riolo RL, editors, Genetic programming 1996: Proc. of the First Annual Conference: 166-174, MIT Press, MA.
[13] Palit AK, Popovic D (2000), Intelligent processing of time series using neuro-fuzzy adaptive Genetic approach, Proc. of IEEE-ICIT Conference, Goa, India, ISBN: 0-7803-3932-0, vol. 1: 141-146.
[14] Panchariya PC, Palit AK, Sharma AL, Popovic D (2004) Rule extraction, complexity reduction and evolutionary optimization, International Journal of Knowledge-Based and Intelligent Engineering Systems, 8(4): 189-203.
[15] Rechenberg I (1973) Evolutionsstrategie: Optimierung Technischer Systeme nach Prinzipien der Biologischen Evolution. Frommann-Holzboog Verlag, Stuttgart.
[16] Roubos H, Setnes M (2001), Compact and transparent fuzzy models and classifiers through iterative complexity reduction, IEEE Trans. on Fuzzy Syst., 9(4): 516-524.
[17] Santini M, Tettamanzi A (2001) Genetic Programming for Financial Time Series Prediction. ECRO GP 2001: 361-370.
[18] Setnes M, Roubos JA (1999) Transparent Fuzzy modeling using Fuzzy Clustering and GAs, in proceedings of NAFIPS'99, pp. 198-202, New York, June, 1999.
[19] Storn and Price (1995): Differential Evolution- A simple and efficient adaptive scheme for global optimization over continuous spaces, TR-95-012, ICSI, March 1995. http://http.icsi.berkeley.edu/~storn/litera.html
[20] Storn R, Price K.(1996), Minimizing the real functions of the ICEC'96 contest Differential Evolution, Int. Conf. on Evol. Comp., Nagoya, Japan.
[21] Storn, R (1995) Constrained optimization, Dr. Dobb's journal, May, pp. 119-123.
[22] Tang KS, Man KF, Kwong S and He Q (1996) Genetic algorithms and their applications, IEEE Signal Processing Magazine, November, pp. 22-36.
[23] Voigt, H.M.(1992), Fuzzy evolutionary algorithms, Technical Report TR-92-038 at ICSI, ftp.icsi.berkeley.edu.
[24] Wang LX (1994) Fuzzy Systems and Control, Design and Stability Analysis, PTR Prentice Hall, Englewood Cliffs, New Jersey.
[25] Yoshihara I, Aoyama T, and Yasunaga M (2000) Genetic programming based modeling method for time series prediction with parameter optimization and node alternation. In: Proc. of Congress on Evolutionary Computation CEC00: 1475-1481.
[26] Zang B, Ohm P, and Mühlenbein H (1997) Evolutionary induction of sparse neural trees. Evolutionary computation 5(2): 213-236.

Part III

Hybrid Computational Technologies

6

Neuro-fuzzy Approach

6.1 Motivation for Technology Merging

Contemporary intelligent technologies have various characteristic features that can be used to implement systems that mimic the behaviour of human beings. For example, expert systems are capable of reasoning about the facts and situations using the rules out of a specific domain, *etc.* The outstanding feature of neural networks is their capability of learning, which can help in building artificial systems for pattern recognition, classification, *etc.* Fuzzy logic systems, again, are capable of interpreting the imprecise data that can be helpful in making possible decisions. On the other hand, genetic algorithms provide implementation of random, parallel solution search procedures within a large search space. Therefore, in fact, the complementary features of individual categories of intelligent technologies make them ideal for isolated use in solving some specific problems, but not well suited for solving other kinds of intelligent problem. For example, the black-box modelling approach through neural networks is evidently well suited for process modelling or for intelligent control, but less suitable for decision making. On the other hand, the fuzzy logic systems can easily handle imprecise data, and explain their decisions in the context of the available facts in linguistic form; however, they cannot automatically acquire the linguistic rules to make those decisions. Such capabilities and restrictions of individual intelligent technologies have actually been a central driving force behind their fusion for creation of ***hybrid intelligent systems*** capable of solving many complex problems.

The permanent growing interest in intelligent technology merging, particularly in merging of neural and fuzzy technology, the two technologies that complement each other (Bezdek, 1993), to create neuro-fuzzy or fuzzy-neural structures, has largely extended the capabilities of both technologies in hybrid intelligent systems. The advantages of neural networks in learning and adaptation and those of fuzzy logic systems in dealing with the issues of human-like reasoning on a linguistic level, transparency and interpretability of the generated model, and handling of uncertain or imprecise data, enable building of higher level intelligent systems. The synergism of integrating neural networks with fuzzy logic technology into a hybrid functional system with low-level learning and high-level reasoning transforms the burden of the tedious design problems of the fuzzy logic decision systems to the learning of connectionist neural networks. In this way the approximation capability and the overall performance of the resulting system are enhanced.

A number of different schemes and architectures of this hybrid system have been proposed, such as ***fuzzy-logic-based neurons*** (Pedrycz, 1995), ***fuzzy neurons*** (Gupta, 1994), ***neural networks with fuzzy weights*** (Buckley and Hayashi, 1994), ***neuro-fuzzy adaptive models*** (Brown and Harris, 1994), *etc*. The proposed architectures have been successful in solving various engineering and real-world problems, such as in applications like system identification and modelling, process control, systems diagnosis, cognitive simulation, classification, pattern recognition, image processing, engineering design, financial trading, signal processing, time series prediction and forecasting, *etc*.

6.2 Neuro-fuzzy Modelling

There are several methods for implementing the neuro-fuzzy modelling technique. An early merging approach was to replace the input-output signals or the weights in neural networks by membership values of fuzzy sets, along with the application of fuzzy neurons (Mitra and Hayashi, 2000). Several authors have proposed an internal structure for fuzzy neurons (Gupta, 1994; Buckley and Hayashi, 1995), as presented in the following section.

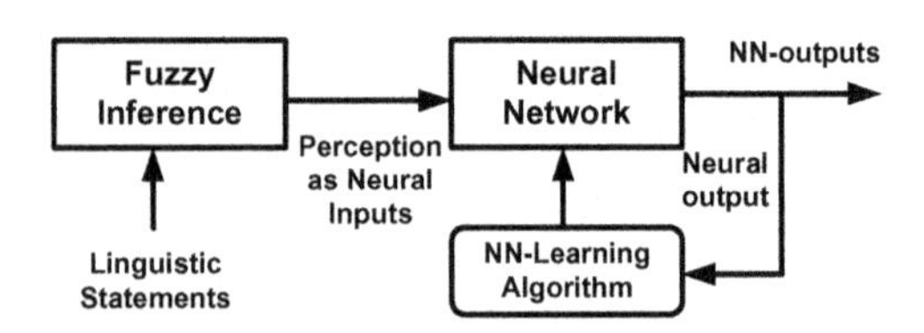

Figure 6.1. (a) Fuzzy-neural system (first model)

In general, neuro-fuzzy hybridization is done in two ways (Mitra and Hayashi, 2000):

- a neural network equipped with the capability of handling fuzzy information processing, termed a fuzzy-neural network (FNN)
- a fuzzy system augmented by neural networks to enhance some of its characteristics, like flexibility, speed, and adaptability, termed a neural-fuzzy system (NFS).

Neural networks with fuzzy neurons are also termed FNN, because they are also capable of processing fuzzy information. A neural-fuzzy system (NFS), on the other hand, is designed to realize the process of fuzzy reasoning, where the connection weights of the network correspond to the parameters of fuzzy reasoning (Nauck *et al.*, 1997).

Gupta (1994) has presented two additional models for fuzzy neural systems. The first model (Figure 6.1(a)) consists of a fuzzy inference block, followed by a neural network block, consisting of a multilayer feedforward neural network, the input of which is fed by the inference block (Fuller, 1995). The neural network used can be adapted and adequately trained with training samples to yield the desired outputs.

In the second model (Figure 6.1(b)), the neural network block drives the fuzzy inference system to generate the corresponding decisions. Hence, the first model takes linguistic inputs and generates the numerical outputs, whereas the second model takes numerical inputs and generates the linguistic outputs.

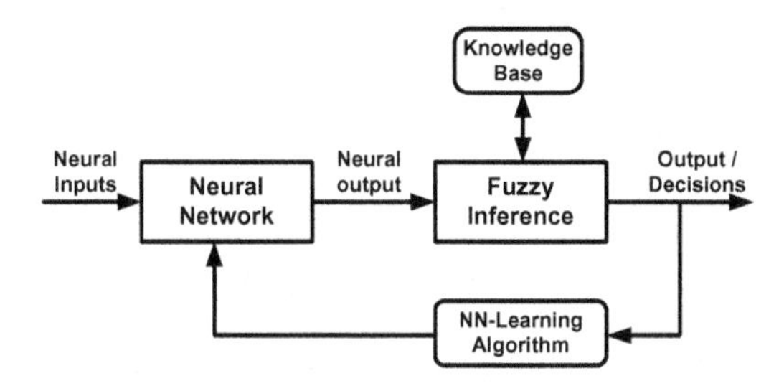

Figure 6.1. (b) Fuzzy-neural system (second model)

Alternatively, the second approach is to use fuzzy membership functions to pre-process or post-process signals with neural networks as shown in Figure 6.2. A fuzzy inference system can encode an expert's knowledge directly and easily using rules with linguistic labels (Kulkarni, 2001).

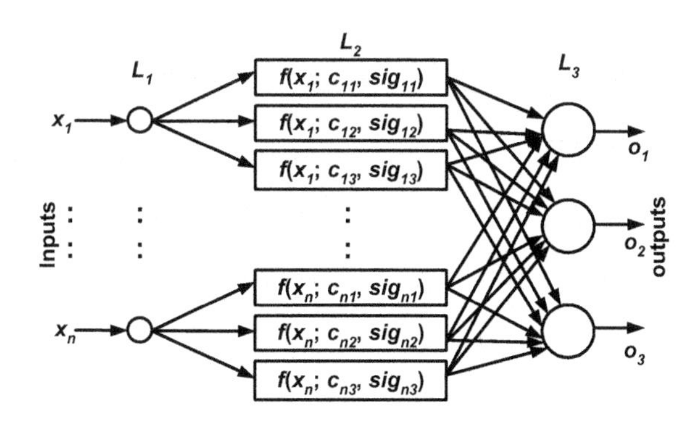

Figure 6.2. Fuzzy-neural model with tuneable membership function

In practice, for optimal tuning of membership functions of the fuzzy logic part of a neuro-fuzzy system, a reliable skill is required. The incorporated neural network part of the same system can, using its learning capability, perform on-line tuning of membership functions and gradually improve the performance of the entire hybrid system. This concept, which became very popular in engineering applications, was originally proposed and extended to multidimensional membership functions by Takagi and Hayashi (1991).

Lin and Lee (1991) proposed a neural-network-based model for fuzzy logic control consisting of a feedforward neural network, the input nodes of which are fed by input signals and its output nodes delivering the output and decision signals. Nodes in the hidden layers of the system implement the membership functions and the fuzzy rules, making up a fuzzy inference system with distributed representation and learning algorithms of the neural network. Parameters representing membership functions are determined using any suitable network training algorithm. Pal and Mitra (1992) proposed a similar model in which inputs are fed to a preprocessor block, which performs the same functions as that in the above fuzzy inference system. The output of the preprocessor delivers the fuzzy membership function values. For each input variable term, linguistic labels such as *low*, *medium*, and *high* are used. If input consists of n variables, then the preprocessor block yields $m \times n$ outputs, where m represents the number of term values used in the model. The output of the preprocessor block is then fed to a multilayer perceptron model that implements the inference engine. The model was successfully used for classifying vowels in English alphabets. Kulkarni (1998), again, developed a similar model and successfully used it for multi-spectral image analysis. Some authors have designed neuro-fuzzy systems incorporating some processing stages implemented with neural networks and some with a fuzzy inference system. In another design, a neural-network-based tree classifier was used. Finally, Kosko (1992) suggested some remarkable neuro-fuzzy models for fuzzy associative memory (FAM).

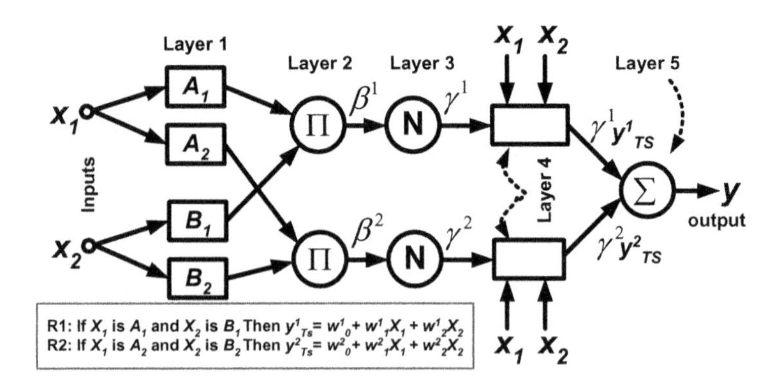

Figure 6.3. ANFIS architecture with Takagi-Sugeno-type fuzzy model with two rules

The neuro-fuzzy model ANFIS (adaptive-network-based fuzzy inference system) of Jang (1993), presented in Figure 6.3, incorporates a five-layer network to implement a Takagi-Sugeno-type fuzzy system. The proposed model has a relatively complex architecture for a large number of inputs, and it can process a large number of fuzzy rules. It uses the least mean square training algorithm in the forward computation to determine the linear consequents of the Takagi-Sugeno rules, while for the optimal tuning of an antecedent membership function backpropagation is used (Kim and Kim, 1997).

The neuro-fuzzy model of Chak *et al.* (1998) can locate the fuzzy rules and optimize their membership functions by competitive learning and a Kalman filter algorithm. The key feature is that a high-dimensional fuzzy system can be implemented with fewer rules than that required by a conventional Sugeno-type model. This is because the input space partitions are unevenly distributed, thus enabling a real-time network implementation.

The approach of Nie (1997) concerns the development of a ***multivariable fuzzy model*** from numerical data using a self-organizing counterpropagation network. Both supervised and unsupervised learning algorithms are used for network training. Knowledge can be extracted from the data in the form of a set of rules. This rule base is then utilized by a fuzzy reasoning model. The rule base of the system, which is supposed to be relatively simple, is updated on-line in an adaptive way (in terms of connection weights) in response to the incoming data.

Cho and Wang (1996) developed an adaptive fuzzy system to extract the IF-THEN rules from sampled data through learning using a radial basis functions network. Different types of consequent, such as constants, first-order linear functions, and fuzzy variables are modelled, thereby enabling the network to handle arbitrary fuzzy inference schemes. There is not an initial rule base, and neither does one need to specify in advance the number of rules required to be identified by the system. Fuzzy rules are generated (when needed) by employing basis function units.

Wang and Mendel (1992a) described a fuzzy system by series of basis functions, which are algebraic superpositions of membership functions. Each such basis function corresponds to one fuzzy logic rule. An orthogonal least squares training algorithm is utilized to determine the significant fuzzy logic rules (structure learning) and associated parameters (parameter learning) from input-output training pairs. Owing to the possibility of acquiring and interpreting the linguistic IF-THEN rules by human experts, the fuzzy basis function network provides a framework for combining both numerical and linguistic information in a uniform manner.

Zhang and Morris (1999) used a recurrent neuro-fuzzy network to build long-term prediction models for nonlinear processes. Process knowledge is initially used to partition the process operation into several local fuzzy operating regions and also to set up the initial fuzzification layer weights. Membership functions of fuzzy operating regions are refined through training, enabling the local models to learn. The global model output is obtained by ***centre-of-gravity defuzzification*** involving the local models.

6.2.1 Fuzzy Neurons

The perceptron or processing unit described in Chapter 3, which employs multiplication, addition, and the sigmoid activation function to produce the nonlinear output from the applied input, is generally known as a *simple neural network*. However, if their architectures are extended by adding other mathematical operations, such as triangular-norm, a triangular-co-norm, *etc*., to combine the incoming signals to the neuron, the extended networks give rise to a hybrid neural network based on fuzzy arithmetic operations. The fuzzy neural network architecture is practically based on such a processing element known as *fuzzy neuron* (Fuller,1995).

6.2.1.1 AND Fuzzy Neuron

Consider a perceptron-like structure as shown in Figure 6.4 with n input neurons acting as fan out elements (*i.e.* having the same output values as their inputs) and with one output neuron. The outputs x_i of the input-layer neurons are multiplied by the connecting weights w_i and, thereafter, fed to the output-layer neuron. If, however, the input signals x_i and the weights w_i are combined by an ***S-norm***, *i.e.* the ***triangular-conorm***

$$p_i = S(w_i, x_i), \quad i = 1, 2, \cdots, n. \tag{6.1}$$

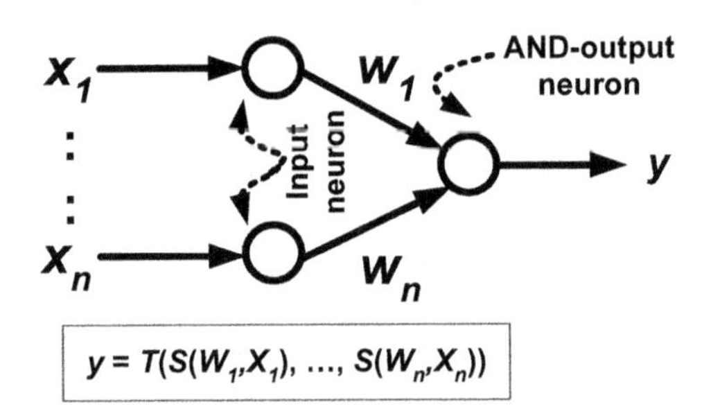

Figure 6.4. AND fuzzy neuron

and the input information p_i is further aggregated by a ***T-norm***, *i.e.* ***triangular norm***, to yield the final output of the neuron as

$$\begin{aligned} y &= AND(p_1, p_2, \dots, p_n) = T(p_1, p_2, \dots, p_n) \\ &= T(S(w_1, x_1), S(w_2, x_2), \dots, S(w_n, x_n)). \end{aligned} \tag{6.2}$$

then the configuration in Figure 6.4 will represent the implementation of an ***AND fuzzy neuron*** under the condition that the *T-norm* represents a ***min operator*** and the *S-norm* represents a ***max operator***. Then the *min-max* composition

$$y = \min\{\max(w_1, x_1), \dots, \max(w_n, x_n)\}. \tag{6.3}$$

can be realized by the AND fuzzy neuron.

6.2.1.2 OR Fuzzy Neuron

If a similar configuration to Figure 6.4 is used, but the signals x_i and the weights w_i are combined by a *triangular-norm* (*T-norm*)

$$p_i = T(w_i, x_i), \quad i = 1, 2, \cdots, n. \tag{6.4}$$

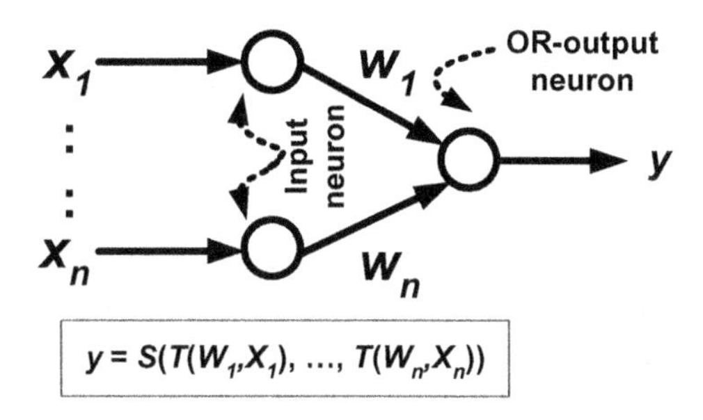

Figure 6.5. OR fuzzy neuron

and, thereafter, the input information p_i is further aggregated by a *triangular conorm* (*t-conorm* or *s-norm*) to yield the final output of the neuron as follows:

$$\begin{aligned} y &= OR(p_1, p_2, \ldots, p_n) = S(p_1, p_2, \ldots, p_n) \\ &= S(T(w_1, x_1), T(w_2, x_2), \ldots, T(w_n, x_n)). \end{aligned} \tag{6.5}$$

So, if the *t*-norm or the T = min operator and the *t-co norm* or the *s-norm* S = *max operator*, then the *max-min* composition can be realized by the *OR fuzzy neuron* as follows:

$$y = \max\{\min(w_1, x_1), \ldots, \min(w_n, x_n)\}. \tag{6.6}$$

Both fuzzy neurons realize logic operations on the membership values. The role of the connections is to differentiate between particular levels of impact that the individual inputs might have on the result of aggregation. We note that: (i) the higher the value of w_i the stronger is the impact of x_i on the output y of an *OR neuron*; (ii) the lower the value of w_i the stronger is the impact of x_i on the output y of an *AND neuron*.

The range of the output value y for the *AND neuron* is computed by letting all x_i equal to zero or one. By virtue of the monotonic property of the *triangular norms*, we obtain

$$y \in [T(w_1, \ldots, w_n), 1], \tag{6.7}$$

and for the OR neuron one derives the boundaries as

$$y \in \left[0, S\left(w_1, \ldots, w_n\right)\right]. \tag{6.8}$$

Similar to *AND* and *OR fuzzy neurons*, several other *fuzzy neurons*, such as *implication-OR*, *Kwan and Cai's fuzzy neuron*, *etc.* have been proposed (Fuller, 1995).

6.3 Neuro-fuzzy System Selection for Forecasting

The most common approach to numerical-data-driven neuro-fuzzy modelling is to use a Takagi-Sugeno-type fuzzy model along with differentiable operators and continuously differentiable membership functions (*e.g.* Gaussian function) for building the fuzzy inference mechanism, and the weighted average defuzzifier for defuzzification of output data. The corresponding output inference can then be represented in a multilayer feedforward network structure, such as the one depicted in Figure 6.6. In principle, the neuro-fuzzy network's architecture (Figure 6.6) is identical to the architecture of ANFIS, as shown in Figure 6.3.

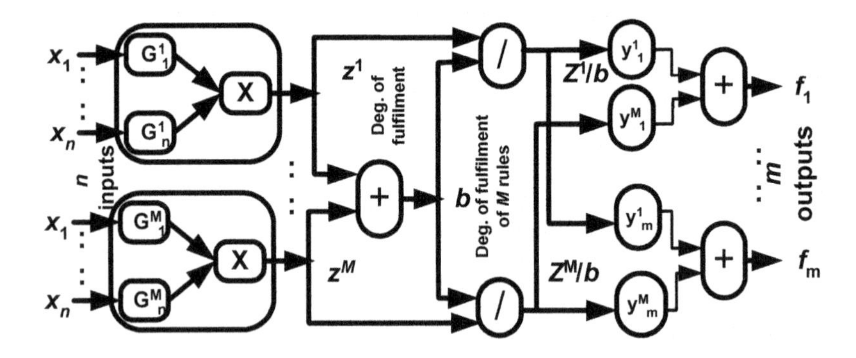

Figure 6.6. Fuzzy system as a multi-input multi-output feedforward neural network

The neuro-fuzzy model presented in Figure 6.6 is based on Gaussian membership functions. It uses Takagi-Sugeno-type fuzzy rules, product inference, and weighted average defuzzification. The nodes in the first layer calculate the degree of membership of the numerical input values in the antecedent fuzzy sets. The product nodes ($\times$) in the rectangular blocks (rounded corners) represent the antecedent conjunction operator and the output of this node is the corresponding degree of fulfilment $\left(z^l;\ l=1,\ 2,\ 3,\cdots,M\right)$ or firing strength of the rule. The division nodes $(/)$, together with summation nodes (+), help implement the normalized degree of fulfilment (z^l/b) of the corresponding rule, which, after multiplication with the corresponding Takagi-Sugeno rule consequent (y_m^l), is used as input to the summation block (+) at the final output layer. The output of this summation node is the final defuzzified output value, which, being crisp in nature, is directly compatible with the real world data. Once the fuzzy system of the above choice is represented as a feedforward network, the algorithm used for its training is less relevant.

A similar fuzzy model with singleton rule consequents, trained with standard backpropagation algorithm, was used by Wang and Mendel (1992b) for identification of various nonlinear plants.

Forecasting of time series is primarily based on numerical input-output data. To demonstrate this for neuro-fuzzy networks, a Takagi-Sugeno-type model, *i.e.* with linear rules consequent (and also a singleton model as a special case), is selected (Palit and Popovic, 1999; Palit and Babuška, 2001). Here, the number of membership functions to be implemented for fuzzy partitioning of input universes of discourse happens to be equal to the number of *a priori* selected fuzzy rules. To accelerate the convergence speed of the training algorithm and to avoid other inconveniences, the *Levenberg-Marquardt training algorithm* (described in the Section 6.4.2.3) or the *adaptive genetic algorithm* (AGA) can also be used.

In Chapter 4 it was shown that in forecasting of various nonlinear time series the fuzzy logic approach with automatically generated fuzzy rules (Wang and Mendel, 1992c; Palit and Popovic, 1999) works reasonably well. However, it was emphasized that the performance of fuzzy logic systems depends greatly on a set of well-consistent fuzzy rules and on the number of fuzzy membership functions implemented, along with their extent of overlapping. Therefore, determination of the optimum overlapping values of adjacent membership functions is very important in the sense that overlapping values too large or too small may deteriorate the forecasting accuracy. In the absence of firm guiding rules for optimum selection of overlapping, this selection mechanism was rather seen more as an art than as a science, mainly relying on a trial-and-error approach. Alternatively, very time-consuming heuristic approaches, such as the ***evolutionary computation*** or the ***genetic algorithms*** (Setnes and Roubos, 2000), can be used for this purpose.

Fuzzy logic systems encode numerical crisp values using linguistic labels, so it is difficult and time consuming to design and fine tune the membership functions related to such labels. However, neural networks' learning ability can automate this process. The combination of both fuzzy logic and neural network implementations can thus facilitate development of hybrid forecasters.

As an example we will consider the neural-networks-like architecture of the neuro-fuzzy system (Figure 6.6) and the training algorithm selected will fine tune the randomly generated system parameters. The great advantage of this scheme is that, apart from the user-selected number of fuzzy rules to be implemented, all other fuzzy parameters are automatically set by the training algorithm, so that the user does not need to bother about the optimal settings of fuzzy region overlappings and the like. Therefore, the approach to be described here is often referred to as an ***adaptive neuro-fuzzy approach*** and the related fuzzy logic system as an ***adaptive fuzzy logic system*** (Wang, 1994).

6.4 Takagi-Sugeno-type Neuro-fuzzy Network

In the recent years much attention has been paid to deriving an effective data-driven neuro-fuzzy model because of its numerous advantages. For example, ANFIS-based (neuro-fuzzy) modelling was initially developed by Jang (1993) and Jang and Sun (1995), and later on widely applied in engineering. Similarly, singleton-rule-based and data-driven multi-input single output neuro-fuzzy modelling was initially developed by Wang and Mendel (1992b) and used for solving a variety of systems identification and control problems. A similar neuro-fuzzy network, with an improved training algorithm, was later developed and applied by Palit and Popovic (1999, 2000a, 2002b) and Palit and Babuška (2001) for time series forecasting. Because of its advantages compared with ANFIS, at least as far as model accuracy and the training time are concerned, this similar model, but with multi-input and multi-output structure, will be used in this chapter as a neuro-fuzzy forecaster. The advantages of this approach, where an explicitly Takagi-Sugeno-type multi-input multi-output fuzzy model is used, will be demonstrated on simulation examples of benchmark problems. Furthermore, the type of network selected can be regarded as a generalization or upgraded version of both a singleton-consequent-type multi-input single-output neuro-fuzzy network and the Takagi-Sugeno-type multiple input single output neuro-fuzzy network of Palit and Babuška (2001).

To avoid the fine tuning difficulties of initially chosen random membership functions, an efficient training algorithm for modelling of various nonlinear dynamics of multi-input multi-output systems is proposed that relies on a Takagi-Sugeno-type neuro-fuzzy network. The algorithm is further used for training the neuro-fuzzy network with the available data of a nonlinear electrical load time series. Thereafter, the trained network is used as a neuro-fuzzy model to predict the future value of electrical load data. In order to verify its prediction capability with other standard methods, some benchmark problems, such as Mackey-Glass chaotic time series and second-order nonlinear plant modelling, are considered.

Furthermore, the neuro-fuzzy approach described here attempts to exploit the merits of both neural-network and fuzzy-logic-based modelling techniques. For example, the fuzzy models are based on fuzzy IF-THEN rules and are, to a certain degree, transparent to interpretation and analysis, whereas the neural-networks-based black-box model has a unique learning ability.

In the following, the Takagi-Sugeno-type multiple-input multiple-output neuro-fuzzy system is constructed by multilayer feedforward network representation of the fuzzy logic system, as described in Section 6.4.1, and its training algorithm is described in Section 6.4.2. Thereafter, some comparisons between the radial basis function network and the proposed neuro-fuzzy network are made, followed by similar comparisons of the training algorithm for neural networks and neuro-fuzzy networks. Neuro-fuzzy modelling and time series forecasting are subsequently described and then, finally, some engineering examples are presented.

6.4.1 Neural Network Representation of Fuzzy Logic Systems

The fuzzy logic system considered here for constructing neuro-fuzzy structures is based on a Takagi-Sugeno-type fuzzy model with Gaussian membership functions. It uses product inference rules and a weighted-average defuzzifier defined as

$$f_j(x^p) = \sum_{l=1}^{M} y_j^l z^l \Big/ \sum_{l=1}^{M} z^l, \tag{6.9a}$$

where $j = 1, 2, 3, ..., m$; $l = 1, 2, 3, ..., M$;

$$y_j^l = \left(\theta_{0j}^l + \sum_{i=1}^{n} \theta_{ij}^l x_i\right), \text{ with } i = 1, 2, 3, \cdots, n; \tag{6.9b}$$

and $j = 1, 2, 3, ..., m$; $l = 1, 2, 3, ..., M$;

$$z^l = \prod_{i=1}^{n} \mu_{G_i^l}(x_i) \; ; \; \mu_{G_i^l}(x_i) = \exp\left\{ -\left(x_i - c_i^l\right)^2 \Big/ \sigma_i^{l\,2} \right\} \tag{6.9c}$$

with $i = 1, 2, 3, ..., n$.

Here, we assume that $c_i^l \in U_i$, $\sigma_i^l > 0$, and $y_j^l \in V_j$, where U_i, and V_j are the input and output universes of discourse respectively. The corresponding lth rule from the above fuzzy logic system can be written as

R_l: *If* x_1 *is* G_1^l *and* x_2 *is* G_2^l *and ... and* x_n *is* G_n^l

$$\text{Then } y_j^l = \theta_{0j}^l + \theta_{1j}^l x_1 + \theta_{2j}^l x_2 + \cdots + \theta_{nj}^l x_n \tag{6.10}$$

where, x_i with $i = 1, 2, \ldots, n$; are the n system inputs, whereas f_j, with $j = 1, 2, \ldots, m$; are its m system outputs, and G_i^l, with $i = 1, 2, \ldots, n$; and $l = 1, 2, \ldots, M$; are the Gaussian membership functions of the form (6.9c) with the corresponding mean and variance parameters c_i^l and σ_i^l respectively and with y_j^l as the output consequent of the lth rule.

It is to be noted that the Gaussian membership functions (G_i^l) actually represent linguistic terms such as *low*, *medium*, *high*, *etc*. The rules (6.10), as specified above, are known as Takagi-Sugeno rules.

In the fuzzy logic system (6.9a) – (6.9c) the Gaussian membership function is deliberately chosen because the same membership function is continuously differentiable at all points. This is an essential requirement to apply the gradient-method-based training algorithm. Furthermore, it is also important to note that the fuzzy logic system (6.9a) – (6.9c) is capable of uniformly approximating any nonlinear function to any degree of accuracy over a universe of discourse $U \subset \mathbb{R}^n$ (Wang, 1994).

By carefully observing the functional forms (6.9a) – (6.9c), it can be seen that the above fuzzy logic system can be represented as a three-layer multi-input, multi-output feedforward network as shown in Figure 6.6. Because of the neuro implementation of the Takagi-Sugeno type fuzzy logic system, Figure 6.6 actually represents a Takagi-Sugeno-type of multi-input, multi-output neuro-fuzzy network, where, instead of the connection weights and the biases as in backpropagation neural networks, we have the mean (c_i^l) and the variance (σ_i^l) parameters of Gaussian membership functions, along with (θ_{0j}^l, θ_{ij}^l), *i.e.* y_j^l from the rules consequent, as the equivalent adjustable parameters of the network.

If the adjustable parameters of the neuro-fuzzy network are suitably selected, then the above fuzzy logic system can correctly approximate any nonlinear system based on given input–output data pairs.

6.4.2 Training Algorithm for Neuro-fuzzy Network

The fuzzy logic system, once represented as the equivalent multi-input, multi-output feedforward network (Figure 6.6), can generally be trained using any suitable training algorithm, such as the standard backpropagation algorithm (Palit *et al.*, 2002) that is generally used for neural networks training. However, because of its relatively slow speed of convergence, this algorithm needs to be further improved. Alternatively, a more efficient second-order training algorithm, such as the Levenberg-Marquardt algorithm described in the Section 6.4.2.3, can also be used.

6.4.2.1 Backpropagation Training of Takagi-Sugeno-type Neuro-fuzzy Network

Let a set of N input-output data pairs $\left(x^p, d_j^{\,p}\right)$, with p = 1, 2, 3, ..., N; and $x^p \equiv \left(x_1^p, x_2^p, \cdots, x_n^p\right) \in U \subset \mathbb{R}^n$, and $d_j^{\,p} \in V_j \subset \mathbb{R}^m$ is given. The objective is to determine a fuzzy logic system $f_j\left(x^p\right)$ in the form of (6.9a) – (6.9c), such that the performance function S, defined as

$$S_j = 0.5 \cdot \sum_{p=1}^{N} \left(e_j^p\right)^2 = 0.5 \cdot E_j^T E_j \tag{6.11a}$$

and
$$S = \sum_{j=1}^{m} S_j = \left(S_1 + S_2 + \cdots + S_m\right), \tag{6.11b}$$

is minimized, where E_j is the column vector of errors $e_j^p = \left\{f_j(x^p) - d_j^{\,p}\right\}$, and p = *1, 2, ..., N*; for the jth output from the fuzzy logic system. In addition, we also assume that the number of fuzzy rules and also the number of membership functions (to be implemented) M are given. In this way the problem is reduced to the adjustment of y_j^l, *i.e.* the parameters (θ_{0j}^l, θ_{ij}^l) from the rules consequent and the mean (c_i^l) and variance (σ_i^l) parameters of the Gaussian membership functions, so that the performance function (6.11a) is minimized. For convenience, we replace $f_j(x^p)$, d_j^p, and e_j^p in the above definition of error by f_j, d_j, and e_j respectively, so that the individual error becomes $e_j = (f_j - d_j)$.

We recall that the *steepest descent rule* used for training of neuro-fuzzy networks is based on the recursive expressions

$$\theta_{0j}^l(k+1) = \theta_{0j}^l(k) - \eta \cdot \left(\partial S / \partial \theta_{0j}^l\right) \tag{6.12a}$$

$$\theta_{ij}^l(k+1) = \theta_{ij}^l(k) - \eta \cdot \left(\partial S / \partial \theta_{ij}^l\right) \tag{6.12b}$$

$$c_i^l(k+1) = c_i^l(k) - \eta \cdot \left(\partial S / \partial c_i^l\right) \tag{6.12c}$$

$$\sigma_i^l(k+1) = \sigma_i^l(k) - \eta \cdot \left(\partial S / \partial \sigma_i^l\right) \tag{6.12d}$$

where S is the performance function (6.11b) at the *kth* iteration step and $\theta_{0j}^l(k)$, $\theta_{ij}^l(k)$, $c_i^l(k)$, and $\sigma_i^l(k)$ are the free parameters of the network at the same iteration step, the starting values of which are, in general, randomly selected.

In addition, η is the constant step size or learning rate (usually $\eta \ll 1$), $i = 1, 2, \ldots, n$ (with n as the number of inputs to the neuro-fuzzy network); $j = 1, 2, \ldots, m$ (with m as the number of outputs from the neuro-fuzzy network); and $l = 1, 2, 3, \ldots, M$ (with M as the number of Gaussian membership functions selected, as well as the number of fuzzy rules to be implemented).

From Figure 6.6, it is evident that the network output f_j and hence the performance function S_j and, therefore, finally S depends on θ_{0j}^l and θ_{ij}^l only through y_j^l. Similarly, the network output f_j and, thereby, the performance functions S_j and S depend on c_i^l and σ_i^l only through z^l, where, f_j, y_j^l, b, and z^l are represented by

$$f_j = \sum_{l=1}^{M} y_j^l \cdot h^l \tag{6.13a}$$

$$y_j^l = \theta_{0j}^l + \theta_{1j}^l x_1 + \theta_{2j}^l x_2 + \cdots + \theta_{nj}^l x_n \tag{6.13b}$$

$$h^l = \left(z^l / b\right), \text{ and } \quad b = \sum_{l=1}^{M} z^l \tag{6.13c}$$

$$z^l = \prod_{i=1}^{n} \exp\left(-\left(\frac{x_i - c_i^l}{\sigma_i^l}\right)^2\right) \tag{6.13d}$$

Therefore, the corresponding chain rules

$$\left(\partial S/\partial \theta_{0j}^{l}\right)=\left(\partial S_{j}/\partial f_{j}\right)\cdot\left(\partial f_{j}/\partial y_{j}^{l}\right)\cdot\left(\partial y_{j}^{l}/\partial \theta_{0j}^{l}\right) \tag{6.14a}$$

$$\left(\partial S/\partial \theta_{ij}^{l}\right)=\left(\partial S_{j}/\partial f_{j}\right)\cdot\left(\partial f_{j}/\partial y_{j}^{l}\right)\cdot\left(\partial y_{j}^{l}/\partial \theta_{ij}^{l}\right) \tag{6.14b}$$

$$\begin{aligned}\left(\partial S/\partial c_{i}^{l}\right)&=\left(\partial\left(\sum_{j=1}^{m} S_{j}\right)\Big/\partial z^{l}\right)\cdot\left(\partial z^{l}/\partial c_{i}^{l}\right)\\&=\left(\sum_{j=1}^{m}\left(\partial S_{j}/\partial f_{j}\right)\cdot\left(\partial f_{j}/\partial z^{l}\right)\right)\cdot\left(\partial z^{l}/\partial c_{i}^{l}\right)\end{aligned} \tag{6.14c}$$

$$\begin{aligned}\left(\partial S/\partial \sigma_{i}^{l}\right)&=\left(\partial\left(\sum_{j=1}^{m} S_{j}\right)\Big/\partial z^{l}\right)\cdot\left(\partial z^{l}/\partial \sigma_{i}^{l}\right)\\&=\left(\sum_{j=1}^{m}\left(\partial S_{j}/\partial f_{j}\right)\cdot\left(\partial f_{j}/\partial z^{l}\right)\right)\cdot\left(\partial z^{l}/\partial \sigma_{i}^{l}\right)\end{aligned} \tag{6.14d}$$

can finally be written as

$$\left(\partial S/\partial \theta_{0j}^{l}\right)=\left(f_{j}-d_{j}\right)\cdot\left(z^{l}/b\right) \tag{6.15a}$$

$$\left(\partial S/\partial \theta_{ij}^{l}\right)=\left(f_{j}-d_{j}\right)\cdot\left(z^{l}/b\right)\cdot x_{i} \tag{6.15b}$$

$$\left(\partial S/\partial c_{i}^{l}\right)=A\cdot\left\{2\cdot\left(z^{l}/b\right)\cdot\left(x_{i}-c_{i}^{l}\right)\Big/\left(\sigma_{i}^{l}\right)^{2}\right\} \tag{6.15c}$$

$$\left(\partial S/\partial \sigma_{i}^{l}\right)=A\cdot\left\{2\cdot\left(z^{l}/b\right)\cdot\left(x_{i}-c_{i}^{l}\right)^{2}\Big/\left(\sigma_{i}^{l}\right)^{3}\right\} \tag{6.15d}$$

where,

$$\begin{aligned}A&=\left(\sum_{j=1}^{m}\left(y_{j}^{l}-f_{j}\right)\cdot\left(f_{j}-d_{j}\right)\right)\\&=\left(\sum_{j=1}^{m}\left(\left(\theta_{0j}^{l}+\sum_{i=1}^{n}\theta_{ij}^{l}\cdot x_{i}\right)-f_{j}\right)\cdot\left(f_{j}-d_{j}\right)\right)\end{aligned} \tag{6.15e}$$

Using the above results, the final update rules for the networks free parameters can be written as

$$\theta_{0j}^{l}(k+1)=\theta_{0j}^{l}(k)-\eta\cdot\left(f_{j}-d_{j}\right)\cdot\left(z^{l}/b\right) \tag{6.16a}$$

$$\theta_{ij}^{l}(k+1)=\theta_{ij}^{l}(k)-\eta\cdot\left(f_{j}-d_{j}\right)\cdot\left(z^{l}/b\right)\cdot x_{i} \tag{6.16b}$$

$$c_{i}^{l}(k+1)=c_{i}^{l}(k)-\eta.\left\{2\cdot A\cdot h^{l}\cdot\left(x_{i}-c_{i}^{l}\right)\Big/\left(\sigma_{i}^{l}\right)^{2}\right\} \tag{6.16c}$$

$$\sigma_{i}^{l}(k+1)=\sigma_{i}^{l}(k)-\eta\cdot\left\{2.A\cdot h^{l}\cdot\left(x_{i}-c_{i}^{l}\right)^{2}\Big/\left(\sigma_{i}^{l}\right)^{3}\right\}, \tag{6.16d}$$

where $h^l = (z^l/b)$, with $b = \sum_{l=1}^{M} z^l$ and h^l is the normalized degree of fulfilment (firing strength) of lth rule.

Equations (6.11a) – (6.16d) represent the *backpropagation training algorithm* (BPA) for Takagi-Sugeno-type multi-input multi-output neuro-fuzzy networks or the equivalent fuzzy logic system of form (6.9a) – (6.9c) with linear fuzzy rules consequent part as

$$y^l_j = \theta^l_{0j} + \theta^l_{1j} x_1 + \theta^l_{2j} x_2 + \theta^l_{3j} x_3 + \cdots + \theta^l_{nj} x_n \, .$$

In the above Takagi-Sugeno-type fuzzy rules (linear) consequent, if the coefficients $\theta^l_{ij} = 0$, for i = 1, 2, 3, ..., n; l =1, 2, 3, ..., M; and m = 1, then the equivalent neuro-fuzzy network is identical with the multi-input, single-output neuro-fuzzy network described by Wang and Mendel (1992b) and Palit and Popovic (1999 and 2000a). The resulting fuzzy logic system can be seen as a special case of both the Mamdani- and -Takagi-Sugeno-type systems, where the rule consequent is a singleton (constant number) fuzzy set. However, if $\theta^l_{ij} \neq 0$, for i = 1, 2, 3, ..., n; l= 1, 2, 3, ..., M; and for m = 1, then the resulting fuzzy logic system is identical with Takagi-Sugeno type multi-input and single-output neuro-fuzzy network, as described by Palit and Babuška (2001).

It is generally known that the backpropagation algorithm based on steepest descent rule, in order to avoid the possible oscillations in the final phase of the training, uses a relatively low learning rate $\eta << 1$. Therefore, the backpropagation training usually requires a large number of recursive steps or epochs. The acceleration of the training process with classical backpropagation, however, is achievable if the adaptive version of the learning rate or the momentum version of the steepest descent rule is used:

$$\theta^l_{0j}(k+1) = \theta^l_{0j}(k) - \eta \cdot (1-mo) \cdot \left\{ \left(f_j - d_j\right)\left(z^l/b\right) \right\} + mo \cdot \Delta\theta^l_{0j}(k-1) \tag{6.17a}$$

$$\theta^l_{ij}(k+1) = \theta^l_{ij}(k) - \eta \cdot (1-mo) \cdot \left\{ \left(f_j - d_j\right)\left(z^l/b\right) \right\} \cdot x_i + mo \cdot \Delta\theta^l_{ij}(k-1) \tag{6.17b}$$

$$c^l_i(k+1) = c^l_i(k) - \eta \cdot (1-mo) \cdot \left\{ A \cdot 2 \cdot z^l \cdot \left(x_i - c^l_i\right) / \left(\sigma^l_i\right)^2 \right\} + mo \cdot \Delta c^l_i(k-1) \tag{6.17c}$$

$$\sigma^l_i(k+1) = \sigma^l_i(k) - \eta \cdot (1-mo) \cdot \left\{ A \cdot 2 \cdot z^l \cdot \left(x_i - c^l_i\right)^2 / \left(\sigma^l_i\right)^3 \right\} + mo \cdot \Delta\sigma^l_i(k-1) \tag{6.17d}$$

where,

$$\Delta\mathbf{w}(k-1) = \mathbf{w}(k) - \mathbf{w}(k-1), \tag{6.17e}$$

and **w** represents the networks free parameter vector in general. The momentum constant is usually less than one. Therefore, we can write $mo < 1$.

6.4.2.2 Improved Backpropagation Training Algorithm

To improve the training performance of the proposed neuro-fuzzy network, we have modified the momentum version of the backpropagation algorithm by adding to it the modified error index term/modified performance index term (6.18a), as proposed by Xiaosong *et al.* (1995).

$$S_m(\mathbf{w}) = 0.5 \cdot \gamma \cdot \sum_{r=1}^{N} \left(e_r(\mathbf{w}) - e_{avg}\right)^2, \tag{6.18a}$$

where,

$$e_{avg} = \frac{1}{N} \cdot \sum_{r=1}^{N} e_r(\mathbf{w}). \tag{6.18b}$$

and e_{avg} is the average error. Thus, the new error index (new performance index) is finally defined as

$$S_{new}(\mathbf{w}) = S(\mathbf{w}) + S_m(\mathbf{w}), \tag{6.19}$$

where, $S(\mathbf{w})$ is the unmodified performance index as defined in (6.11b). From this, the corresponding gradient can be defined as

$$\nabla S(\mathbf{w}) = \sum_{r=1}^{N} \left(e_r(\mathbf{w})\right) \cdot \left(\partial e_r(\mathbf{w}) / \partial w\right) \tag{6.20a}$$

$$\nabla S_m(\mathbf{w}) = \sum_{r=1}^{N} \left(\gamma \cdot \left(e_r(\mathbf{w}) - e_{avg}\right)\right) \cdot \left(\partial e_r(\mathbf{w}) / \partial w\right) \tag{6.20b}$$

$$\nabla S_{new}(\mathbf{w}) = \sum_{r=1}^{N} \left(e_r(\mathbf{w}) + \gamma \cdot \left(e_r(\mathbf{w}) - e_{avg}\right)\right) \cdot \left(\partial e_r(\mathbf{w}) / \partial w\right), \tag{6.20c}$$

where the constant term (gama) $\gamma < 1$ has to be chosen appropriately.

With the modified error index extension as per Equation (6.20c) we need only to add a new vector term $\gamma \cdot \left(\mathbf{e}(\mathbf{w}) - e_{avg}\right)$ with the original error vector $\mathbf{e}(\mathbf{w})$. Theoretical justification of the improved training performance of the network by the use of a modified error index term has been described in Xiaosong *et al.* (1995).

6.4.2.3 Levenberg-Marquardt Training Algorithm

Training experiments with a neuro-fuzzy network using the momentum version of backpropagation algorithm, as well as its modified error index extension form, have shown that, with the first 200 training (four inputs- one output) data sets of a Mackey-Glass chaotic series, backpropagation algorithm usually requires several hundred epochs to bring the SSE value down to the desired error goal (see Palit and Popvic, 1999). This calls for an alternative, much faster training algorithm. Hence, to accelerate the convergence speed of neuro-fuzzy network training, the *Levenberg-Marquardt algorithm* (LMA) was proposed.

Although being an approximation to Newton's method, based on a Hessian matrix, the Levenberg-Marquardt algorithm can still implement the second-order training speed without direct computation of the Hessian matrix (Hagan and Menhaj, 1994). This is achieved in the following way.

Suppose that a function $V(\mathbf{w})$ is to be minimized with respect to the network's free-parameter vector $\mathbf{w}$ using Newton's method. The update of $\mathbf{w}$ to be used here is

$$\Delta\mathbf{w} = -\left[\nabla^2 V(\mathbf{w})\right]^{-1} \cdot \nabla V(\mathbf{w}) \tag{6.21a}$$

$$\mathbf{w}(k+1) = \mathbf{w}(k) + \Delta\mathbf{w} \tag{6.21b}$$

where $\nabla^2 V(\mathbf{w})$ is the Hessian matrix and $\nabla V(\mathbf{w})$ is the gradient of $V(\mathbf{w})$. If the function $V(\mathbf{w})$ is taken to be the sum squared error function, *i.e.*

$$V(\mathbf{w}) = 0.5 \cdot \sum_{r=1}^{N} e_r^2(\mathbf{w}) \tag{6.22}$$

then the gradient $\nabla V(\mathbf{w})$ and the Hessian matrix $\nabla^2 V(\mathbf{w})$ are generally defined using the Jacobian matrix $J(\mathbf{w})$ as

$$\nabla V(\mathbf{w}) = J^T(\mathbf{w}) \cdot e(\mathbf{w}) \tag{6.23a}$$

$$\nabla^2 V(\mathbf{w}) = J^T(\mathbf{w}) \cdot J(\mathbf{w}) + \sum_{r=1}^{N} e_r(\mathbf{w}) \cdot \nabla^2 e_r(\mathbf{w}), \tag{6.23b}$$

where

$$J(\mathbf{w}) = \begin{bmatrix} \frac{\partial e_1(\mathbf{w})}{\partial w_1} & \frac{\partial e_1(\mathbf{w})}{\partial w_2} & \cdots & \frac{\partial e_1(\mathbf{w})}{\partial w_{N_p}} \\ \frac{\partial e_2(\mathbf{w})}{\partial w_1} & \frac{\partial e_2(\mathbf{w})}{\partial w_2} & \cdots & \frac{\partial e_2(\mathbf{w})}{\partial w_{N_p}} \\ \vdots & \vdots & & \vdots \\ \frac{\partial e_N(\mathbf{w})}{\partial w_1} & \frac{\partial e_N(\mathbf{w})}{\partial w_2} & \cdots & \frac{\partial e_N(\mathbf{w})}{\partial w_{N_p}} \end{bmatrix} \tag{6.23c}$$

and $w = \left[w_1, \, w_2, \, \cdots, w_{N_p} \right]$ is the parameter vector of network. From (6.23c) it is seen that the dimension of the Jacobian matrix is $(N \times N_p)$, N and N_p being the number of training samples and the number of adjustable network parameters respectively. For the Gauss-Newton method the second term in (6.23b) is assumed to be zero, so that the update according to (6.21a) becomes

$$\Delta \mathbf{w} = -\left[J^T(\mathbf{w}) \cdot J(\mathbf{w})\right]^{-1} \cdot J^T(\mathbf{w}) \cdot e(\mathbf{w}). \qquad (6.24a).$$

The Levenberg-Marquardt modification of the Gauss-Newton method is

$$\Delta \mathbf{w} = -\left[J^T(\mathbf{w}) \cdot J(\mathbf{w}) + \mu \cdot I\right]^{-1} \cdot J^T(\mathbf{w}) \cdot e(\mathbf{w}) \qquad (6.24b).$$

in which I is the $(N_p \times N_p)$ identity matrix and the parameter μ is multiplied by some constant factor μ_{inc} whenever an iteration step increases the value of $V(\mathbf{w})$, and divided by μ_{dec} whenever a step reduces the value of $V(\mathbf{w})$. Hence, the update according to (6.21b) is

$$\mathbf{w}(k+1) = \mathbf{w}(k) - \left[J^T(\mathbf{w}) \cdot J(\mathbf{w}) + \mu \cdot I\right]^{-1} \cdot J^T(\mathbf{w}) \cdot e(\mathbf{w}). \tag{6.24c}$$

Note that for large μ the algorithm becomes the steepest descent gradient algorithm with step size $(1/\mu)$, whereas for small μ, *i.e.* $\mu \approx 0$, it becomes the Gauss-Newton algorithm. Usually, $\mu_{inc} = \mu_{dec}$. However, in our program we have selected two different values for them. In order to get even faster convergence, a small momentum term mo = 0.098 was also added, so that the final update becomes

$$\begin{aligned} \mathbf{w}(k+1) = \mathbf{w}(k) - \left[J^T(\mathbf{w}) \cdot J(\mathbf{w}) + \mu \cdot I\right]^{-1} \cdot J^T(\mathbf{w}) \cdot e(\mathbf{w}) \\ + \, mo \cdot \left(\mathbf{w}(k) - \mathbf{w}(k-1)\right) \end{aligned} \tag{6.24d}$$

It is to be noted that the use of a momentum term is quite usual with the classical backpropagation algorithm, whereas this may appear to be unusual with the Levenberg-Marquardt algorithm. However, the latter is justified, as the use of a momentum term in the backpropagation algorithm is primarily to overcome the possible trap at local minima and also to prevent small oscillations during the training of the network; similarly, the use of a small momentum term, as experimentally verified through simulation, also helps to increase network training convergence with the Levenberg-Marquardt algorithm. Furthermore, similar to the backpropagation algorithm, here also the Levenberg-Marquardt algorithm was extended by adding a modified error index term, as proposed by Xiaosong *et al.* (1995), to improve further the training convergence. Therefore, as per (6.20c), the corresponding new gradient can now be expressed or defined using a Jacobian matrix as

$$\nabla S_{new}(\mathbf{w}) = J^T(\mathbf{w}) \cdot \left[\mathbf{e}(\mathbf{w}) + \gamma \cdot \left(\mathbf{e}(\mathbf{w}) - e_{avg}\right)\right], \tag{6.25}$$

where $\mathbf{e(w)}$ represents the column vector of errors, and the constant factor $\gamma << 1$ (for the Levenberg-Marquardt algorithm) has to be chosen appropriately. Equation (6.25) suggests that even with consideration of the modified error index extension of the original performance function the Jacobian matrix remains unaltered and, with the above modification, we need to add only a new error vector term $\gamma \cdot (\mathbf{e}(\mathbf{w}) - e_{avg})$ with the original error vector $\mathbf{e}(\mathbf{w})$ as we did with the back-propagation algorithm.

6.4.2.3.1 Computation of Jacobian Matrix

We now describe a simplified technique to compute, layer by layer, the Jacobian matrix and the related parameters from the backpropagation results. Layer-wise or parameter-wise computation of the Jacobian matrix is permissible because, as stated in Equations (6.26a) and (6.26b), the final contents of the Hessian matrix remain unaltered even if the whole Jacobian is divided into smaller parts. Furthermore, this division of the Jacobian matrix helps to avoid computer memory shortage problem, which is likely to occur for large neural networks.

From

$$\nabla^2 V(w) \approx \left[J^T(w)\right] \cdot \left[J(w)\right] = \left[J_1^T(w), J_2^T(w)\right] \cdot \begin{bmatrix} J_1(w) \\ J_2(w) \end{bmatrix} \tag{6.26a}$$

it follows that

$$\nabla^2 V(w) \approx \left[J_1^T(w) \cdot J_1(w) + J_2^T(w) \cdot J_2(w)\right]. \tag{6.26b}$$

Computation of the Jacobian matrix is in fact the most crucial step in implementing the Levenberg-Marquardt algorithm for neuro-fuzzy networks. For this purpose, the results obtained in the Section 6.4.2.1 will be used, where the derivatives of the sum square error S with respect to the network's adjustable parameters (free-parameters) $\theta_{0j}^l, \theta_{ij}^l, c_i^l$, and σ_i^l for the fuzzy logic system (6.9a) – (6.9c) were already computed and listed in (6.15a) – (6.15e).

Now, considering the singleton consequent part (constant term) of the rules and taking into account Equation (6.15a), we can rewrite the gradient $V(\theta_{0j}^l) \equiv S$ as

$$\nabla V(\theta_{0j}^l) \equiv (\partial S / \partial \theta_{0j}^l) = \{z^l / b\} \cdot (f_j - d_j), \tag{6.27}$$

where f_j is the actual output vector from the jth output node of the Takagi-Sugeno-type multiple input multiple output neuro-fuzzy network and d_j is the corresponding desired output vector at the jth output node for a given set of input-output training data. Taking into account Equation (6.27) and comparing it with (6.23a), where the gradient is expressed using the transpose of the Jacobian matrix multiplied by the network's error vector, *i.e.*

$$\nabla V(\mathbf{w}) = J^T(\mathbf{w}) \cdot e(\mathbf{w}), \tag{6.28}$$

where $\mathbf{w}$ is the free parameter of the network, the transpose of the Jacobian matrix $J^T(\theta_{0j}^l)$ and the Jacobian matrix $J(\theta_{0j}^l)$ for the free parameter θ_{0j}^l of the neuro-fuzzy network can be defined by

$$J^T(\theta_{0j}^l) = (z^l / b) \tag{6.29a}$$

$$J(\theta_{0j}^l) \equiv [J^T(\theta_{0j}^l)]^T = [z^l / b]^T. \tag{6.29b}$$

This is because the prediction error at the jth output node of the Takagi-Sugeno-type neuro-fuzzy network is

$$e_j \equiv (f_j - d_j). \tag{6.30}$$

However, if we consider the normalized prediction error of the network at the jth output node, instead of the original prediction error at the jth output node, then by applying a similar technique, the transposition of the Jacobian matrix $J^T(\theta_{0j}^l)$ and the Jacobian matrix $J(\theta_{0j}^l)$ itself for the free parameter θ_{0j}^l will be

$$J^T(\theta_{0j}^l) = (z^l) \tag{6.31a}$$

$$J(\theta_{0j}^l) \equiv [J^T(\theta_{0j}^l)]^T = [z^l]^T, \tag{6.31b}$$

this is because the normalized prediction error at the *j*th output node of the multi-input multi-output neuro-fuzzy network is:

$$e_j\,(\text{normalized}) \equiv \left(f_j - d_j\right)/b\,. \tag{6.32}$$

In the above equation, z^l is a matrix of size $(M \times N)$ that contains the degree of fulfilment (firing strength) of each fuzzy rule computed for a given set of training samples, where *M* is the number of fuzzy rules (and also the number of Gaussian membership functions implemented for fuzzy partition of input universes of discourse) and *N* is the number of training samples (input-output data samples).

Adopting a similar technique and taking into account Equation (6.28), the original prediction error (6.30) and Equation (6.15b), which computes the derivative of *S* with respect to θ_{ij}^l, we can get the transposition of the Jacobian matrix and its further transposition, *i.e.* the Jacobian matrix itself, for the network's free-parameter θ_{ij}^l using

$$J^T\left(\theta_{ij}^l\right) = \left(z^l/b\right)\cdot x_i \tag{6.33a}$$

$$J\left(\theta_{ij}^l\right) \equiv \left[J^T\left(\theta_{ij}^l\right)\right]^T = \left[\left(z^l/b\right)\cdot x_i\right]^T. \tag{6.33b}$$

Also, instead of the original prediction error, if here we consider the normalized prediction error of Equation (6.32) and, as usual, Equations (6.28) and (6.15b), then we can get the transposed Jacobian matrix and the Jacobian matrix itself for the same parameter θ_{ij}^l as

$$J^T\left(\theta_{ij}^l\right) = \left(z^l\cdot x_i\right) \tag{6.34a}$$

$$J\left(\theta_{ij}^l\right) \equiv \left[J^T\left(\theta_{ij}^l\right)\right]^T = \left[z^l\cdot x_i\right]^T \tag{6.34b}$$

Finally, to compute the Jacobian matrices and their transpositions for the remaining free parameters of the network, *i.e.* for parameters c_i^l, and σ_i^l, we also use a similar technique, whereby Equation (6.15e), which computes the term *A*, has to be reorganized.

Let us denote

$$D_j \equiv \left(y^l{}_j - f_j\right). \tag{6.35}$$

Using Equations (6.30) and (6.35) we can rewrite (6.15e) as

$$A \equiv \sum_{j=1}^{m} \left(D_j \cdot e_j\right) = \left(D_1 \cdot e_1 + D_2 \cdot e_2 + \cdots + D_m \cdot e_m\right) \tag{6.36}$$

Our objective is to find suitable terms D_{eqv} and e_{eqv} such that their product is equal to

$$A \equiv D_{eqv} \cdot e_{eqv} = \left(D_1 \cdot e_1 + D_2 \cdot e_2 + \cdots + D_m \cdot e_m\right) \tag{6.37}$$

where the term e_{eqv} is such that it contributes the same amount of sum squared error value S of equation (6.11b) as that can be obtained jointly by all the $e_j \equiv \left(f_j - d_j\right)$ from the multiple-input multiple-output network. Therefore,

$$e_{eqv}^{p} = \sqrt{\left(e_1^{p^2} + e_2^{p^2} + \cdots + e_m^{p^2}\right)}, \tag{6.38}$$

where, $p = 1, 2, 3, \ldots, N$; corresponding to N training samples. This results in

$$D_{eqv} = A \cdot \left(e_{eqv}\right)^{-1} \tag{6.39a}$$

This can be written in matrix form using the pseudo inverse as

$$\mathbf{D}_{eqv} = \mathbf{A} \cdot \left(\mathbf{E}_{eqv}\right)^T \cdot \left(\mathbf{E}_{eqv} \cdot \mathbf{E}_{eqv}^T\right)^{-1} \tag{6.39b}$$

where $\mathbf{E}_{eqv}$ is the equivalent error vector of size $(N \times 1)$ containing $\left(e_{eqv}^{p}\right)$ as its elements for all (N) training samples. Similarly, $\mathbf{D}_{eqv}$ and $\mathbf{A}$ are matrices of size $(M \times N)$ and $(M \times 1)$ respectively. Once the matrix $\mathbf{D}_{eqv}$ and the equivalent error vector $\mathbf{E}_{eqv}$ are known, we can replace matrix $\mathbf{A}$ with their product. Therefore,

$$\mathbf{A} = \mathbf{D}_{eqv} \cdot \mathbf{E}_{eqv} \tag{6.40a}$$

or, equivalently as,

$$A = D_{eqv} \cdot e_{eqv} \tag{6.40b}$$

can be calculated. In the case of a multiple-input single-output neuro-fuzzy network, *i.e.* for $m = 1$ and $A = D_1 \cdot e_1$, $D_{eqv} = D_1$ and $e_{eqv} = e_1$ hold. This means that, in this case, Equations (6.37) – (6.40b) need not be computed.

However, for the multiple-input multiple-output case, where $m \geq 2$, using (6.37) we can write Equations (6.15c) and (6.15d) as

$$\left(\partial S/\partial c_i^l\right) = \left(D_{\text{eqv}} \cdot e_{\text{eqv}}\right) \cdot \left\{ 2 \cdot \left(z^l/b\right) \cdot \left(x_i - c_i^l\right) / \left(\sigma_i^l\right)^2 \right\} \tag{6.41a}$$

$$\left(\partial S/\partial \sigma_i^l\right) = \left(D_{\text{eqv}} \cdot e_{\text{eqv}}\right) \cdot \left\{ 2 \cdot \left(z^l/b\right) \cdot \left(x_i - c_i^l\right)^2 / \left(\sigma_i^l\right)^3 \right\} \tag{6.41b}$$

Now, following the previous technique and realizing that $\left(e_{\text{eqv}}/b\right)$ can be considered as the normalized equivalent error and, in addition, taking into account Equation (6.28) and comparing it respectively with (6.41a) and (6.41b), transposed Jacobian matrix and the Jacobians $J^T\left(c_i^l\right)$, $J\left(c_i^l\right)$ and $J^T\left(\sigma_i^l\right)$, $J\left(\sigma_i^l\right)$ for the network free parameters c_i^l and σ_i^l can be computed as:

$$J^T\left(c_i^l\right) = \left\{ 2 \cdot D_{eqv} \cdot z^l \cdot \left(x_i - c_i^l\right) / \left(\sigma_i^l\right)^2 \right\} \tag{6.42a}$$

$$J\left(c_i^l\right) = \left[J^T\left(c_i^l\right)\right]^T = \left[2 \cdot D_{eqv} \cdot z^l \cdot \left(x_i - c_i^l\right) / \left(\sigma_i^l\right)^2\right]^T \tag{6.42b}$$

$$J^T\left(\sigma_i^l\right) = \left\{2 \cdot D_{eqv} \cdot z^l \cdot \left(x_i - c_i^l\right)^2 / \left(\sigma_i^l\right)^3\right\} \tag{6.42c}$$

$$J\left(\sigma_i^l\right) = \left[J^T\left(\sigma_i^l\right)\right]^T = \left[2 \cdot D_{eqv} \cdot z^l \cdot \left(x_i - c_i^l\right)^2 / \left(\sigma_i^l\right)^3\right]^T \tag{6.42d}$$

The above equations describe the Jacobian matrices and their transpositions for the Takagi-Sugeno-type fuzzy logic systems with the adjustable free parameters c_i^l and σ_i^l when normalized (equivalent) error is considered.

If, however, instead of normalized (equivalent) error only the equivalent error is considered, then the Jacobian matrices and their transpositions will be the same, except that in the right-hand sides of Equations (6.42a) – (6.42c) the term z^l has to be replaced by normalized degree of fulfilment of the lth rule $h^l = \left(z^l/b\right)$, where $b = \sum_{l=1}^{M} z^l$ represents the sum of degree of fulfilment of all rules.

It is to be noted that, while computing the Jacobian matrices, care has to be taken so that the dimensions of the Jacobians match correctly with $\left(N \times N_p\right)$, where N is the number of training data sets and N_p the number of adjustable parameters in the network's layer considered. In all our simulation experiments with neuro-fuzzy networks the *normalized prediction error* has been considered for the computation of Jacobian matrices for the network's free parameters θ_{0j}^l and θ_{ij}^l, so that Equations (6.31a), (6.31b) and Equations (6.34a), (6.34b) delivered the corresponding transposed Jacobian matrices and their Jacobians respectively. In contrast, *normalized equivalent error* has been considered for the computation of transposed Jacobian matrices and their Jacobians respectively for the mean and variance parameters c_i^l and σ_i^l of the Gaussian membership functions; therefore, Equations (6.42a) – (6.42c) delivered the corresponding transpositions of the Jacobian matrices and the Jacobian matrices themselves for the Takagi-Sugeno-type multi-input, multi-output neuro-fuzzy network's free parameter and gave the Levenberg-Marquardt algorithm better convergence in most experiments.

6.4.2.4 Adaptive Learning Rate and Oscillation Control

The proposed backpropagation training algorithm and the Levenberg-Marquardt training algorithm, both with the modified error index extension as performance function and with the added small momentum term, have proven to be very efficient, and faster in training the Takagi-Sugeno-type neuro-fuzzy networks than the standard back-propagation algorithm. But still, the performance function of the network (if left without any proper care) is not always guaranteed to reduce, in every epoch, towards the desired error goal. As a consequence, the training can proceed in the opposite direction, giving rise to a continuous increase of performance function or to its oscillation. This prolongs the training time or makes the training impossible. To avoid this, three sets of adjustable parameters are recommended to be stored for the backpropagation algorithm and two sets for the Levenberg-Marquardt algorithm. The stored sets are then used in the following way.

In the case of the backpropagation algorithm, if two consecutive new sets of adjustable parameters reduce the network performance function, then in the following epochs the same sets are used and the learning rate in the next step is increased slightly by a factor of 1.1. In the opposite case, *i.e.* if the performance function with the new sets of parameters tends to increase beyond a given limit - say *WF* (*wildness factor* of oscillation) times the current value of the performance function – then the new sets are discarded and training proceeds with the old sets of adjustable parameters. Thereafter, a new direction of training is sought with the old sets of parameters and with lower values of the learning rate parameter, *e.g.* 0.8 or 0.9 times the old learning rate.

In the case of the Levenberg-Marquardt algorithm, if the following epoch reduces the value of the performance function, then the training proceeds with a new set of parameters and the μ value is reduced by a preassigned factor $(1/\mu_{dec})$. In the opposite case, *i.e.* if the next epoch tends to increase this performance value beyond the given limits (*WF* times of current value of performance function) or remains the same, then the μ value is increased by another preassigned factor (μ_{inc}) but the new set of adjustable parameters is discarded and training proceeds with the old set of parameters. In this way, in every epoch the value of the performance function is either decreased steadily or at least maintained within the given limit values.

6.5 Comparison of Radial Basis Function Network and Neuro-fuzzy Network

There are considerable similarities, as well as dissimilarities, between the RBF-type neural network and neuro-fuzzy network. In this section we present a few comparisons between them.

A radial basis function network can be considered as a three-layer network consisting of an input layer, a hidden layer and an output layer (see Chapter 3 for details). The hidden layer performs the nonlinear transformation, so that the input space is mapped into a new space. The output layer then combines the outputs of the hidden layer linearly. The structure of an RBF network with an input vector $\mathbf{x} \in \mathbb{R}^n$ and output $y \in \mathbb{R}$ is shown in Chapter 3. The output from such a network can be written as

$$y(\mathbf{x}) = \sum_{i=1}^{N} w_i R_i(\mathbf{x}),$$

where w_i are the weights and $R_i(\mathbf{x})$ is the nonlinear activation function of the hidden-layer neurons.

The fuzzy logic system considered in Equations (6.9a) – (6.9c) can be rewritten as

$$f_j(x^p) = \sum_{l=1}^{M} y_j^l \cdot h^l \quad \text{where} \quad h^l = z^l \Big/ \sum_{l=1}^{M} z^l,$$
$$\text{and} \quad j = 1, 2, 3, \cdots, m; \quad l = 1, 2, 3, \cdots, M.$$

noting that, when using the definition of the radial basis function the normalized degree of fulfilment of the lth rule, *i.e.* $h^l \equiv h^l(x_i)$, is similar to an RBF. Therefore, the fuzzy logic system can also be represented as an RBF neural network model. However, the following points have to be carefully noted:

- Functions in the form of (6.9a) are just one kind of fuzzy logic system with a particular choice of fuzzy inference engine with *product inference rules*, a fuzzifier, and a weighted-average defuzzifier. If another choice is made, such as the *mean-of-maxima (MOM) defuzzifier*, then the fuzzy logic system will be quite different from the RBF network. Therefore, an RBF network in fact is a special case of the fuzzy logic system.

- The membership functions of the fuzzy logic system can take various geometric forms (such as Gaussian, triangular, trapezoidal, bell-shaped, *etc.*). They can also be non-homogeneous (*i.e.* the membership functions that divide the input or output universe of discourse may not all be of the same functional form), whereas the RBF network takes a lesser number of functional forms, like a Gaussian function, and are usually homogeneous. This is due to the different justifications of the neuro-fuzzy network and the RBF networks. The fuzzy logic systems are justified from the human reasoning point of view and, therefore, the membership functions can have any suitable form within the range [0, 1], appropriate to representing the knowledge of a human expert through IF-THEN rules. On the other hand, RBF networks are based on biological motivations. Therefore, it is difficult to justify the use of many different kinds of non-homogeneous basis functions in a single RBF network.

One of the fundamental differences between a neuro-fuzzy network and an RBF network is that the former takes the linguistic information explicitly into consideration and makes use of it in a systematic manner, whereas the latter does not. Furthermore, while using the neuro-fuzzy network, besides the generated model accuracy we are also concerned about the transparency of the model, whereas for the RBF network, and also for other types of neural network, we are only concerned about the model accuracy (black-box modelling).

6.6 Comparison of Neural Network and Neuro-fuzzy Network Training

We would now like to compare the back-propagation training algorithms for the multi-layer perceptron networks and neuro-fuzzy networks described in this chapter. The training algorithms are similar in the following sense:

- Their basic operation, *i.e.* forward computation and backward training, is the same, and in order to minimize the sum squared error between the actual output and the desired output of the network, both of them use either the same gradient method or the second-derivative-based recursive algorithm, *i.e.* the approximate Hessian matrix.

- Both of them are universal approximators and, therefore, well qualified to solve any nonlinear mapping to any degree of accuracy within the universe of discourse.

However, they differ distinctly in the following:

- The parameters (weights and biases) of the neural networks have no clear physical meaning or interpretation (black-box modelling), which makes the selection of their initial values difficult; thus, they are chosen rather randomly. On the other hand, the parameters of the neuro-fuzzy networks have clear physical meaning (membership functions), so that if the sufficient knowledge about the system to be modelled by the neuro-fuzzy networks is available, then a good initial parameter setting procedure can be developed.

- Besides numerical information, linguistic information can also be incorporated into neuro-fuzzy systems.

6.7 Modelling and Identification of Nonlinear Dynamics

We would now like to illustrate the efficiency of the neuro-fuzzy approach proposed in Section 6.4.1 on some forecasting examples.

6.7.1 Short-term Forecasting of Electrical Load

This application concerns the forecasting the electrical load demand, based on a time series that predicts the values at time $(t + L)$ using the available observation data up to the time point t. For modelling purposes the time series data $\mathbf{X} = \{X_1, X_2, X_3, \ldots, X_q\}$ have been rearranged in input-output form **XIO**. The neuro-fuzzy predictor to be developed for time series modelling and forecasting is supposed to operate with four inputs (*i.e.* $n = 4$) and with three outputs (*i.e.* $m = 3$). Taking both the sampling interval and the lead time of forecast to be one time unit, then for each $t \geq 4$ the input data have to be represented as a four-dimensional vector and the output data as a three-dimensional vector

$$\mathrm{XI} = [X(t\text{-}3), X(t\text{-}2), X(t\text{-}1), X(t)],$$
$$\mathrm{XO} = [X(t+1), X(t+2), X(t+3)]$$

Furthermore, in order to have sequential output in each row, the values of t should run as 4, 7, 10, 13, …, $(q\text{-}3)$. The corresponding **XIO** matrix will then look like (6.43), in which the first four columns represent the four inputs of the network and the last three columns represent its output.

$$\mathrm{XIO} = \begin{bmatrix} X_1, & X_2, & X_3, & X_4 \Rightarrow X_5, & X_6, & X_7 \\ X_4, & X_5, & X_6, & X_7 \Rightarrow X_8, & X_9, & X_{10} \\ \vdots & \vdots & \vdots & \vdots \quad\;\; \vdots & \vdots & \vdots \\ X_{q-6}, & X_{q-5}, & X_{q-4}, & X_{q-3} \Rightarrow X_{q-2}, & X_{q-1}, & X_q \end{bmatrix} \tag{6.43}$$

In the selected forecasting example, 1163 input-output data were generated, from which only the first 500 input-output data sets, *i.e.* the first 500 rows from the XIO matrix, were used for the multi-input multi-output neuro-fuzzy network training. The remaining 663 rows of the XIO matrix were used for verification of the forecasting results. The training and forecasting performances achieved with the neuro-fuzzy network are illustrated in Figures 6.7(a) – (d) and in Tables 6.1(a) and 6.1(b) respectively.

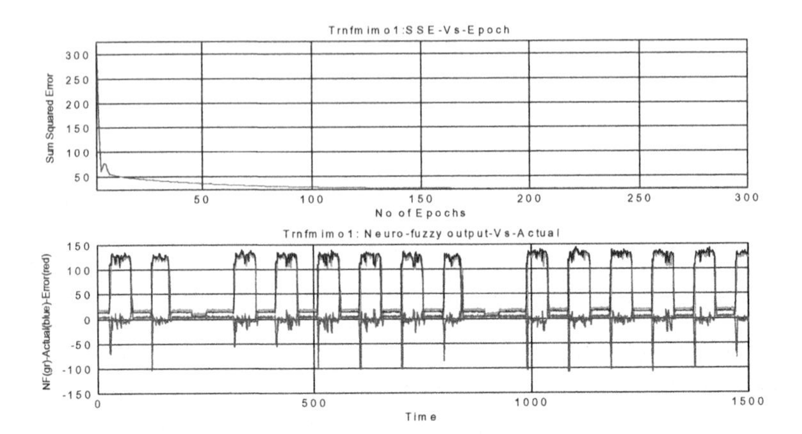

Figure 6.7(a). Training performance of Takagi-Sugeno-type multi-input multi-output neuro-fuzzy network with n = 4 inputs, m = 3 outputs, M = 15 fuzzy rules and 15 GMFs for short-term forecasting of electrical load time series when trained with proposed backpropagation algorithm. Backpropagation algorithm training parameters: η = 0.0005, γ = 0.5, mo = 0.5, maximum epoch = 300, training (pre-scaled) data = 1 to 500 rows of XIO matrix, initial SSE = 324.6016 (with random starting parameter), final SSE – 23.8580, data scaling factor = 0.01.

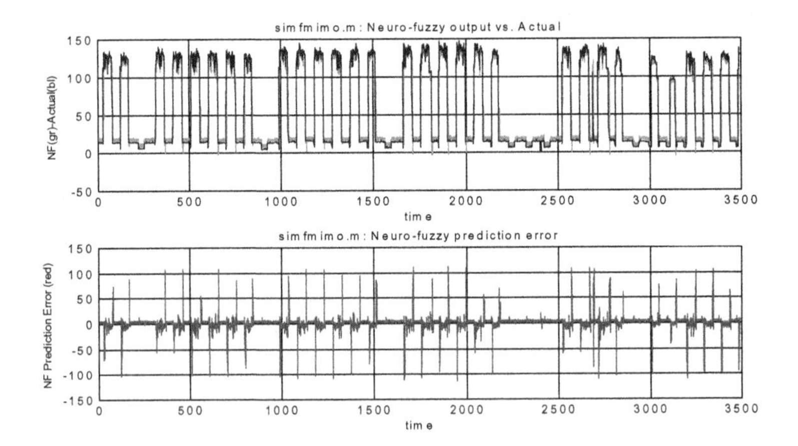

Figure 6.7(b). Forecasting performance of Takagi-Sugeno-type multi-input multi-output neuro-fuzzy network with n = 4 inputs, m = 3 outputs, M = 15 fuzzy rules and 15 GMFs for short-term forecasting of electrical load when trained with proposed backpropagation algorithm. Data 1 to 1500 correspond to training data and data 1501 to 3489 (*i.e.* row 501 to 1163 from XIO matrix) represent the forecasting performance.

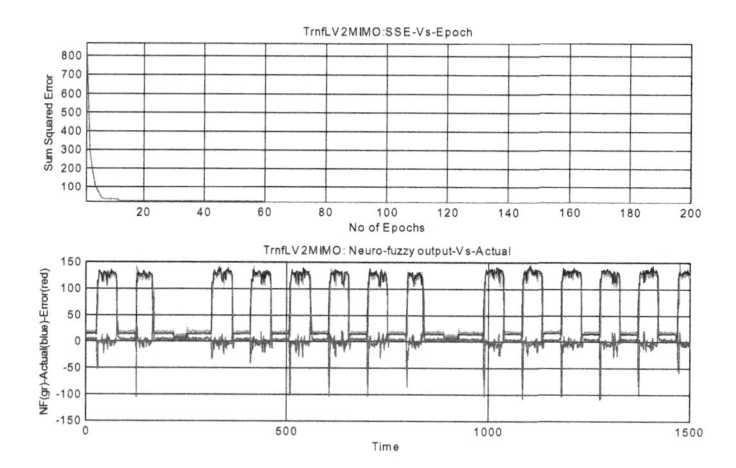

Figure 6.7(c). Training performance of the Takagi-Sugeno-type multi-input multi-output neuro-fuzzy network with $n = 4$ inputs, $m = 3$ outputs, $M = 15$ fuzzy rules and 15 GMFs for short-term forecasting of electrical load time series with the proposed Levenberg-Marquardt algorithm. Training parameters of Levenberg-Marquardt algorithm: $\mu = 0.001$, $\gamma = 0.1$, $mo = 0.1$, maximum epoch = 200, training (pre-scaled) data = 1 to 500 rows of XIO matrix, initial SSE = 868.9336 (with random starting parameter of neuro-fuzzy network), final SSE = 22.5777, data scaling factor = 0.01.

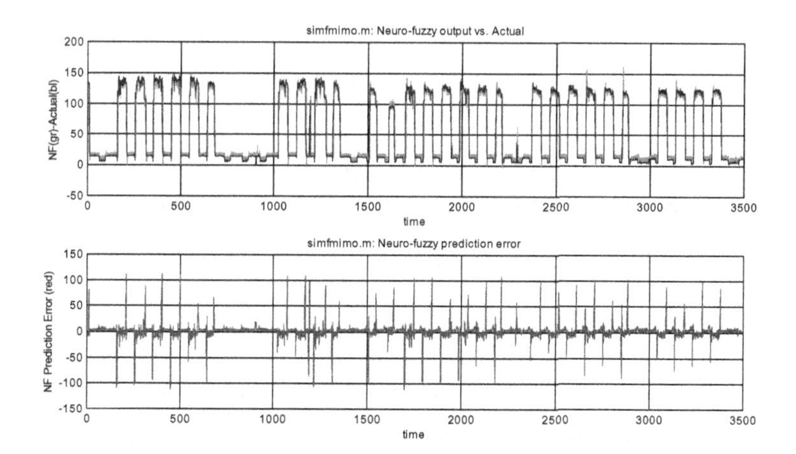

Figure 6.7(d). Forecasting performance of the Takagi-Sugeno-type multi-input multi-output neuro-fuzzy network with $n = 4$ inputs, $m = 3$ outputs, $M = 15$ fuzzy rules and 15 GMFs for short-term forecasting of electrical load after the training with the proposed Levenberg-Marquardt algorithm. Note that in both Figures 6.7(b) and 6.7(d) data from 1 to 1500 correspond to training data and data from 1501 to 3489 represent the forecasting performance with validation data set. It is important to note that data within the time points 2200 to 2510 are different from the training data. Still the Takagi-Sugeno-type multi-input

multi-output neuro-fuzzy network can predict this data region with reasonably high accuracy.

Table 6.1(a). Training and forecasting performance of Takagi-Sugeno-type multi-input multi-output neuro-fuzzy network with proposed backpropagation algorithm for electrical load time series.

Sl. No.	Final SSE with pre-scaled data (scale factor = 0.001)	Final SSE, MSE, RMSE with original (nonscaled) data
1.	SSE = 23.8580 SSE1 = 3.0077, SSE2 = 7.2863, SSE3 = 13.5640 (with training data 1 to 1500) (After training with backpropagation algorithm)	SSE = 2.3858e+005 (with training data 1 to 1500) MSE1 = 30.0772, MSE2 = 72.8630, MSE3 = 135.6401; RMSE1 = 5.4843, RMSE2 = 8.5360, RMSE3 = 11.6465
2.	SSE = 53.5633; SSE1 = 6.8169, SSE2 = 16.6395, SSE3 = 30.1069, (with training and validation data points 1 to 3489)	SSE = 5.3563e+005 (with training and validation data points 1 to 3489)

Figure 6.7(a) and Table 6.1(a) demonstrate that the proposed backpropagation algorithm brings the sum squared error as the performance function smoothly from its initial value of 324.6016 down to 23.8580 in 300 epochs, whereas Figure 6.7(c) and Table 6.1(b) demonstrate the training performance with the proposed Levenberg-Marquardt algorithm. In the latter case the performance function is brought down to 22.5777 from its initial value of 868.9336 within just 200 epochs, indicating the much higher convergence speed of the proposed Levenberg-Marquardt algorithm in comparison with the backpropagation algorithm. Furthermore, the sum square error plots in both Figure 6.7(a) and Figure 6.7(c) show that the training does not exhibit much oscillation. The results illustrated in Figure 6.7(b) and Figure 6.7(d) and also in Table 6.1(a) and Table 6.1(b) clearly show the excellent training and forecasting performance of the Takagi-Sugeno-type multiple-input, multiple-output neuro-fuzzy network with the proposed training algorithms.

Table 6.1(b). Training and forecasting performance of Takagi-Sugeno-type multi-input multi-output neuro-fuzzy network with proposed Levenberg-Marquardt algorithm for electrical load time series.

Sl. No.	Final SSE with pre-scaled data (scale factor = 0.001)	Final SSE, MSE, and RMSE with original (nonscaled) data
1.	SSE = 22.5777 SSE1 = 2.6365 SSE2 = 6.7828 SSE3 = 13.1584 (with training data 1 to 1500) (After training with Levenberg-Marquardt algorithm)	SSE = 2.2578e+005 MSE1 = 26.3650 MSE2= 67.8278 MSE3= 131.5837 RMSE1= 5.1347 RMSE2= 8.2358 RMSE3=11.471
2.	SSE = 42.3026 SSE1 = 5.0096 SSE2 = 11.7879 SSE3 = 25.5051 (with training and validation data points 1 to 3489)	SSE = 4.2303e+005 (with training and validation data points 1 to 3489)

Note that in the Table 6.1(a) and Table 6.1(b) SSE1, SSE2, and SSE3 indicate the sum squared error values at the output nodes 1, 2 and 3 respectively of the Takagi-Sugeno-type multi-input multi-output neuro-fuzzy network as formulated in (Equation 6.11a), and SSE indicates the cumulative sum of the sum square error values contributed by all three output nodes of the multi-input multi-output neuro-fuzzy network as formulated in (Equation 6.11b).

6.7.2 Prediction of Chaotic Time Series

In the next application example the proposed neuro-fuzzy algorithm has been tested for modelling and forecasting the Mackey-Glass chaotic time series, generated by solving the Mackey-Glass time delay differential equation (6.44) (MATLAB, 1998).

$$(dx/dt) = \left(0.2 \cdot x(t-\delta) / \left(1 + x^{10}(t-\delta)\right)\right) - 0.1 \cdot x(t), \tag{6.44}$$

for $x(0) = 1.2$, $\delta = 17$, and $x(t) = 0$, for $t < 0$.

The equation describes the arterial CO_2 concentration in the case of normal and abnormal respiration and belongs to a class of time-delayed differential equations that are capable of generating chaotic behaviour. It is a well-known benchmark problem in fuzzy logic and neural network research communities. Like in the previous example for forecasting purposes the time series data $X = \{X_1, X_2, X_3, \ldots, X_q\}$ were rearranged in a multi-input single-output (XIO)-like structure. For modelling and forecasting of the given time series the respective neuro-fuzzy predictor that has to be developed is taken to have four inputs (n = 4) and one output (m = 1). In addition, both the sampling interval and the lead time of forecast is supposed to be six time units, so that for each $t > 18$ the input data represents a four-dimensional vector

$$\mathrm{XI}(t\text{-}18) = [X(t\text{-}18), X(t\text{-}12), X(t\text{-}6), X(t)],$$

and the output data a scalar value

$$\mathrm{XO}(t\text{-}18) = [X(t\text{+}6)].$$

In the forecasting example considered, using Equation (6.44) and neglecting the first 100 transient data from the chaotic series, in addition 1000 input-output data were generated for the XIO matrix. Out of 1000 generated input-output data, only the first 200 data sets were used for network training, and the remaining 800 data were used for verification of forecasting results.

The training and forecasting performances achieved with the implemented neuro-fuzzy network and with stored seven fuzzy rules are illustrated in Figure 6.8(a) and Figure 6.8(b) and listed in Table 6.2(a) and also compared with other standard models in Table 6.2(b). The items listed in serial numbers 1 to 12 of Table 6.2(b) were taken from Kim and Kim (1997), whereas serial number 13 is taken from Park *et al.* (1999). The results clearly confirm excellent training and forecasting performance of the Takagi-Sugeno-type neuro-fuzzy network for Mackey-Glass chaotic time series.

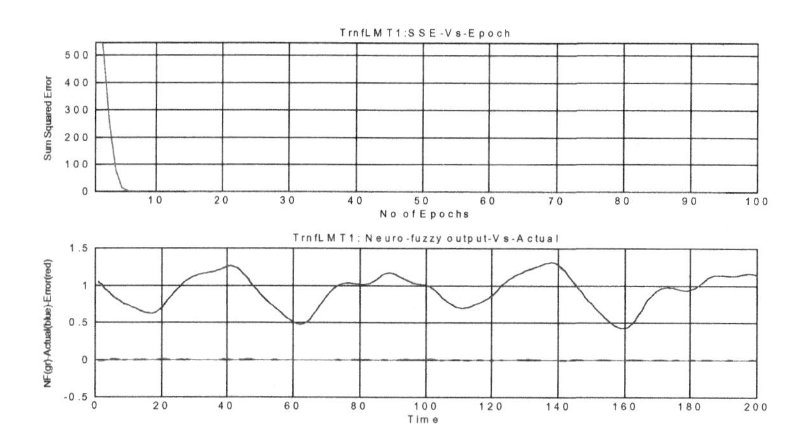

Figure 6.8(a). Training performance of Levenberg-Marquardt algorithm for Takagi-Sugeno-type of multi-input single-output neuro-fuzzy network (using seven fuzzy rules and seven GMFs) with Mackey-Glass chaotic time series data. Parameters of Levenberg-Marquardt algorithm: $\mu = 10$, $\gamma = 0.001$, $mo = 0.098$, $WF = 1.01$.

Table 6.2(a). Training and forecasting performance of Takagi-Sugeno-type of multi-input single-output neuro-fuzzy network (with $M = 7$ fuzzy rules) with proposed Levenberg-Marquardt algorithm for Mackey-Glass chaotic time series (SSE = sum square error, MSE = mean square error, MAE = mean absolute error, RMSE = root mean square error)

Sl. No.	Input data	SSE, MSE achieved	RMSE, MAE achieved
1.	1–200 (Training in 95 epochs)	SSE = 0.0026 MSE = 2.5571e–005	RMSE = 0.0051 MAE = 0.0039
2.	201–500 (Forecasting)	SSE = 0.0047 MSE = 3.1120e–005	RMSE = 0.0056 MAE = 0.0043
3.	501–1000 (Forecasting)	SSE = 0.0071	RMSE = 0.0053
4.	201–1000 (Forecasting)	SSE = 0.0118 MSE = 2.9427e–005	RMSE = 0.0054 MAE = 0.0042

SSE $= 0.5 \cdot \sum_{r=1}^{N} (e_r)^2$, MSE $= \sum_{r=1}^{N} (e_r)^2 / N$, RMSE $= \sqrt{\sum_{r=1}^{N} (e_r)^2 / N}$, and MAE $= \sum_{r=1}^{N} abs(e_r) / N$, where e_r is the error due to rth data sample and N is the number of data samples.

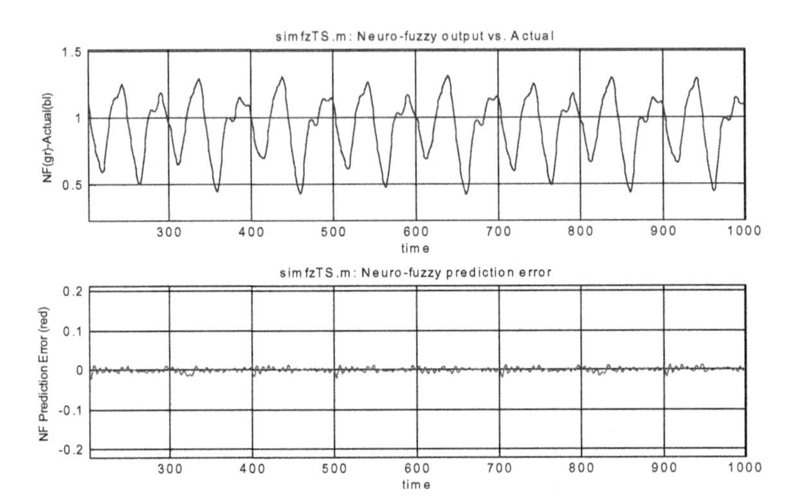

Figure 6.8(b). Performance of a Takagi-Sugeno-type multi-input single-output neuro-fuzzy network in forecasting the Mackey-Glass chaotic time series. Figure 6.8(a) and Figure 6.8(b) demonstrate the excellent training and forecasting performance of the Takagi-Sugeno-type multi-input single-output neuro-fuzzy network respectively for the Mackey-Glass chaotic time series. It is to be noted that the neuro-fuzzy network considered for this problem has only four inputs and one output and uses only seven Gaussian membership functions for (fuzzy) partitioning of input universes of discourse and seven fuzzy rules for neuro-fuzzy modelling.

Table 6.2(b). Comparison of training and prediction performance of fuzzy and other model with selected Takagi-Sugeno-type multi-input single-output neuro-fuzzy network with proposed Levenberg-Marquardt training algorithm for Mackey-Glass chaotic time series.

Sl. No.	Method	Training / forecasting with Input data	RMSE prediction error
1.	Kim and Kim (coarse partition)	500	0.050809 (5 partitions) 0.044957 (7 partitions) 0.038011 (9 partitions)
2.	Kim and Kim (after fine tuning)	500	0.049206 (5 partitions) 0.042275 (7 partitions) 0.037873 (9 partitions)
3.	Kim and Kim (genetic-fuzzy predictor ensemble)	500	0.026431
4.	Lee and Kim	500	0.0816
5.	Wang (product operator)	500	0.0907
6.	Min operator	500	0.0904
7.	Jang (ANFIS)	500	0.0070 (16 rules)
8.	Auto regression model	500	0.19
9.	Cascade correlation neural network	500	0.06
10.	Backpropagation neural network	500	0.02
11.	Sixth-order polynomial	500	0.04
12.	Linear prediction model	500	0.55
13.	FPNN (quadratic polynomial fuzzy inference)	500	0.0012 (16 rules)
14.	Takagi-Sugeno-type multi-input single-output neuro-fuzzy network (proposed work)	500 (forecasting)	0.0053 (7 rules, 7 GMFs, non-optimized)

RMSE = $\sqrt{\sum_{r=1}^{N}(e_r)^2/N}$, e_r is the error due to rth training sample and N is the number of training / predicted data samples.

6.7.3 Modelling and Prediction of Wang Data

This example deals with the modelling of a second-order nonlinear plant

$$y(k) = g(y(k-1), y(k-2)) + u(k) \tag{6.45a}$$

studied by Wang and Yen (1998, 1999a, and 1999b) and by Setnes and Roubos (2000 and 2001), with

$$g(y(k-1), y(k-2)) = \frac{y(k-1)y(k-2)(y(k-1)-0.5)}{1+y^2(k-1)y^2(k-2)} \tag{6.45b}$$

The goal is to approximate the nonlinear component $g(y(k-1), y(k-2))$ of the plant with a suitable fuzzy model. Wang and Yen (1999) generated 400 simulated data points from the plant model (6.45a) and (6.45b). 200 samples of identification data were obtained with a random input signal $u(k)$ uniformly distributed in [-1.5, 1.5], followed by 200 samples of evaluation data obtained by using a sinusoid input signal $u(k) = \sin(2\pi k/25)$, as shown in Figure 6.9(a). This example was also used by Setnes and Roubos (2000 and 2001) and a comparison with the results of Wang and Yen (1998, 1999a, and 1999b) was made. Here, we also apply the proposed Takagi-Sugeno-type neuro-fuzzy modelling scheme on the original Wang-data and show the results for linear rules consequents and compare the results with others described in the above references.

In order to apply the Takagi-Sugeno-type neuro-fuzzy modelling scheme the original Wang data (which is available to us in the form of an XIO matrix of size 400 × 3 that contains the first two columns as inputs and the third column as the desired output) was scaled and normalized down to the range [0, 1] for convenience. In the following, since our objective is to approximate the nonlinear component $g(y(k-1), y(k-2))$ of the plant, the same is treated as the desired output from the neuro-fuzzy network, whereas $u(k)$ and $y(k)$ have been considered as two inputs to the neuro-fuzzy network. The scaling and normalization were performed separately on each column of the XIO matrix, *i.e.* $\text{XIO} = [u, y, g]$, and the three column vectors $u = [u_1, u_2, \cdots, u_N]^T$, $y = [y_1, y_2, \cdots, y_N]^T$ and $g = [g_1, g_2, \cdots, g_N]^T$, each contains N data points. The scaled and normalized vector

$$\begin{aligned} u_{\text{nsc}} &= K_0\left[(u_1 - u_{\text{min}}), (u_2 - u_{\text{min}}), \cdots, (u_N - u_{\text{min}})\right]^T + u_{lo}, \\ K_0 &= (u_{hi} - u_{lo})/(u_{\max} - u_{\min}) \end{aligned} \tag{6.46}$$

is then computed where u_{max} and u_{min} are the maximum and minimum values of the u vector, and $u_{hi} = 1$ and $u_{lo} = 0$ are the desired highest and lowest values of the scaled or normalized u_{nsc} vector.

Once the scaling/normalization is performed, the scaled/normalized data are fed to the neuro-fuzzy network with $n = 2$ inputs and $m = 1$ output for training. Once the network is trained, its final parameter values are stored and the network is used for prediction. In this experiment, the first 200 data samples were used for training and the remaining 200 data samples were used for evaluation. The training performance of the network is illustrated in Figure 6.9(b) and also listed in Table 6.3(a). It is also illustrated that using only $M = 10$ fuzzy rules (*first model*) and also 10 Gaussian membership functions implemented for fuzzy partition of the input universe of discourse the proposed training algorithm could bring the network performance index (SSE) down to 3.0836×10^{-4} or equivalently MSE to 3.0836×10^{-6} from their initial values 45.338 in only 999 epochs. This is equivalent to achieving an actual SSE = 0.0012 or an actual MSE = 1.1866×10^{-5} when computed back on the original data.

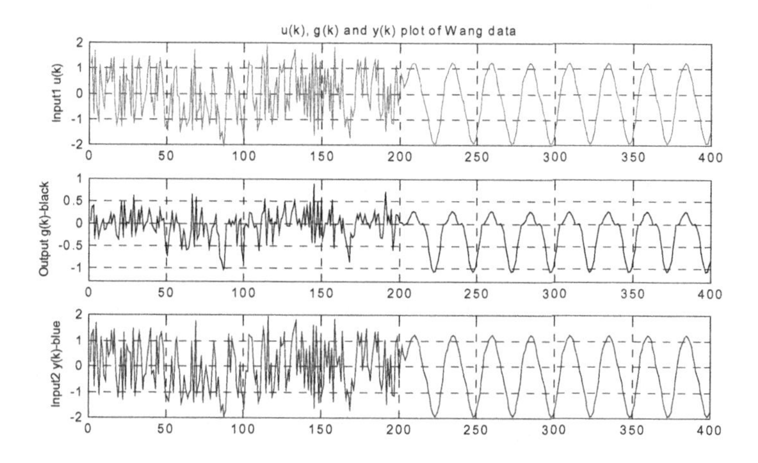

Figure 6.9(a). Plot of first input $u(k)$ (top), output $g(k)$ (middle) and second input $y(k)$ (bottom) of non-scaled Wang data (second-order nonlinear plant).

The corresponding evaluation performance of the trained network, as shown in Figure 6.9(c) and also listed in Table 6.3(a), illustrates that using the scaled or normalized evaluation data set from 201 to 400, the SSE value of 5.5319×10^{-4}, or equivalently MSE value of 5.5319×10^{-6}, were obtained. The above results further correspond to an actual SSE value of 0.0021, or equivalently to an actual MSE value of 2.1268×10^{-5}, which were computed back on the original evaluation data set. Evidently, the evaluation performance (actual MSE value), reported in Table 6.3(b), is at least 10 times better than that achieved by Setnes and Roubos (2000), Roubos and Setnes (2001) and much better than that achieved by Yen and Wang (1998, 1999a, 1999b).

Table 6.3(a). Training (999 epochs) and evaluation performance of Takagi-Sugeno-type multi-input single-output neuro-fuzzy network with proposed Levenberg-Marquardt algorithm for Wang data (second-order nonlinear plant data). Tuning parameter values of Levenberg-Marquardt algorithm for first model, *i.e.* $M = 10$, GMFs* = 10: $\mu = 10$, $\gamma = 0.01$, $mo = 0.098$, $WF = 1.05$; for second model, *i.e.* $M = 5$, GMFs* = 5: $\mu = 10$, $\gamma = 0.01$, $mo = 0.098$, $WF = 1.05$

Sl. No.	Input data	SSE & MSE (with pre-scaled and non-scaled actual data)	RMSE & MAE (with pre-scaled and non-scaled actual data)
1	1–200 Training data (*first model*)	SSE_train = 3.0836e–004 MSE_train = 3.0836e–006 Equivalently actual SSE_train = 0.0012 MSE_ train= 1.1866e–005	RMSE_train = 0.0018 MAE_train = 0.0012 Equivalently actual RMSE_train = 0.0034 MAE_train = 0.0024
2	201–400 Evaluation data (*first model*)	SSE_test = 5.5319e–004 MSE_test = 5.5319e–006 Equivalently actual SSE_test = 0.0021, MSE_test = 2.1268e–005	RMSE_test = 0.0024 MAE_test = 0.0015 Equivalently actual RMSE_test = 0.0046 MAE_test = 0.0030
3	1–200 Training data (*second model*)	SSE_train = 0.0135 MSE_train = 1.3491e–004 Equivalently actual SSE_train = 0.0519 MSE_train = 5.1866e–004	RMSE_train = 0.0116 MAE_train = 0.0087 Equivalently actual RMSE_train = 0.0228 MAE_train = 0.0170
4	201–400 Evaluation data (*second model*)	SSE_test = 0.0203 MSE_test = 2.0289e–004 Equivalently actual SSE_test = 0.0780 MSE_test = 7.8002e–004	RMSE_test = 0.0142 MAE_test = 0.0104 Equivalently actual RMSE_test = 0.0279 MAE_test = 0.0204

GMFs* = Gaussian membership functions

The same experiment was also carried out for $M = 5$ (*second model*), which exhibited the following training performance with the first 1 to 200 normalized and scaled training data: SSE and MSE values of 0.0135 and 1.3491×10^{-4} respectively, which correspond to the actual SSE and MSE values of 0.0519 and 5.1866×10^{-4} respectively. In addition, as listed below, the testing or evaluation performance of the Wang data with 201 to 400 rows, for five fuzzy rules and five Gaussian membership functions has produced SSE and MSE values of 0.0203 and 2.0289×10^{-4} respectively. These results further correspond to actual SSE and MSE values of 0.0780 and 7.8002×10^{-4} respectively, which are computed back from original (non-scaled) evaluation data.

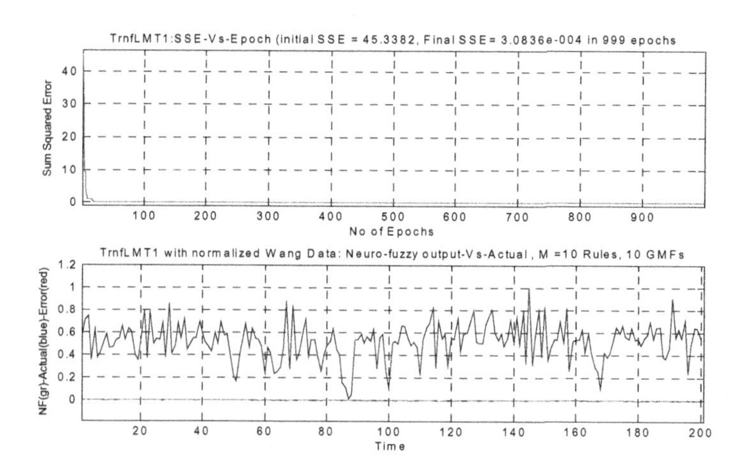

Figure 6.9(b). Performance of Takagi-Sugeno-type multi-input single-output neuro-fuzzy network with M = 10 rules (first model) for normalized Wang data when trained with proposed Levenberg-Marquardt algorithm.

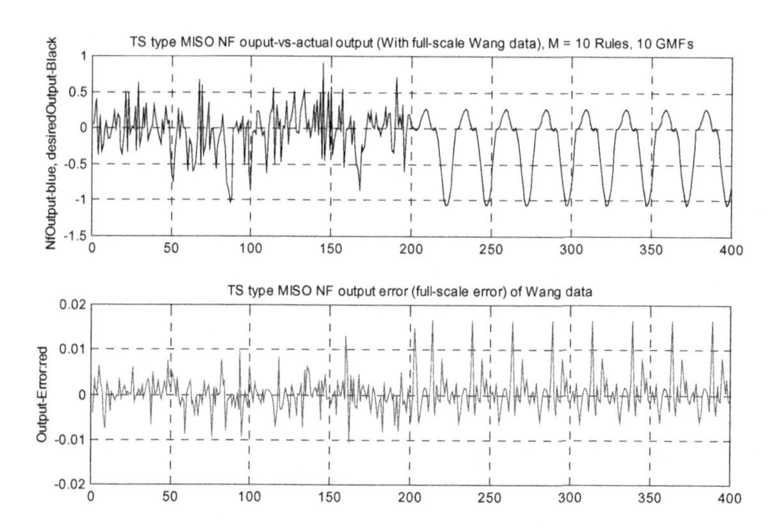

Figure 6.9(c). Prediction performance of Takagi-Sugeno-type multi-input single-output neuro-fuzzy network with M= 10 rules (first model) for non-scaled Wang data after training with proposed Levenberg-Marquardt algorithm.

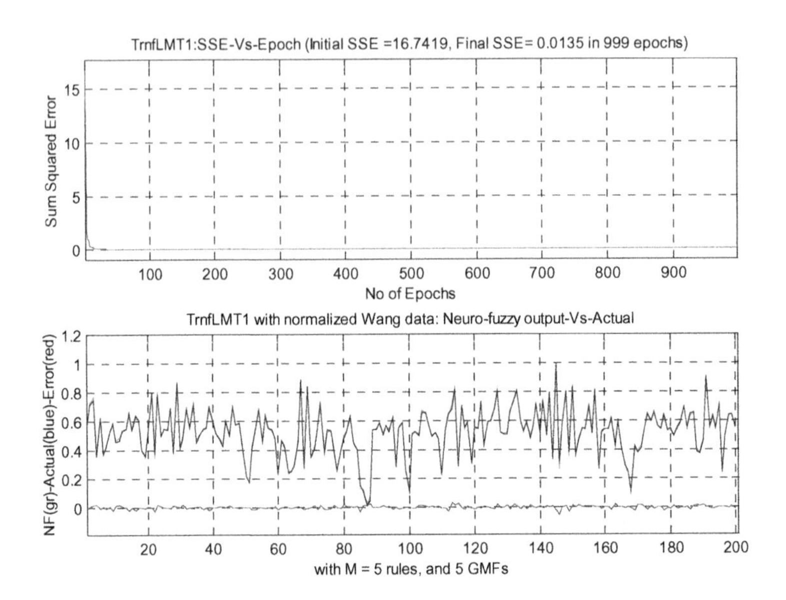

Figure 6.9(d). Performance of Takagi-Sugeno-type multi-input single-output neuro-fuzzy network with $M - 5$ rules (second model) and five GMFs for normalized Wang data when trained with proposed Levenberg-Marquardt algorithm.

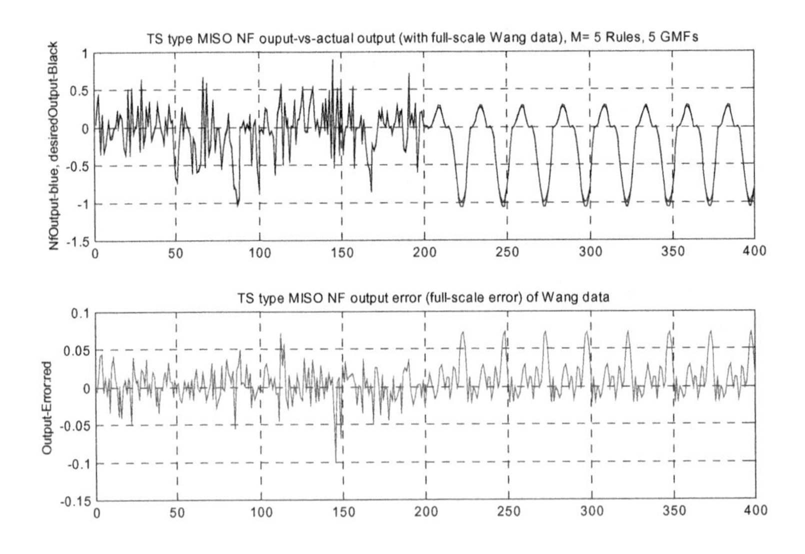

Figure 6.9(e). Prediction performance of Takagi-Sugeno-type multi-input single-output neuro-fuzzy network with $M = 5$ rules (second model) for non-scaled Wang data after training with proposed Levenberg-Marquardt algorithm.

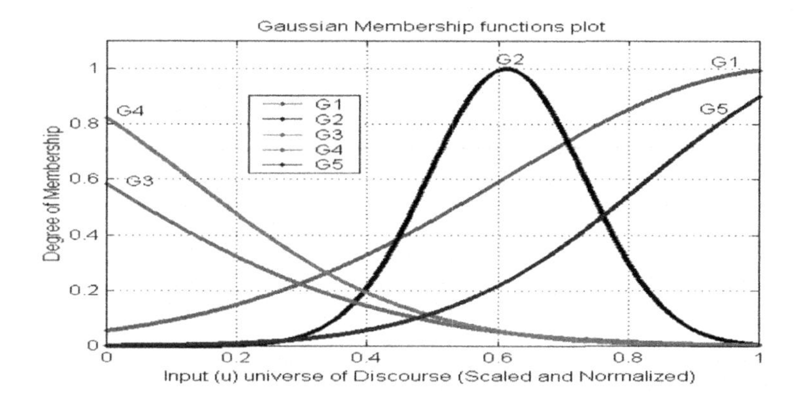

Figure 6.9(f). Finally tuned five Gaussian membership functions plot for fuzzy partition of input u (scaled/normalized) universe of discourse for Wang data. X-axis (input universe of discourse, scaled/normalized), Y-axis (degree of membership).

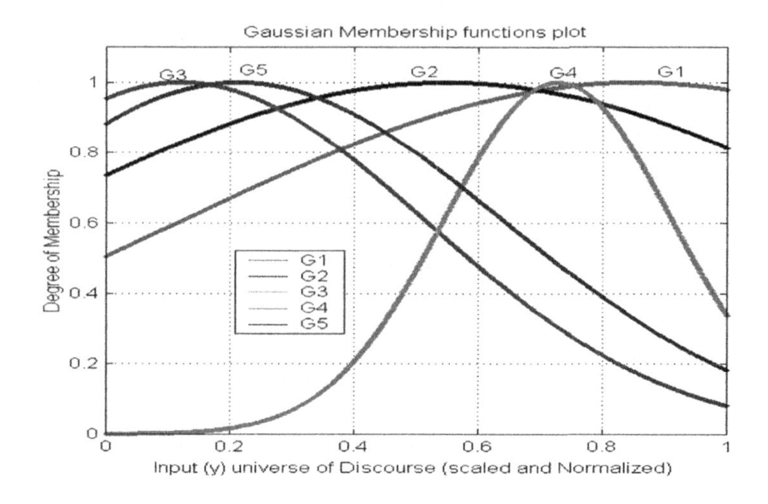

Figure 6.9(g). Finally tuned five Gaussian membership functions plot for fuzzy partition of input y (scaled/normalized) universe of discourse for Wang data. X-axis (input universe of discourse – scaled/normalized), Y-axis (degree of membership). Note that fuzzy membership functions in Figure 6.9(f) and Figure 6.9(g) are largely overlapping. Accuracy and transparency of the model are expected to be further improved if the similar fuzzy sets (for *e.g.* G4 and G3 in Figure 6.9(f) and in Figure 6.9(g) G3 and G5 are highly similar fuzzy sets) are merged and further fine-tuned using genetic algorithm or evolutionary computation.

Table 6.3(b). Comparison of training and evaluation performances of other fuzzy model and Takagi-Sugeno-type multi-input single-output neuro-fuzzy networks trained with the proposed Levenberg-Marquardt algorithm for Wang data (second-order nonlinear plant data)

Method	No. of rules	No. of fuzzy sets	Rules conseq.	MSE training	MSE eval.
Wang and Yen (1999)	40 (initial)	40 Gauss. (2D)	Singleton	3.3e–4	6.9e–4
	28 (optimized)	28 Gauss. (2D)	Singleton	3.3e–4	6.0e–4
Yen and Wang (1998)	36 (initial)	12 B-splines	Singleton	2.8e–5	5.1e–3
	23 (optimized)	12 B-splines	Singleton	3.2e–5	1.9e–3
	36 (initial)	12 B-splines	Linear	1.9e–6	2.9e–3
	24 (optimized)	12 B-splines	Linear	2.0e–6	6.4e–3
Yen and Wang (1999)	25 (initial)	25 Gauss. (2D)	Singleton	2.3e–4	4.1e–4
	20 (optimized)	20 Gauss. (2D)	Singleton	6.8e–4	2.4e–4
Setnes and Roubos (2000)	7 (initial)	14 Triangular	Singleton	1.6e–2	1.2e–3
	7 (optimized)	14 Triangular	Singleton	3.0e–3	4.9e–4
	5 (initial)	10 Triangular	Linear	5.8e–3	2.5e–3
	5 (optimized)	8 Triangular	Linear	7.5e–4	3.5e–4
	4 (optimized)	4 Triangular	Linear	1.2e–3	4.7e–4
Roubos and Setnes (2001)	5 (initial)	10 Triangular	Linear	4.9e–3	2.9e–3
	5 (optimized)	10 Triangular	Linear	1.4e–3	5.9e–4
	5 (optimized)	5 Triangular	Linear	8.3e–4	3.5e–4
Proposed neuro-fuzzy TS model	10 (initial, non-optimized)	10 Gaussian	Linear	1.1866e–5	2.1268e–5
	5 (initial, non-optimized)	5 Gaussian	Linear	5.1866e–4	7.8003e–4

The plots of the finally tuned GMFs that made the fuzzy partitions of universes of discourse of normalized input $u(k)$ and input $y(k)$ are shown in Figures 6.3(f) and 6.3(g) respectively. The figures also show that there is further scope for improving the accuracy, transparency and interpretability of neuro-fuzzy model obtained through similarity measures and genetic-algorithm-based optimizations. These issues, namely model transparency and interpretability, will be the main subject of discussion in Chapter 7. The results obtained in this example also, in general, summarize the excellent prediction performance of Takagi-Sugeno-type multi-input single-output neuro-fuzzy networks when trained with the proposed Levenberg-Marquardt Algorithm.

6.8 Other Engineering Application Examples

In the following, some engineering application examples are given in which the systematic neuro-fuzzy modelling approach has been used to solve the problem of

- material property prediction
- pyrometer reading correction in temperature measurement of wafers, based on prediction of wafer emissivity changes in a rapid thermal processing system, such as chemical vapour deposition and rapid thermal oxidation

- monitoring of tool wear.

6.8.1 Application of Neuro-fuzzy Modelling to Material Property Prediction

Chen and Linkens (2001) have proposed a systematic neuro-fuzzy modelling framework with application to mechanical property prediction in hot-rolled steel. Their methodology includes three main phases:

- the initial fuzzy model, which consists of generation of fuzzy rules by a self-organizing network
- the second phase, which includes the selection of important input variables on the basis of the initial fuzzy model and also the assessment of the optimum number of fuzzy rules (hidden neurons in the RBF network) and the corresponding receptive fields determination via the fuzzy *c*-means clustering algorithm
- third phase, dedicated to the model optimization, including parameter learning and structure simplification on the basis of backpropagation learning and the similarity analysis of fuzzy membership functions.

Thereafter, the neuro-fuzzy model developed is used to predict the tensile stress, yield stress, and the like in materials engineering.

In materials engineering, property prediction models for materials are important for design and development. This has for many years been an important subject of research for steel. Much of this work has concentrated on the generation of structure - property relationships based on linear regression models (Pickering, 1978), (Hodgson, 1996), developed only for some specific class of steels and specific processing routes. Recently, some improved, neural-networks-based models have been developed for prediction of mechanical properties of hot-rolled steels (Hodgson, 1996), (Chen *et al.*, 1998), and (Bakshi and Chatterjee, 1998). Using complex nonlinear mapping, the models provide more accurate prediction than traditional linear regression models. But the drawback here is that the development of these kinds of model is usually highly problem specific and time consuming, so that the development of a fast, efficient, and systematic data-driven modelling framework for material property prediction is still needed.

The problem of modelling of hot-rolled metal materials can be broadly stated as follows. Given a certain material which undergoes a specified set of manufacturing processes, what are the final properties of this material? Typical final properties, in which metallurgical engineers are interested, are the mechanical properties such as, tensile strength (*TS*), yield stress, elongation, *etc.* Chen *et al.* (2001) have developed a neuro-fuzzy model for the prediction of the composition-microstructure-property relations of a wide range of hot-rolled steels. More than 600 experimental data from carbon-manganese (C-Mn) steels and niobium micro-alloyed steels have been used to train and test the neuro-fuzzy model, which relates the chemical compositions and microstructure with the mechanical properties.

In the experimental data set, they have considered 13 chemical compositions, two microstructure variables, and measured tensile stress values, which corresponds to a system with 15 possible input variables and with one output variable. Several performance indices (RMSE and MAE), and the correlation coefficient between the measured and the model predicted tensile stress were used to evaluate the performance of the fuzzy models developed. Property prediction results for different types of steel are summarized below.

6.8.1.1 Property Prediction for C-Mn Steels

Using the proposed input selection paradigm, five inputs (the carbon, silicon, manganese, nitrogen contents and the ferrite grain size $D^{-1/2}$ ($mm^{-1/2}$), were selected from the 15 possible input variables. Three hundred and fifty-eight industrial data were used, with 50% of them for training and the remaining 50% for model testing. After partition validation and parameter learning, the final fuzzy models of the Mamdani type consisting of six rules were obtained. The rule-based fuzzy model was represented by six fuzzy rules. From the fuzzy model generated, Chen *et al.* (2001) used linguistic hedges to derive the corresponding linguistic model.

The fuzzy model with linguistic hedges finally generated used six Mamdani-type fuzzy rules, such as one described below:

Rule-1: IF Carbon is large and Silicon is medium and Manganese is large and Nitrogen is medium and $D^{-1/2}$ is more or less medium, THEN Tensile Stress is large

Using the above model, Chen *et al.* (2001) obtained good prediction results that gave RMSE = 12.44 and 16.85 and MAE = 9.46 and 13.15 for model training and testing respectively.

According to their simulation result, the out-of-10% error-band prediction patterns for the testing data is 2.2%. It was claimed that the fuzzy model generated gave good prediction and generalization capability.

6.8.1.2 Property Prediction for C-Mn-Nb Steels

In another experiment of Chen *et al.* (2001), for property prediction for C-Mn-Nb steels, more than 600 measured data, including the previously used 358 C-Mn data, were used to build the fuzzy model. Three hundred and fifteen data were selected for training and the remaining 314 data were used for testing. Using their proposed fuzzy modelling approach, six out of 15 variables were selected as the inputs (C, Si, Mn, N, Nb, $D^{-1/2}$) with tensile stress as output. A six-rule fuzzy model was developed after structure identification and parameter training. The property prediction resulted in RMSE = 15.48 and 19.74 and MAE = 12.11 and 14.46 for training and testing, respectively. Furthermore, the out-of-10% error-band patterns for the testing data were found to be only 3%.

6.8.2 Correction of Pyrometer Reading

As a second engineering application, we describe here the prediction capability of a self-constructing ***neural-fuzzy inference network*** (SONFIN) proposed by Lai and Lin (1999) for pyrometer reading correction in wafer temperature measurement, based on emissivity changes. The motivation for this was that, because of several distinct advantages of rapid thermal processing (RTP) over other batch processing, such as significant reduction in thermal budget and better control over the processing environment, rapid thermal processing has been extensively used in high-density integrated circuit manufacturing on single wafers.

Wafer temperature measurement and control are two critical issues here. Currently, a single-wavelength pyrometer is used as a non-contact temperature sensor. However, for applications where the characteristics of the surface change with the time, the wafer emissivity also varies simultaneously. This can lead to temperature errors in excess of 50 degree Celsius in a few seconds. Various methods were suggested to overcome this problem, such as use of a dual-wavelength pyrometer, model-based emissivity correction, *etc.* A global mathematical model for the rapid thermal process, which includes the temperature sensor along with a control loop and lamp system, was developed and simulated by Lai and Lin (1999). In the same model, emissivity changes during oxidation are calculated according to reflections, refraction within thin dielectric films on a silicon substrate. The oxide thickness as a function of oxidation time at various temperatures, is simulated by a linear parabolic model. Using the basic heat transfer law, a pyrometer model to simulate the temperature sensor in the rapid thermal process is derived and, thereafter, a neural-fuzzy network is used to learn and predict the variations of oxidation growth rate of the film under different process temperatures. Based on this neural-fuzzy prediction and an already available optical model the emissivity of the wafer can be correctly computed.

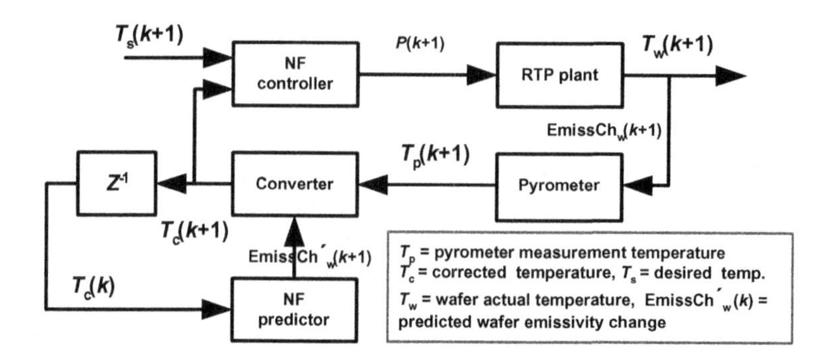

Figure 6.10. Block diagram of the neural-fuzzy method to predict wafer emissivity variation and to correct the pyrometer readings

Another neural-fuzzy network was used by Lai and Lin (1999) to control the temperature of an RTP system by using the inverse model of the RTP system to achieve two control objectives: trajectory following and temperature uniformity on the wafer. Figure 6.10 shows the block diagram of the neural-fuzzy method to predict wafer emissivity variation and to correct the pyrometer readings. The previous corrected temperature value $T_c(k)$ and the current processing time k are used as the inputs of the neural-fuzzy network to predict the current film thickness, which is further used to compute the emissivity of the wafer $ew'(k+1)$ according to wafer optical model. The converter is then used to correct the pyrometer reading value $T_p(k+1)$ to $T_c(k+1)$.

The neural-fuzzy network used for this purpose was the SONFIN, which has a fuzzy rule-based network possessing neural learning ability. Compared with other existing neural-fuzzy networks, a major characteristic of this network is that no preassignment and design of fuzzy rules are required. The rules are constructed automatically during the training process. Besides, SONFIN can overcome both the difficulty of finding a number of proper rules for the fuzzy logic controllers and the overtuned and slow convergence phenomena of backpropagation neural networks. SONFIN can also optimally determine the consequent of fuzzy IF-THEN rules during its structure learning phase, and it also outperforms the pure neural networks greatly, both in learning speed and accuracy.

6.8.3 Application for Tool Wear Monitoring

In automated manufacturing systems, such as flexible manufacturing systems, one of the most important issues is the detection of tool wear during the cutting process to avoid poor quality in the product or even damage to the workpiece or the machine. It will be shown that a neuro-fuzzy model, based on a prediction technique, can be applied for monitoring tool wear in the drilling process.

The alternating direction of the cutting force leads to vibrations of the machine structure. These vibrations will change owing to the tool wear conditions. Despite the relatively harsh environment in the proximity of the cutting zone, the vibrations can be measured conveniently by accelerometers at a comparably affordable price. Neural networks have, for a long time, been used for classification of various signals. However, because of many limitations, including the slow training performance of neural networks, alternatively a neural network with fuzzy inference has been used because of its much faster learning ability. The latter is nothing but a neuro-fuzzy type of hybrid learning network. Using such a network a new drill condition monitoring method is described, as proposed by Li *et al.* (2000). The method is based on spectral analysis of the vibration signal. The results are used to generate a set of **indices for monitoring**, utilizing the fact that the frequency distribution of vibration changes as the tool wears. The nonlinear relationship between the tool wear condition and these monitoring indices is modelled using a hybrid neuro-fuzzy network. The hybrid network selected in this case has five inputs and five outputs. The inputs to the network are the monitoring indices based on the vibration signal of the drilling process. It is to be noted that the mean value of each frequency band can be used to characterize the different tool conditions. The monitoring indices selected as network inputs are summarized in Table 6.4(a). The content of the Table 6.4(a) is read follows:

$$x_1 = \text{the r.m.s value of the signal in the frequency band } [0, 300] \text{ Hz.}$$

Unlike the inputs of the network, the tool wear condition of the network was divided into five states represented by five fuzzy membership functions (MF), namely initial wear, normal wear, acceptable wear, severe wear and failure. Based on the flank wear of the tool, these conditions are summarized in the Table 6.4(b).

Table 6.4(a). Summary of monitoring indices selected as network inputs

Input terminal of network	Input representation	RMS value of the signal in the frequency band
1	x_1	[0, 300] Hz.
2	x_2	[300, 600] Hz
3	x_3	[600, 1000] Hz
4	x_4	[1000, 1500] Hz
5	x_5	[1500, 2500] Hz

Table 6.4(b). Summary of the conditions for various flank wear

Output terminal of the network	Fuzzy MF	Tool condition	Flank wear
y_1	1	Initial wear	0 < wear < 0.1 mm
y_2	2	Normal wear	0.05 < wear < 0.3 mm
y_3	3	Acceptable wear	0.25 < wear <0.5 mm
y_4	4	Severe wear	0.45< wear <0.6 mm
y_5	5	Failure	wear > 0.6 mm

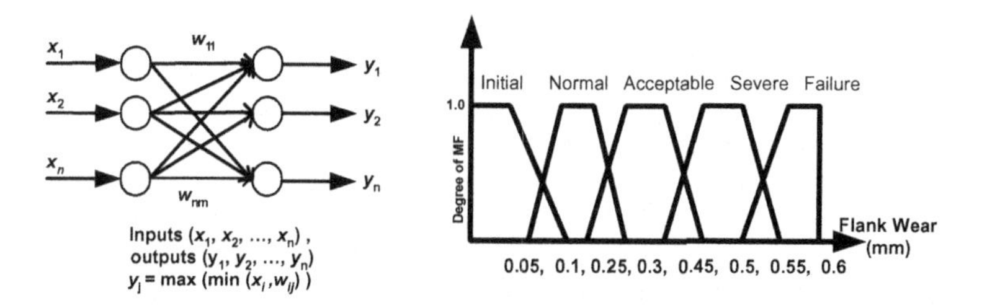

Figure 6.11. Fuzzy neural net topology (left), fuzzy membership functions of drilling conditions (right)

The fuzzy membership functions of drilling conditions based on experimental data and the observed system behaviour are set for output indices of the hybrid network, and are shown in the Figure 6.11. The reason for choosing a trapezoidal membership function is that it is difficult to quantify what exact percentage of tool wear corresponds to a certain linguistic variable. In order to improve the training speed of the hybrid network, the tool wear conditions are coded as follows: initial (1,0,0,0,0); normal (0,1,0,0,0); acceptable (0,0,1,0,0); severe (0,0,0,1,0); and failure

(0,0,0,0,1). If the tool condition is acceptable, then the output values of the hybrid network are (0,0,1,0,0).

Once the hybrid neuro-fuzzy network has learnt the above nonlinear mapping from a given set of training examples consisting of (x, y) values, thereafter, for a new set of monitoring indices (*i.e.* related to the frequency band of the vibration), obtained from the drilling process through accelerometer, charge amplifier, and the signal processing unit, the network will generate or predict a set of y values. The maximum of y_i, namely J, is converted to 1, and the others are converted to 0. For instance, if $J = \max\{y_i \mid i = 1, 2, 3, \ldots, 5) = y_2 = 0.8$, the predicted output of the hybrid network is (0,1,0,0,0). This prediction indicates that the tool wear condition belongs to the normal category. Exploiting the prediction capability of the hybrid network and adopting similar methodologies one can monitor the tool wear in an automated manufacturing system.

6.9 Concluding Remarks

In this chapter a hybrid neuro-fuzzy modelling frame work is proposed. An accelerated training algorithm, based on either the backpropagation or Levenberg-Marquardt algorithm and in combination with a modified error index extension, has also been developed for training Takagi-Sugeno-type multi-input multi-output or multi-input single-output neuro-fuzzy networks. The increased speed of training convergence was experimentally confirmed on some examples of modelling and forecasting of time series. It was observed that the addition of a small modified error index term to the original performance function improves the convergence speed of both standard backpropagation and the Levenberg-Marquardt algorithm significantly.

The trained neuro-fuzzy network itself is found to be powerful for modelling and prediction of dynamics of various nonlinear phenomena. However, the fuzzy rules generated through neuro-fuzzy training are occasionally found to be not transparent enough, in the sense that a clear interpretation of all the tuned fuzzy sets is not possible. This is due to the fact that the membership functions, finally tuned through neuro-fuzzy network training, are frequently very similar to each other or they greatly overlap each other, giving rise to a difficult situation to interpret. To solve this problem and to improve the interpretability of fuzzy rules, set-theoretical ***similarity measures*** should be computed for each pair of fuzzy sets and highly similar fuzzy sets should be merged together into a single set (Setnes, Babuška, Kaymark, 1998) as discussed in detail in Chapter 7. Furthermore, the tuned membership functions building a universal fuzzy set within the universe of discourse should be removed because they do not contribute anything to the rule base. Also, because the parameters of the Gaussian membership functions are unconstrained, it is probable that the fuzzy partition occasionally may not look like the usual fuzzy partition. In such cases, the interpretation of a trained neuro-fuzzy system may also not be possible.

An additional issue is the determination of the optimum number of fuzzy rules and hence, also the determination of optimum number of membership functions. This is essential, because an unnecessarily larger rule base may overfit the noisy data and thereby worsen the prediction ability. For determination of the optimum number of rules and of membership functions, genetic algorithms or, in general, evolutionary computation, should preferably be used as a proper support tool.

It should finally be underlined that, after the completion of backpropagation or Levenberg-Marquardt training, if the final (linear/singleton) rules consequent parameters are determined by applying the least squares error estimator using only the tuned GMF parameters of the network, then the accuracy of the model could occasionally be increased further. Furthermore, the simulation results have shown that the Levenberg-Marquardt algorithm, based on Jacobian matrices computed using normalized prediction error or normalized equivalent error (Section 6.4.2.3.1), though computationally very heavy, often leads to a better training performance and to a faster convergence when applied to the Takagi-Sugeno type of neuro-fuzzy networks. In the experiments investigated here, the proposed training algorithms (modified backpropagation/Levenberg-Marquardt algorithm) proved to be efficient enough for neuro-fuzzy modelling and for prediction of electrical load time series, chaotic time series, *etc.* Furthermore, some recently published additional engineering examples confirm the versatility and possible other applications of neuro-fuzzy networks in different fields of engineering.

References

[1] Bakshi BR and Chatterjee R (1998) Unification of neural and statistical methods as applied to materials structure-property mapping, J. Alloys Compounds, 279(1): 39–46.

[2] Bezdek JC (1993) Editorial-fuzzy models: What are they and why, IEEE Trans. on Fuzzy Systems, vol. 1, pp. 1–5.

[3] Brown M and Harris C (1994) Neuro-fuzzy adaptive modelling and control, Prentice Hall, New York.

[4] Buckley JJ and Hayashi Y (1994) Fuzzy neural networks, In: Fuzzy Sets, Neural Networks and Soft Computing, edited by Yager R and Zadeh L, Van Nostrand Reinhold, New York.

[5] Chak CK, Feng G and Ma J (1998) An adaptive fuzzy-neural network for MIMO system model approximation in high-dimensional spaces, IEEE Trans. on System, Man and Cybernetics, 28: 436–446.

[6] Chen M and Linkens DA (1998) A fast fuzzy modeling approach using clustering neural networks, In Proc. IEEE world congress on Intell. Computat. 2: 1406–1411.

[7] Chen M Linkens DA (2001), A systematic neuro-fuzzy modelling framework with application to material property prediction, IEEE Trans. on SMC, B 31(5): 781–790.

[8] Cho KB and Wang BH (1996) Radial basis function based adaptive fuzzy systems and their application to system identification and prediction, Fuzzy Sets System., 83: 325–339.

[9] Fuller R (1995) Neural-fuzzy systems, Abo Akademi.

[10] Gupta MM (1994) Fuzzy neural networks: Theory and Applications, Proceedings of SPIE, vol. 2353, pp. 303–325.

[11] Gustafson DE, Kessel WC (1979) Fuzzy clustering with fuzzy covariance matrix, Proc. of the IEEE CDC, San Diego, 761–766.

[12] Hagan MT, Menhaj MB (1994) Training feedforward networks with the Marquardt algorithm, IEEE Trans. on Neural Networks, 5(6): 989–993.

[13] Hodgson PD (1996) Microstructure modeling for property prediction and control, J. of Materials Process Technology, 60: 27–33.

[14] Jang JSR (1993) ANFIS: Adaptive Network Based Fuzzy Inference System, IEEE Trans. on SMC., 23(3): 665–685

[15] Jang JSR, Sun CT (1995) Neuro-fuzzy modelling and control, Proc. of IEEE, 83: 378–406.

[16] Kim D, Kim C (1997) Forecasting time series with genetic fuzzy predictor ensemble, IEEE Trans. on Fuzzy Systems, 5(4): 523–535.

[17] Kosko B (1992) Neural networks and fuzzy systems, Prentice Hall, Englewood Cliffs, New Jersey.

[18] Kulkarni AD (1998) Neural-fuzzy models for multi-spectral image analysis, Internat. J. of Applied Intelligence, 8: 173–187

[19] Kulkarni AD (2001) Computer vision and fuzzy-neural systems, Upper Saddle River, New Jersey: Prentice Hall PTR.

[20] Lai JH and Lin CT (1999) Application of neural fuzzy network to pyrometer correction and temperature control in rapid thermal processing, IEEE Trans. Fuzzy Systems, 7(2):160–174.

[21] Lee SH, Kim I (1994) Time series analysis using fuzzy learning, Proc. of Intern. Conf. on Neural Information Processing, Seoul, Korea, 6: 1577–1582.

[22] Li X, Dong S, Venuvinod PK (2000) Hybrid Learning for tool wear monitoring, Int. J. Adv. Manufacturing Technology, 16: 303–307.

[23] Lin CT and Lee CSG (1991) Neural networks based fuzzy logic and control systems, IEEE Trans. On Computers, vol. 40, pp. 1320–1336.

[24] MATLAB (1998) Fuzzy logic toolbox, user's guide, The Math Works Inc., vers. 5.2

[25] Mitra S, Hayashi Y (2000) Neuro-fuzzy rule generation: survey in soft computing framework, IEEE Trans. on Neural Networks, 11(3): 748–768.

[26] Nauck D, Klawonn F and Kruse R (1997) Foundations of neuro-fuzzy systems, Wiley, Chichester, U.K.

[27] Nie J (1997) Nonlinear time-series forecasting : A fuzzy-neural approach, Neurocomputing, 16(1997): 63–76.

[28] Pal SK and Mitra S (1992) Multilayer perceptron, fuzzy sets and classification, IEEE Trans. On Neural Networks, 2(5): 683–697.

[29] Palit AK and Babuška R (2001) Efficient training algorithm for Takagi-Sugeno type neuro-fuzzy network, Proc. of FUZZ-IEEE, Melbourne, Australia, vol. 3: 1367–1371.

[30] Palit AK and Popovic D (1999) Forecasting chaotic time series using neuro-fuzzy approach, Proc. of IEEE-IJCNN, Washington DC, USA, vol. 3: 1538–1543.

[31] Palit AK and Popovic D (1999) Fuzzy logic based automatic rule generation and forecasting of time series, Proc. of FUZZ-IEEE, Seoul, Korea, vol. 1: 360–365.

[32] Palit AK and Popovic D (2000) Intelligent processing of time series using neuro-fuzzy adaptive genetic approach, Proc. of IEEE-ICIT, Goa, India, vol. 1:141–146.

[33] Palit AK and Popovic D (2000) Nonlinear combination of forecasts using artificial neural network, fuzzy logic and neuro-fuzzy Approaches, Proc. of FUZZ-IEEE, San Antonio, Texas, USA, vol. 2: 566–571.

[34] Palit AK, Doeding G, Anheier W, Popovic D (2002) Backpropagation based training algorithm for Takagi-Sugeno type MIMO neuro-fuzzy network to forecast electrical load time series, Proc. of FUZZ-IEEE, Honolulu, Hawai, USA. vol. 1: 86–91.

[35] Park HS, Oh SK, Ahn TC and Pedrycz W (1999) A study on multi-layer based fuzzy polynomial inference system based on an extended GMDH algorithm, Proc. of FUZZ-IEEE, Seoul, Korea, vol. 1: 354–359

[36] Pedrycz W (1995) Fuzzy sets engineering, CRC Press, Boca Raton, Florida.

[37] Pickering FB (1978) Physical metallurgy and the design of steels, Applied Science, London, U.K.

[38] Roubos H, Setnes M (2001) Compact and transparent fuzzy models and classifiers through iterative complexity reduction, IEEE Trans. on Fuzzy System, 9(4): 516–524
[39] Setnes M, Babuška R, Kaymark U, *et al.*, (1998) Similarity measures in fuzzy rule base simplification, IEEE trans. on SMC., B-28: 376–386
[40] Setnes M, Roubos JA (2000) GA-fuzzy modelling and classification: complexity and performance, IEEE Trans. on Fuzzy Systems, 8(5): 509–522
[41] Takagi and Hayashi (1991) NN-driven fuzzy reasoning, Internat. J. of Approximate Reasoning, 5(3): 191–212.
[42] Wang L and Yen J (1999) Extracting fuzzy rules for system modelling using a hybrid of genetic algorithms and Kalman filter, Fuzzy Sets System, 101: 353–362
[43] Wang LX (1994) Adaptive fuzzy systems and control: design and stability analysis, Englewood Cliffs, New Jersey: Prentice Hall.
[44] Wang LX and Mendel JM (1992a) Fuzzy basis functions, universal approximation, and orthogonal least squares learning, IEEE Trans. on Neural Network, 3: 807 – 814.
[45] Wang LX and Mendel JM (1992b) Back-propagation fuzzy system as nonlinear dynamic system identifiers, Proc. of FUZZ-IEEE, vol. 2: 1409–1418.
[46] Wang LX and Mendel JM (1992c) Generating fuzzy rules by learning from examples, IEEE Trans. on SMC, 22(6): 1414–1427.
[47] Xiaosong D, Popovic D, Schulz-Ekloff G (1995) Oscillation resisting in the learning of backpropagation neural networks, Proc. of 3rd IFAC/IFIP workshop on algorithm and architectures for real-time control, Ostend, Belgium.
[48] Yen J and Wang L (1998) Application of statistical information criteria for optimal fuzzy model construction, IEEE Trans. on Fuzzy System, 6(3): 362–371.
[49] Yen J and Wang L (1999) Simplifying fuzzy rule-based models using orthogonal transformation methods, IEEE Trans. on SMC, 29(1): 13–24.
[50] Zhang J and Morris AJ (1999) Recurrent neuro-fuzzy networks for nonlinear process modelling, IEEE Trans. on Neural Networks, 10: 313–326.

7

Transparent Fuzzy/Neuro-fuzzy Modelling

7.1 Introduction

Fuzzy logic is a methodology widely applied in model building of dynamic systems for implementation of advanced control systems. Fuzzy models are developed using the universal approximation capability of fuzzy logic systems. Such models differ from other types of model built using ***non-symbolic methodology***, mainly because they can represent knowledge in a transparent manner using fuzzy IF-THEN rules which are understandable to the human expert who can directly operate on them. This provides the direct man-machine communication.

Fuzzy models are generally built by extracting and encoding expert knowledge into the IF-THEN rules with the linguistic arguments, in this way generating a transparent knowledge appropriate for its easy inspection, modification, and maintenance by human experts. However, the process of knowledge acquisition and building of adequate IF-THEN rules are not trivial tasks, because the experts are not always available and their knowledge is often incomplete, episodic and time varying. This was the motivation for switching model building approach from the seminal ideas of knowledge acquisition described above to a data-driven approach. Unfortunately, many of newly developed algorithms for data-driven fuzzy modelling aim at good numerical approximation and pay little attention to the transparency and computational load of the resulting rule base. In this chapter we will therefore present a rule base simplification method that can be used - along with arbitrary fuzzy modelling methods - for obtaining transparent and compact fuzzy models from data. The efficiency of the approach will be demonstrated on the example of nonlinear plant modelling and prediction of its future output value.

7.2 Model Transparency and Compactness

Fuzzy models are often referred to as ***white-box models***, in contrast to the neural-networks-based models which are considered as ***black-box models***. This is because fuzzy models are, to some extent, transparent to interpretation and analysis, implying that the model's output can be justified through developed IF-THEN linguistic rules. However, the transparency of a fuzzy model cannot be achieved automatically, unless some special measure is taken *a priori*. This is especially, true for the automated data-driven fuzzy modelling technique, where the fuzzy models generated are not at all or to a restricted degree transparent to interpretation.

A system can be described by a few fuzzy rules using distinct, *i.e.* non-overlapping, interpretable fuzzy sets. It can, of course, also be described by a few fuzzy rules, but with a large number of highly overlapping fuzzy sets that hardly allow for any interpretation. Alternatively, if a system is described by a large number of rules but with a few (or many) distinct and non-overlapping fuzzy sets, then the fuzzy models generated in such a case could also be unclear or close to non-interpretable because of the large number of rules. This situation can occur practically when the fuzzy rules are generated using the Wang and Mendel (1992a) approach, or by its modification as proposed by Palit and Popovic (1999a), presented in Chapter 4. In both rule-generation approaches a large number of input-output data pairs (or training samples) generate a large number of rules, even though fuzzy domains are partitioned by large (or small) numbers of distinct and non-overlapping /partially overlapping fuzzy sets such as Small(N), Small(N-1), ..., Small(1), Centre (CE), Big(1), ..., Big(N), *etc.* The reason for loss of interpretability in the above case is mainly because the large number of rules fire simultaneously for an unknown input condition (within the fuzzy domain) to infer the corresponding output decision. Therefore, the corresponding output decision cannot be easily justified by human reasoning.

Yet, in contrast to the above case when a fuzzy model is developed using expert knowledge, the model designer usually takes care that neither the number of rules nor the fuzzy sets, which are used to partition the domains, are large at all, besides maintaining the proper distinguishability of applied fuzzy sets for domain partition. On the other hand, when automated data-driven techniques are applied to build fuzzy models from data, a certain degree of redundancy, and thus unnecessary complexity, cannot be avoided.

In the following, we present a rule base simplification and reduction method proposed by Setnes *et al.* (1998a and 1998b) and Setnes (2001) that seeks to simplify an already available rule base by reducing redundant information present in the form of similar fuzzy sets. Similar fuzzy sets are overlapping fuzzy sets that describe almost the same region in the domain of some model variable. In such cases, the model uses more fuzzy sets than necessary, since these fuzzy sets represent more or less the same concept. We intend to use the concept of set theoretic ***similarity measure***, as extensively used by Setnes *et al.* (1998a, 1998b), that helps to identify the similar fuzzy sets, and to replace these similar fuzzy sets by a common fuzzy set representative of those original fuzzy sets. If the redundancy in the model is very high, then merging the similar fuzzy sets might result in identical rules that can be subsequently removed from the rule base, leading to a reduction of the number of rules too. Also, the number of dimensions (features) in the model's premise can be reduced in the case of partition similarity.

In the rule base simplification method presented here, initially the set-theoretic similarity between two fuzzy sets is defined, based on which the similarity between the same sets can be numerically calculated. If the calculated similarity measure is larger than a threshold value (say 0.7) predefined by the fuzzy model designer, then the similar fuzzy sets are merged together, resulting in a unique fuzzy set representative of both fuzzy sets. By selecting different values of similarity threshold from the same initial (non-transparent/non-interpretable) rule base, several final (transparent/interpretable) fuzzy models can be generated in which the degree of acceptability of the final model is a trade-off between the three model competitive issues: modelling accuracy, transparency, and compactness.

Setnes *et al.*(1998a, 1998b) have pointed out that several methods have been proposed for optimizing the size of the rule base. However, the fuzzy set-theoretic similarity-based rule base simplification method differs from other fuzzy rule base reduction methods mainly in the way that its main objective is *to reduce the number of fuzzy sets* used in the model and *not the number of rules*. Furthermore, the method can favourably be combined with any data-driven modelling tools, such as fuzzy clustering, or even the neuro-fuzzy approach of Palit and Babuška (2001) and genetic algorithms in order to obtain a tool for transparent, yet reasonably accurate and compact fuzzy modelling (Setnes and Roubos, 2000; Roubos and Setnes, 2001).

In what follows, we will briefly discuss the transparent modelling procedure followed by a general data-driven modelling scheme in which fuzzy set-theoretic similarity-driven simplification is included. The concepts of similarity and redundancy to be described here are illustrated through a similarity-driven rule base simplification method, applied to the example of forecasting a nonlinear time series using a fuzzy model.

7.3 Fuzzy Modelling with Enhanced Transparency

In the fuzzy modelling scheme presented below, our objective is to achieve a good approximation accuracy and model transparency in a data-driven fuzzy modelling approach. In order to make the model transparent and computationally more efficient, an initial fuzzy model is extracted from observation data. In order to remove the unnecessary redundancy in the knowledge learnt from the data, the principle of ***set-theoretic similarity-driven fuzzy rule base simplification*** will be used.

7.3.1 Redundancy in Numerical Data-driven Modelling

In the recent past a variety of numerical data-driven fuzzy modelling tools have been developed for automated building of data-driven models (Roubos *et al.*, 2001; Setnes, 2001). Usually, when building a fuzzy model, the model premise space is partitioned by means of fuzzy sets. However, rule-based models obtained from numerical data can contain unnecessary redundancy in the form of highly overlapping and compatible membership functions. Also, when modelling approaches such as fuzzy clusterings are applied, this redundancy is predominant because the rules defined in the multidimensional premise are overlapping in one or more dimensions. As a result, more membership functions will be required to describe the same concept adequately in the final rule base.

Another common fuzzy modelling approach, such as neuro-fuzzy approach proposed by (Wang and Mendel, 1992), and it's modification by (Palit and Popović, 1999b), and (Palit and Babuška, 2001), is based on parameter adaptation. In this approach, an initial partition of the input space is usually given by randomly generated fuzzy sets or by a number of equidistant symmetrical fuzzy sets defined for all the premise variables of the system. This partition can be seen as a uniform grid in the premise space. Thereafter, the parameters of the membership functions are adapted using the steepest descent method (backpropagation algorithm) (Wang and Mendel, 1992b; Palit and Popović, 1999b) or by it's superior form, such as Levenberg-Marquradt algorithms (Palit and Popović, 1999b), (Palit and Babuška, 2001). An undesired effect of adaptation is that antecedent Gaussian fuzzy sets can move closer to each other and may end up in overlapping positions. Also, some sets may grow to cover the whole space (***universal fuzzy set***), or diminish to non-influential ***singletons***. As illustrated in Figure 7.1, an initially transparent fuzzy model may become unreadable after parameter adaptation.

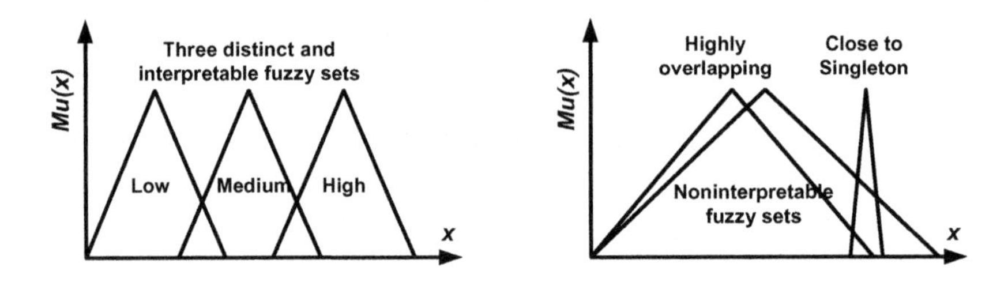

Figure 7.1. Fuzzy sets before adaptation (left) and after adaptation (right)

Undesired redundancy in the form of similarity between fuzzy sets can manifest itself in three different ways:

- Similarity of a particular fuzzy set A with other fuzzy sets in the model.
- Similarity of a fuzzy set A to the universal fuzzy set U: $\mu_A(x) \approx 1, \forall x \in X$.
- Similarity of a fuzzy set A to a singleton fuzzy set such that, $\mu_A(x) = 1,\ if\ x = x_0$; and $\mu_A(x) = 0, \forall x \neq x_0, x \in X$.

As similar fuzzy sets represent compatible concepts in the rule base, a model with many similar fuzzy sets becomes redundant, unnecessarily complex and computationally less efficient. Linguistic interpretation of such a model is also difficult, as it is not trivial to assign qualitatively meaningful labels to highly similar fuzzy sets. As an illustration of the latter, consider the Figure 7.1 (right), where it can be seen that the first two triangular fuzzy sets after adaptation become highly overlapping and they approximately represent the same concept. Consequently, assigning them any meaningful label, such as low or medium, is no longer appropriate. Furthermore, some of the fuzzy sets extracted from numerical data may be similar to the universal set U. Such fuzzy sets are irrelevant because, for all the elements within the universe of discourse, they have degree of membership approximately equal to 1, which fails to categorize the data. The opposite effect is similarity to a singleton fuzzy set (see Figure 7.1 (right)). In this case, a particular data point has degree of membership equal to 1 and for all other data points it gives zero degrees of membership. If a rule has one or more such fuzzy sets in it's premise, then it may never fire, and thus the rule does not contribute to the output model. However, it should be noted that such a rule may represent an exception in the overall model behaviour and, therefore, deserves a special care as it's removal may force one to neglect the exceptionality in the model behaviour.

7.3.2 Compact and Transparent Modelling Scheme

We will now turn our attention to the application problem of similarity-driven simplification to enhance the transparency and compactness of a fuzzy rule base. In order to reduce the redundancy of fuzzy models obtained from data, this simplification can naturally be combined with a data-driven modelling tool, which results in a transparent fuzzy model scheme. This is the approach followed in nonlinear time series modelling for the purpose of forecasting it's future values. As such, a data-driven modelling tool, either of the ***fuzzy clustering*** or the ***neuro-fuzzy*** method, can be considered. However, other methods, such as Wang and Mendel's (1992a) approach, or it's modification by Palit and Popovic (1999a) for rule base generation or fuzzy modelling, can also be considered. Setnes *et al.* (1998a) considered a similarity-driven simplification in combination with fuzzy-neural networks, and Setnes and Roubos (2000), and Roubos and Setnes (2001) considered the genetic-fuzzy approach for second-order nonlinear plant modelling using Wang data (Wang and Yen, 1999), the principal steps of which for a transparent modelling scheme are described below.

Step 1: Model Structure Selection

- The relevant input and output variables that are used for fuzzy model building are determined. Here, the structure selection for dynamic systems means translation of the identification problem into the equivalent regression problem that can be solved in a static manner (Babuška, 1996). Frequently, a reasonable choice of model structure can be made by the user, based on prior knowledge about the process. For the time series forecasting problem considered in this chapter, four input variables and one output variable are considered, so that the input data is a vector of size (1×4) and output is a scalar.

Step 2: Data Clustering

- The fuzzy clustering is usually used to discover the substructures in the product space of the available observations, where each cluster defines a fuzzy region in which the system can be approximated locally by a corresponding submodel. The location and the parameters of the submodels are derived from the clusters of the data. By applying ***cluster validity measures*** (Bezdek and Pal, 1998; Gath and Geva, 1989) such as ***Xie-Beni's index*** (Xie and Beni, 1991) or ***compatible cluster merging*** (Kaymak and Babuška, 1995); (Setnes and Kaymak, 1998) and (Setnes, 1999), an appropriate number of clusters can be found. Alternatively, Yao *et al.* (2000) have proposed an entropy-based simple fuzzy clustering algorithm where the number of clusters is automatically determined by the clustering algorithm itself. In the recent publications of Panchariya *et al.* (2003a, 2003b, 2004a, 2004b) a distance-based simple clustering algorithm has been developed that uses an almost similar idea for the determination of the number of clusters.

Step 3: Initial Fuzzy Model

- For a rule-based fuzzy model derived from the fuzzy partition matrix and the cluster prototypes, the rules themselves, the membership functions, and other model parameters, such as rules consequent parameters, are automatically extracted. The extraction procedure used depends on the type of fuzzy model to be built. In our case, fuzzy models of the type Takagi-Sugeno are considered.

Step 4: Similarity Based Simplification

- In order to upgrade or improve the transparency and the computational issues, the initial fuzzy model is simplified in this step. By selecting an acceptable degree of similarity (redundancy) between the fuzzy sets in the model, it is possible to generate models with varying degrees of complexity for different purposes. Thereafter, depending upon the needs, an appropriate model can be selected for validation.

Step 5: Model Evaluation

- The ultimate version of the fuzzy model built undergoes an evaluation process that is decisive for its final acceptance for the given purpose. In addition to the numerical model validation by simulation, the interpretation of the fuzzy model plays an important role in the process of model validation. This includes the analysis of the input space coverage by the rules. If the rule base generated is found to be incomplete, *i.e.* if no rule is available involving an antecedent fuzzy set, then some additional rule is to be provided to complete the rule base. Such, an interpretation is made easier by the simplification in step 4.

Very often, the number of rules, and hence the number of clusters, are not known *a priori*. From the function approximation point of view creation, of too many clusters does not necessarily pose any problem. However, for the inspection of the resulting model this means higher complexity, less transparency, and possibly wrong conclusions about the characteristics of the system.

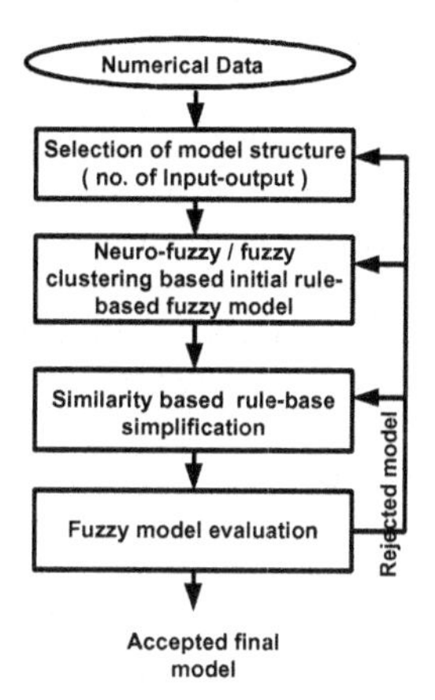

Figure 7.2. Flow chart of transparent fuzzy modelling scheme

Furthermore, in the modelling approach proposed in Figure 7.2, the aggregation of similar fuzzy sets to a certain degree will correct for bias introduced by having too many clusters, making the modelling less sensitive to the determination of the correct number of clusters.

7.4 Similarity Between Fuzzy Sets

The definition of similarity concept between the fuzzy sets depends on their context. The concept of similarity has been defined, in our case, as the degree to which the fuzzy sets are equal. For instance, the fuzzy sets F_1 (slow) and F_2 (fast) in Figure 7.3(a) have exactly the same (triangular) shape, but clearly represent two distinct concepts, because they are representatives of slow and fast speeds respectively.

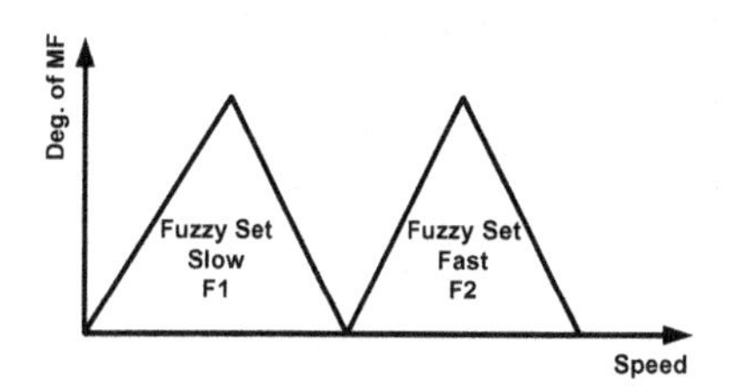

Figure 7.3(a). Dissimilar fuzzy sets

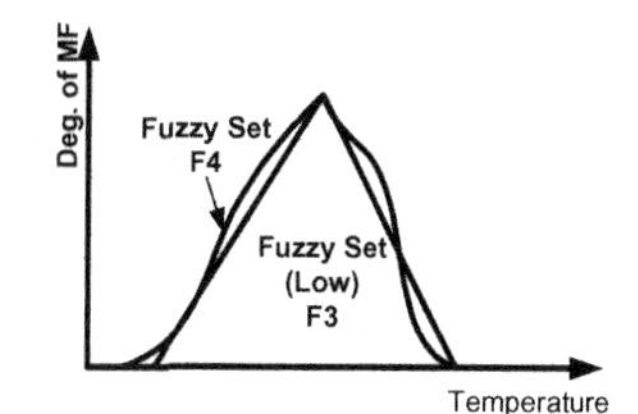

Figure 7.3(b). Similar fuzzy sets

This means that they have a zero degree of equality and are, therefore, considered dissimilar. On the other hand, the two fuzzy sets F_3 and F_4 in Figure 7.3(b), although different in shape, have a high degree of similarity or resemblance. They represent compatible concepts (low temperature) and are considered largely similar.

7.4.1 Similarity Measure

In the method presented here, two fuzzy sets are considered similar if the two overlapping membership functions assign approximately the same values of membership grade to the elements in their universe of discourse. So, the similarity here is the degree to which they can be considered as equal. Equality is a crisp set in the classical definition.

Let us now consider two fuzzy sets F_1 and F_2 with the membership functions $\mu_{F_1}(x)$ and $\mu_{F_2}(x)$ respectively. Then, it holds that the fuzzy sets F_1 and F_2 on X are equal if $\mu_{F_1}(x) = \mu_{F_2}(x)$ and $\forall x \in X$, where X is the universe of discourse. Applying this concept of equality to the fuzzy sets in Figure 7.3, we get that $F_1 \neq F_2$ and $F_3 \neq F_4$, because in both cases their membership functions are different. However, F_3 and F_4 can be said to have high degree of equality, and hence are similar.

As the fuzzy sets allow for gradual transition between full membership and total non-membership, therefore, the ***similarity measure*** S should capture a gradual transition between equality and non-equality

$$s = S(F_1, F_2), \quad s \in [0,1], \tag{7.1}$$

The similarity measure is a function of assigning a similarity value "s" to the pair of fuzzy sets (F_1, F_2) that indicates the degree to which F_1 and F_2 are equal.

7.4.2 Similarity-based Rule Base Simplification

For the purpose of rule base simplification, the fuzzy sets in a rule base that represent a more-or-less compatible concept should be detected by a similarity measure. Therefore, the fuzzy sets, representatives of a compatible concept, should be assigned a high similarity value, whereas more distinct sets should be assigned a lower similarity value. Furthermore, for a correct comparison of similarity values, the similarity measure in any case should be independent of the scaling of the domain on which fuzzy sets are defined. As a consequence, this eliminates the necessity of normalization of the domains.

Now, let F_1 and F_2 be two fuzzy sets on X with the membership functions $\mu_{F_1}(x)$ and $\mu_{F_2}(x)$ respectively. If the four criteria, as listed below, are satisfied by the similarity measure, then it can be used as a suitable candidate for an automated rule base simplification scheme.

1. Two overlapping fuzzy sets should have a similarity value $s > 0$:

$$S(F_1, F_2) > 0 \Leftrightarrow \exists x \in X, \ \mu_{F_1}(x)\mu_{F_2}(x) \neq 0.$$

According to this criterion, two overlapping fuzzy sets F_1 and F_2 should be assigned a non-zero degree of similarity and should not be regarded as a totally non-equal.

2. Only two equal fuzzy sets should have a similarity value $s = 1$:

$$S(F_1, F_2) = 1 \Leftrightarrow \mu_{F_1}(x) = \mu_{F_2}(x), \forall x \in X.$$

This criterion assures that the equality is a special case of similarity, in the same way as the crisp sets can be considered as a special case of fuzzy sets.

3. Non-overlapping fuzzy sets should be totally non-equal, *i.e.* $s = 0$:

$$S(F_1, F_2) = 0 \Leftrightarrow \mu_{F_1}(x)\mu_{F_2}(x) = 0, \forall x \in X.$$

This assures that dissimilar (non-overlapping) fuzzy sets are excluded from the set of similar fuzzy sets. Various degrees of similarity between distinct fuzzy sets are related to the distance between them, and can be quantified by a distance measure.

4. Similarity between two fuzzy sets should not be influenced by scaling or shifting the domain on which they are defined:

$$S(F_1', F_2') = S(F_1, F_2), \quad \mu_{F_1'}(l + kx) = \mu_{F_1}(x),$$
$$\mu_{F_2'}(l + kx) = \mu_{F_2}(x), \ k, l \in \mathbb{R}, k > 0.$$

This criterion is required for a fair comparison of similarities in the rule base, as a similarity measure that satisfies this criterion is not influenced by the numerical values of the domain variables.

Many methods have been proposed to assess the similarity or compatibility of fuzzy concepts. A comparative analysis of different measures using human subjects was reported by Zwick *et al.*(1987) and the mathematical relations between the various measures were studied by Cross (1993). Later, Setnes (1995) investigated the usefulness of various measures for fuzzy modelling.

According to the taxonomy presented by Cross (1993), the compatibility measures can be divided into three broad classes: set-theoretical, logic-based, and distance-based measures. Zwick *et al.* (1987) and Setnes (1995) used the term similarity measures as a general description for methods of comparing fuzzy sets. Unlike in the taxonomy by Cross, the term similarity is not reserved for a subclass of measures, and all measures are divided into two main groups:

- geometric similarity measures

- set-theoretic similarity measures.

Compared with the classification of Cross, the geometric similarity measures are the same as the distance-based compatibility measures, and the set-theoretical similarity measure holds for both the set-theoretic and the logic-based compatibility measures.

The theoretical analysis of similarity has been dominated by the geometric models. These models represent fuzzy sets as points in a metric space and the similarity between the sets is regarded as an inverse of their distance in this metric space.

Denoting now the distance between the fuzzy sets F_1 and F_2 as $D(F_1, F_2)$, the similarity of F_1 and F_2 can be written as

$$S(F_1, F_2) = \frac{1}{1 + D(F_1, F_2)}. \tag{7.2}$$

Examples of geometric similarity measures are the ***generalizations of Hausdorff distance*** to fuzzy sets (Zwick *et al.*, 1987). Another example is similarity transformed from the well-known ***Minkowski class of distance functions***:

$$D_r(F_1, F_2) = \left(\sum_{i=1}^{n} \left| \mu_{F_1}(x_i) - \mu_{F_2}(x_i) \right|^r \right)^{\frac{1}{r}}, \quad r \geq 1$$

The above sum of terms holds when the fuzzy sets F_1 and F_2 are defined on discrete universe of discourse $X = \{x_i | i = 1, 2, \cdots, n\}$, whereas for continuous universes the summation is replaced by integration.

As argued by Zwick *et al.* (1987), geometric similarity measures are best suited for measuring similarity (or dissimilarity) among distinct fuzzy sets, while the set-theoretical measures are most suitable for capturing the similarity among overlapping fuzzy sets. Setnes and Cross (1997) found that geometric measures are quite suitable for ranking of fuzzy numbers. The geometric similarity measures represent similarity as the proximity of fuzzy sets, and not as a measure of equality. The interpretation of similarity as "approximate equality" can be better represented by set-theoretic operations like union and intersection. They also have an advantage over geometrical measures, in that they are not affected by scaling and ordering of the domain (Setnes, 1995). For the similarity-driven simplification, we will use the fuzzy ***Jaccard index***, which is based on the set-theoretical operations of intersection and union, in order to determine the similarity between fuzzy sets.

Considering the two fuzzy sets F_1 and F_2 defined on the discrete domain X by their membership functions, the ***Jaccard index of similarity*** is defined as

$$S(F_1, F_2) = \frac{|F_1 \cap F_2|}{|F_1 \cup F_2|} = \frac{\left|\min\left(\mu_{F_1}(x_i), \mu_{F_2}(x_i)\right)\right|}{\left|\max\left(\mu_{F_1}(x_i), \mu_{F_2}(x_i)\right)\right|}, \tag{7.3}$$

where the cardinality is given by

$$\left|\mu_{F_1}(x_i)\right| = \sum_{i=1}^{n} \mu_{F_1}(x_i).$$

The fuzzy ***Jaccard index*** complies with the four criteria described above, and reflects the idea of gradual transition from equal to completely non-equal fuzzy sets with $S(F_1, F_2) = 0$. This similarity is also used by Chao *et al.* (1996) for training the structure of fuzzy artificial neural networks.

The similarity measure values for Gaussian fuzzy sets with a varying degrees of overlap are shown in Figure 7.4. Figure 7.4(b) shows that degree of similarity between the first Gaussian fuzzy set (GMF) and itself is 1.0, whereas it's degree of similarity with the second fuzzy set from Figure 7.4(a) is only 0.4889, and with the

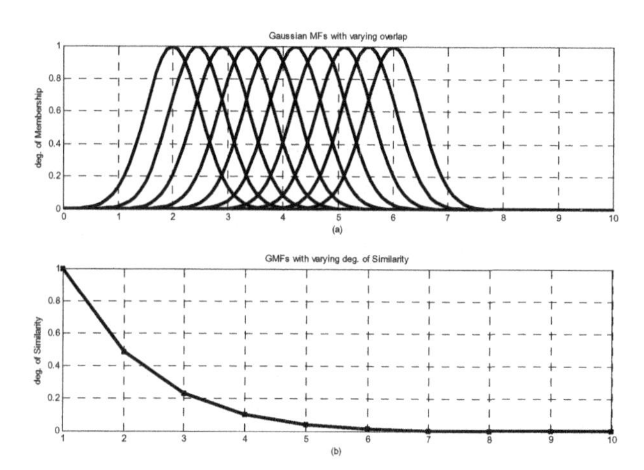

Figure 7.4. Gaussian fuzzy sets and varying degree of similarity

third Gaussian fuzzy set it is 0.2295, and so on. The list of degrees of similarity between the first Gaussian fuzzy set and the fourth set and others are given as follows: 0.1001, 0.0394, 0.0134, 0.0038, 0.0009, 0.0002, 0.0000.

7.5 Simplification of Rule Base

As discussed in Section 7.2, the automated approaches to fuzzy modelling frequently introduce redundancy in terms of several similar fuzzy sets that describe almost the same region in the domain of some model variable. These similarity measures can be used to quantify the similarity between fuzzy sets in the rule base. Two or more similar such fuzzy sets can be merged to create a new set to be stored in the rule base as the representative of the merged sets. In this way, the overall number of fuzzy sets needed to construct the model decreases, which obviously simplifies the rule base. The simplification, however, also results when two or more rules are equal. Here, only one of the equal rules is to be stored in the rule base. Hence, in the approach presented here, there is a difference between ***rule base simplification***, where the primary objective is to simplify the rules by merging similar fuzzy sets that represent almost the similar concept and ***rule base reduction***, which may follow automatically as a result of rule base simplification. Figure 7.5 illustrates the idea of merging similar fuzzy sets, showing both rule base simplification and rule base reduction.

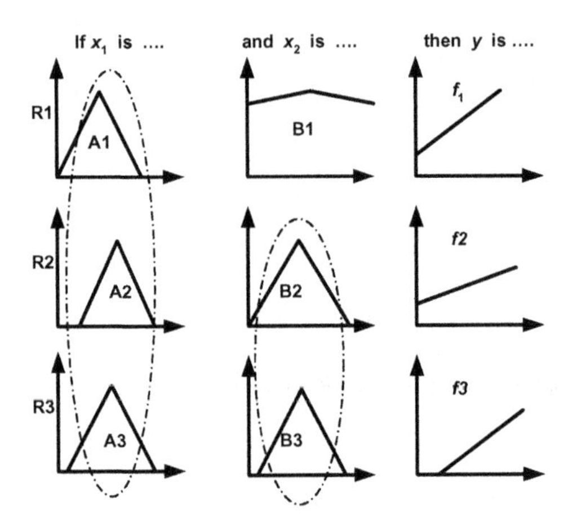

Figure 7.5(a). Similarity-driven rule simplification (A_1, A_2, A_3 are compatible fuzzy sets in Rules 1, 2 and 3; similarly B_2 and B_3 are compatible). Note that fuzzy set B_1 is close to the universal fuzzy set in Rule 1.

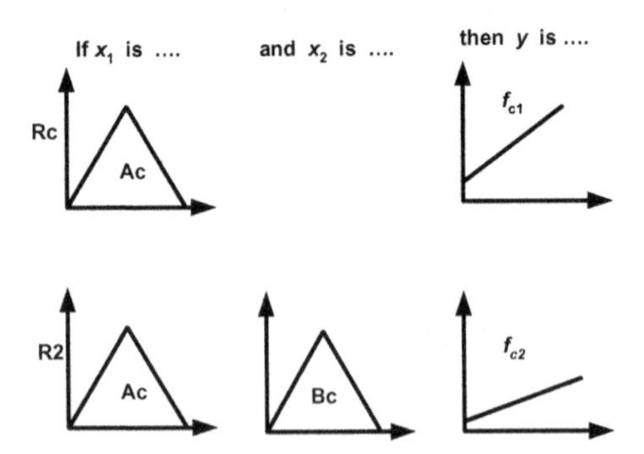

Figure 7.5(b). Similarity-driven rule simplification and rule reduction (after merging compatible fuzzy sets A_1, A_2 and A_3 in Figure 7.5(a) to give common fuzzy set A_c, and similarly merging compatible fuzzy sets B_2 and B_3 in Figure 7.5(a) in Rule 2 and Rule 3 to give common fuzzy set B_c).

In time series modelling or in data-driven identification of dynamic systems, when the similarity of the partitioning as a whole of two or more inputs occurs, another type of redundancy may be encountered. For instance, as illustrated in Figure 7.6, when delayed samples of the same variables are used as input, say $x(k)$ and $x(k\text{-}1)$, they may have a highly similar influence in the model's premise. In this case, the degree of firing of the various rules can be determined by one such input only, which reduces the dimensionality (feature) of the rule base premise.

7.5.1 Merging Similar Fuzzy Sets

In general, when two fuzzy sets are considered to be similar, the rule base can be simplified by

- replacing A by B
- replacing B by A, or
- replacing both A and B by a new fuzzy set C.

When the rule base represents a system model, two important aspects of the simplified rule base are to be considered: the model accuracy and it's coverage of the premise space. Here, owing to the rule base simplification, the uncovered regions should not occur in the premise space. Assuming that the model's accuracy is measured by the sum of squared errors J, the effect of replacing A and B by C should be as small as possible with respect to J. Finding the fuzzy set C best suited to replace A and B becomes a question of evaluating J. Considering the nonlinearity of fuzzy models and the possible interplay between the rule antecedents and the rule consequents, optimizing the fuzzy set C based on J becomes a computationally intensive search problem. In general, if the model is more sensitive to changes in A than to the changes in B, then the fuzzy set A should replace the fuzzy set B, or the common fuzzy set C should resemble A more than B. In particular cases, some additional aspects like model granularity (number of linguistic terms per variable), interpretability or physical relevance may be important.

For a better understanding of merging fuzzy sets, we define a trapezoidal fuzzy set A using parametric membership functions

$$\mu_A(x;a_1,a_2,a_3,a_4);\; a_1 \le a_2 \le a_3 \le a_4\,,$$

$$\mu_A(x;a_1,a_2,a_3,a_4) = \begin{cases} 0, & \text{for } x \le a_1, \text{ or } x \ge a_4 \\ 1, & \text{for } a_2 \le x \le a_3 \\ \mu_A(x) \text{and } \mu_A(x) \in [0,1] & \end{cases} \tag{7.4}$$

One way to merge the fuzzy sets is to take the support of $A \cup B$ as the support of the new fuzzy set C. This guarantees preservation of the coverage of the whole premise space when C replaces A and B in the premise of the rule base. The kernel (cardinality) of C is given by aggregating the parameters describing the kernels of A and B. Thus, merging A and B, defined by $\mu_A(x;a_1,a_2,a_3,a_4)$ and

$\mu_B(x;b_1,b_2,b_3,b_4)$ respectively, gives a fuzzy set C defined by $\mu_C(x;c_1,c_2,c_3,c_4)$, where

$$\begin{aligned} c_1 &= \min(a_1,b_1), & c_2 &= \lambda_1 a_2 + (1-\lambda_1) b_2, \\ c_3 &= \lambda_2 a_3 + (1-\lambda_2) b_3, & c_4 &= \max(a_4,b_4). \end{aligned} \tag{7.5}$$

From the above description one can see that for $a_2 = a_3$ and $b_2 = b_3$, trapezoidal fuzzy sets A and B reduce to two triangular fuzzy sets and for $c_2 = c_3$ fuzzy set C represents the final triangular fuzzy set obtained by merging two triangular fuzzy sets A and B. Following the same discussion, one can also merge two similar Gaussian fuzzy sets G_1 and G_2 represented by the corresponding membership function as

$$\mu_{G_i}(x;c_i,\sigma_i) = \exp\left\{-(x-c_i)^2/\sigma_i^2\right\}, \; i=1,2.$$

Merging of these two fuzzy sets G_1 and G_2 will result in a new fuzzy set G_3 represented also by a Gaussian membership function with mean and variance parameters respectively as

$$c_3 = \lambda_1 c_1 + (1-\lambda_1) c_2, \quad \sigma_3 = \lambda_2 \sigma_1 + (1-\lambda_2) \sigma_2 .$$

The parameters λ_1 and $\lambda_2 \in [0,1]$ determine which of the fuzzy sets G_1, or G_2 has the most influence on the cardinality of G_3. Similarly, in the case of trapezoidal or triangular fuzzy sets the parameters λ_1 and $\lambda_2 \in [0,1]$ determine which of the fuzzy sets A or B has the most influence on the cardinality (kernel) of C.

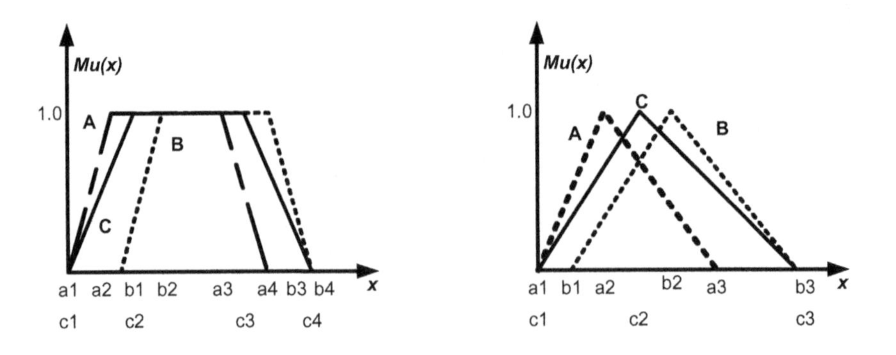

Figure 7.6. Merging of two fuzzy sets, trapezoidal (left) and triangular (right)

In the following, we will suppose that $\lambda_1 = \lambda_2 = 0.5$. This averaging of the kernel gives a trade-off between the contributions of the rules in which the fuzzy sets occur. Figure 7.6 illustrates this method for merging the two fuzzy sets A and B in order to create the fuzzy set C.

7.5.2 Removing Irrelevant Fuzzy Sets

If the rule base contains irrelevant fuzzy sets, *i.e.* if a fuzzy set in the premise of a rule has a membership function $\mu(x) \approx 1, \forall x \in X$, then it is similar to the universal fuzzy set U and can be removed. The similarity of a fuzzy set A to the universal fuzzy set is to be quantified by $S(A,U)$. An example of a fuzzy set quite similar to the universal fuzzy set is illustrated in the Figure 7.7, where the fuzzy set B_1 that is highly similar to a universal fuzzy set can be removed and only A_1 is required in the premise of rule R_1 for distinguishing the associated region in the premise space.

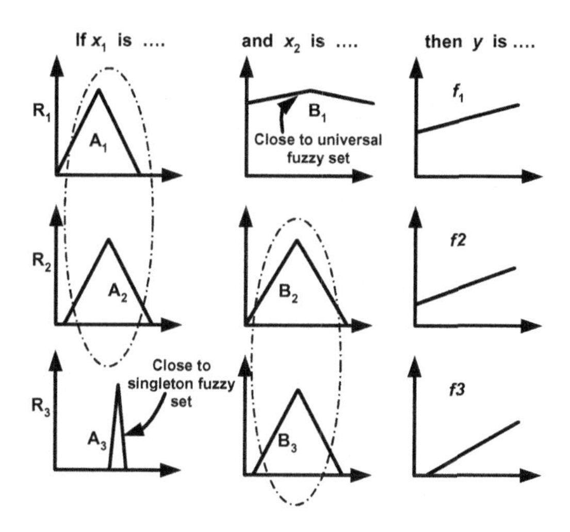

Figure 7.7. Irrelevant set (B_1) in the rule base and non-similar $(A_i \neq B_i)$ domain partition

If the rule base premise consists of all antecedents fuzzy sets similar to the universal set only, then the corresponding rule can be removed from the rule base. The activation of such rules is more or less constant for all inputs within the domain, and the contribution to the output can often be accounted for by re-estimating the consequents of the other rules. The opposite situation may also occur. During parameter adaptation of the fuzzy model, the support of one or more fuzzy sets may become so narrow that they can be almost like fuzzy singletons (see fuzzy set A_3 in rule R_3 of Figure 7.7), prohibiting the rule from firing. Singleton-like fuzzy sets have extremely low similarity to the universal fuzzy set, *i.e.* $S(A, U) = 0$. Rules with such singleton fuzzy sets in their premise are also candidates to be removed from the rule base. In general, one must be careful, as the rules with singleton fuzzy sets may represent exceptions. Interaction from the user is typically needed in such cases to handle such situations. Since our interest is to develop an automated simplification method, these types of rule reduction are not considered here.

7.5.3 Removing Redundant Inputs

Figure 7.7 shows a non-similar partitioning of two input-domains. However, in systems identification and time series modelling, highly similar partitioning of two or more inputs can sometimes occur. An assessment of the similarity S_{pq} between the partitions of a pair of inputs (x_p, x_q) can be obtained by measuring the similarity $S(A_{lp}, A_{lq})$ between all corresponding pairs of fuzzy sets $l = 1, 2, ..., M$, and taking the minimum occurring similarity for each pair of inputs as the partition similarity:

$$S_{pq} = \min_{l} \; S\left(A_{lp}, A_{lq}\right), \; l = 1, 2, ..., M. \tag{7.6}$$

If the partition similarity S_{pq} is above an acceptable threshold value predefined by the user, then one of the two inputs, x_p or x_q, can be removed from the model's premise part. Depending upon the model type and it's performance, *e.g.* in a Takagi-Sugeno fuzzy model, it might still be necessary to keep all variables in the consequent part of the rule base.

7.5.4 Merging Rules

Given a Mamdani-type fuzzy model with k identical rules, if $k \geq 2$, then the rule base simplification will result in the removal of k-1 rules, and thereby reducing rule base. However, if only the premises of the rules (antecedent fuzzy sets) are equal, but not the consequents, then this may indicate a rule conflict situation in the rule base and that has to be solved by assigning a degree to each conflict rule (Wang and Mendel, 1992). In the following, only the fuzzy models of Takagi-Sugeno type are considered.

As in the case of Takagi-Sugeno models, the rule-consequents are not fuzzy; therefore, the similarity concept is applied here only in the premise (antecedents) part of the rules. When the premise parts of $k \geq 2$ Takagi-Sugeno rules are equal, these rules are removed and replaced by one general rule R^g. This general rule has the same premise part as the rules that it replaces. However, the consequent parameters of the general rule are re-estimated taking into account the total influence of all the k-rules in fuzzy inferencing that it replaces. This can be done by weighting R_G with k and letting it's consequent be an average of the consequents of all the k-rules with equal premise parts.

Let Q be a set of indices $l \in 1, 2, \cdots, M$ of the k rules R^l with equal premise parts. These rules are replaced by a single rule R^g with weight k and consequent parameters

$$\theta^g = \frac{1}{k} \sum_{l \in Q}^{k} \theta^l \tag{7.7}$$

where θ^l is a vector of the consequents parameters of rule R^l as described in Chapter 4. The output of the Takagi-Sugeno model can now be calculated as

$$y = \frac{\left(\sum_{l, l \notin Q} \beta^l y^l \right) + \beta^g y^g}{\left(\sum_{l, l \notin Q} \beta^l \right) + k \beta^g} \tag{7.8}$$

For the Takagi-Sugeno model, a substitution of the k-rules with equal common parts by one general rule R^g yields the same input-output mapping. In the above equation, it is assumed that all rules in the initial rule base have a weight $w_l = 1$. A similar expression can be derived for any rule weights.

Another approach is to re-estimate the consequent parameters in the reduced rule base using the training data with the help of the least squares error technique as described in Chapter 4. This requires more computations, but it usually gives a numerically more accurate result than the averaging in the above equation, since it enables the consequents to adapt to the new rule base. However, re-estimation of all rules consequents is the preferred approach using the training samples relying on the least squares error approach.

7.6 Rule Base Simplification Algorithms

Based on the discussions above, an algorithm is now presented for rule base simplification in Takagi-Sugeno models. The same procedure, carried out in three operational steps, can also be used for Mamdani-type fuzzy models.

- **Simplification**, achieved by merging similar fuzzy sets and by removing fuzzy sets similar to the universal set.
- **Dimensionality reduction**, achieved by removing redundant (similar) premise partitions.
- **Rules reduction**, achieved by merging rules whose premise parts have become equal as a result of the two previous steps.

The approach uses the ***Jaccard similarity*** measure (7.3) for determining the similarity between the fuzzy sets in the rule base and requires three threshold values within [0,1], namely the λ for merging fuzzy sets that are mutually similar, γ for removing fuzzy sets similar to the universal fuzzy sets, and η for removing the redundant input partitions. The values of γ and η should be relatively high to ensure that the model's performance will not be deteriorated. As pointed out by Setnes (2000), in many applications the values of $\gamma = 0.8$ and $\eta = 0.8$ have given good results and are used as defaults in the algorithm, but the selection of a suitable threshold λ, which represents the degree to which the user allows for equality between the two fuzzy sets used in the model, depends on the application. The lower the value of λ, the more fuzzy sets are combined, thereby decreasing the term set of the model. In general, one can expect the numerical accuracy of the model to decrease as the λ value decreases.

However, this need not always be the case. If the model is highly redundant or overdetermined, then the numerical accuracy may improve as a result of merging the fuzzy sets and thereby possible reduction in rule base. As a general practice, one may carry out the trial with several values of λ for a particular application with the training samples, and the λ value that gives the best result with the validation data set for a particular application should be finally selected. For instance, in order to explain the operation of a particular system, *e.g.* operator's training or expert's validation, a comprehensible linguistic description is important. In such cases, it is reasonable to trade some accuracy for extra transparency and readability. Consequently, this implies the use of a lower value of λ so that more fuzzy sets can be found to meet this similarity threshold, and which can, in turn, be merged. In contrast to this, an application that aims at prediction or simulation (function approximation) means that one can probably select much higher values of λ, as in this case accuracy is more important. To obtain rules sufficiently distinguishable to describe the system qualitatively, a λ value around 2/3 has been found to give good results in the various experiments of Setnes (2000). Since this part of the simulation requires no additional data acquisition or computationally expensive optimization, the effect of different thresholds can be easily investigated.

The simplification part of the algorithm can be performed in two ways:

- by iterative merging
- using similarity relations.

The main difference lies in the computational effort, and the sensitivity to changes in the threshold λ. Iterative merging requires more computations than similarity relations, but it is more transparent to user interaction. Both approaches are presented below.

7.6.1 Iterative Merging

The algorithm is illustrated in Figure 7.8 and summarized in Algorithm 7.1. The algorithm starts by iteratively merging similar fuzzy sets. In each iteration, the similarities between all pairs of fuzzy sets for each variable are considered, and the pair of fuzzy sets having the highest similarity $S > \lambda$ is merged to create a new fuzzy set. Then, the rule base is updated by substituting this new fuzzy set for the fuzzy sets merged to create it. The algorithm then again evaluates the similarities in the updated rule base. This continues until there are no more fuzzy sets for which $S > \lambda$. Then the fuzzy sets that have similarity $S > \gamma$ to the universal fuzzy set are removed. Thereafter, the rule base premise is checked for redundant inputs. If present, such inputs are removed. The rule base is then checked for rules with equal premise parts. Such rules are merged as discussed in Section 7.5.4. Finally, the rule consequents are re-estimated.

Algorithm 7.1. Algorithm of iterative merging

Given a rule base $R = \{R^l | l = 1, 2, \cdots, M.\}$, with lth rule as R^l: If x_1 is G^l_1 and, ..., and x_n is G^l_n then $y^l = f(x_1, ..., x_n)$, where G^l_i, with inputs $i = 1, 2, ..., n$, are fuzzy sets with membership functions $\mu_{G_i^l} : x_i \rightarrow [0,1]$, select three thresholds $\lambda, \gamma, \eta \in (0,1)$.

Repeat for inputs i =1, 2, ..., n

***Step 1**: Selection of the most similar fuzzy sets*

$$G_i^L = \left\{ G_i^l \middle| S\left(G_i^l, G_i^m\right) = \max_{\substack{p \neq q \\ p,q=1,\ldots,M}} S\left(G_i^p, G_i^q\right) \right\}$$

***Step 2**: Merging of Selected fuzzy sets*

If $S\left(G_i^l, G_i^m\right) > \lambda, \ \left(G_i^l, G_i^m\right) \in G_i^L$:

$G_i^C = Merge(G_i^L), \ \forall G_i^l \in G_i^L$, set $G_i^l = G_i^C$

Until: $S\left(G_i^l, G_i^m\right) < \lambda$.

***Step 3**: Removal of fuzzy sets similar to universal set*
for $i = 1, 2, ..., n$
  for $l = 1, 2, ..., M$

$$S\left(G_i^l, U_i\right) = \left|G_i^l \cap U_i\right| / \left|G_i^l \cup U_i\right|,$$

If $S\left(G_i^l, U_i\right) > \gamma$, remove G^l_i from the antecedent of rule R^l.
  end
end
where $\mu_{U_i}(x_i) = 1, \forall x_i$.

***Step 4**: Removal of redundant inputs*
for $j = 1, 2, ..., n-1$
  for $k = 1, 2, ..., n$

$S_{jk} = \min_l \ S\left(G_j^l, G_k^l\right), l = 1, 2, \cdots, M;$

If $S_{jk} > \eta$, remove x_j from the premise.
  end
end

***Step 5**: Merging of rules with equal premise parts*
for $l = 1, 2, ..., M-1$
  for $m = 1, 2, ..., M$
    if $G_i^l = G_i^m, \forall i$, Merge (R^l, R^m).
  end
end

***Step 6**: Re-estimation of TS rule consequents by LSE method*

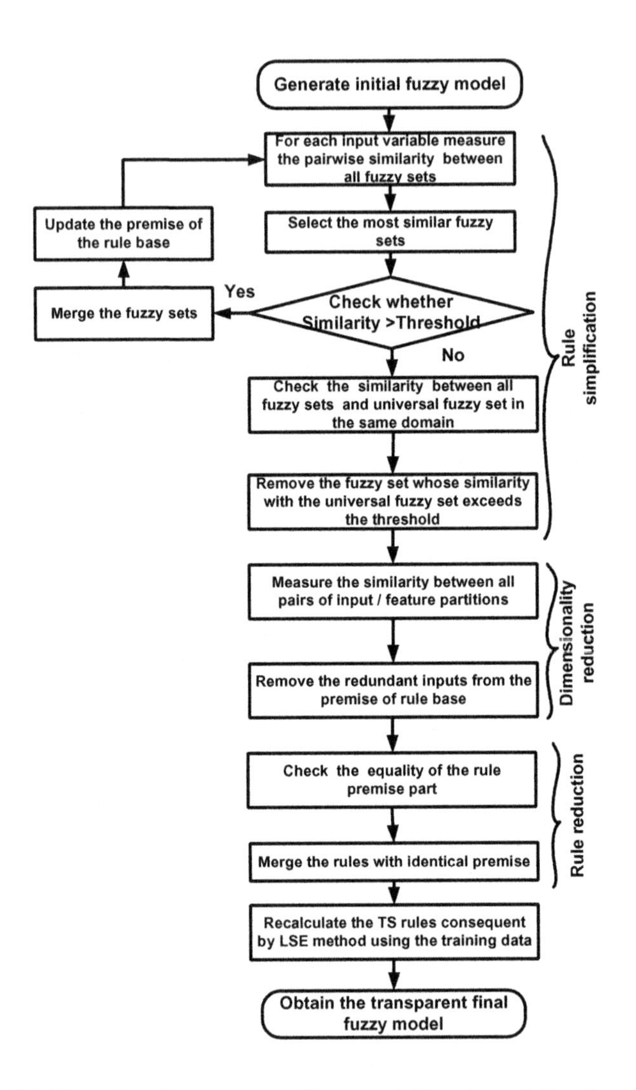

Figure 7.8. Flow chart for transparent fuzzy modelling through iterative merging

7.6.2 Similarity Relations

In this approach all similar fuzzy sets per input are merged in one operation. The fuzzy compatibility relation $C_i = \left[c_{ilm}\right]$ of size $M \times M$ is calculated for each input i = 1, 2, ..., n. The elements of compatibility relation $C_{ilm} = S\left(A_l\left(x_i\right), A_m\left(x_i\right)\right)$ are obtained by the ***Jaccard similarity index*** (7.3). It is to be noted that the ***Jaccard similarity measure*** is not transitive. Thus, it follows that C_i is reflexive and symmetric, but not transitive. In order to obtain a transitive similarity relation S_i, the max-min transitive closure C_{Ti}, of C_i is calculated (Klir and Yuan, 1995):

- $C_i' = \max\left(C_i\left(C_i \;\; T\, C_i\right)\right)$.
- If $C_i' \neq C_i$, set $C_i = C_i'$, and go to previous step.
- Stop: $C_{Ti} = C_i'$, set $S_i = C_{Ti}$.

Here, the *t-norm* used is the *min-operator* and "o*T*" is the *sup-t* composition. The lm^{th} element of the fuzzy similarity relation $S_i = \left[s_{ilm}\right]$, of size $M \times M$, gives the transitive similarity between the concepts represented by the fuzzy sets A_{li} and A_{mi}. The merging of similar fuzzy sets takes place by applying a threshold $\lambda \in (0,1)$ to the similarity relation. Therefore, the similar fuzzy sets are merged, when their similarities are greater than a threshold λ, to produce a fuzzy set representing generalization of the individual concepts represented by the similar fuzzy sets. Thereafter, updated rule base is checked for any fuzzy set which is similar to the universal fuzzy set. The approach is illustrated in Algorithm 7.2 and Example 7.1.

Algorithm 7.2. Algorithm of similarity relations

Given a fuzzy rule base $R = \left\{R^l \middle| l = 1,2,\cdots,M\right\}$, with the lth rule R^l: If x_1 is G^l_1 and, ..., and x_n is G^l_n Then $y^l = f(x_1, ..., x_n)$, where G^l_i, $i = 1, 2,..., n$, are fuzzy sets with membership functions $\mu_{G_i^l} : x_i \rightarrow [0,1]$, select $\lambda, \gamma \in (0,1)$.

Repeat for inputs i =1, 2, ..., n;

Step 1. *Calculate similarity relation:*

$$C_i = \left[c_{ilm}\right];\; l, m = 1, 2,\cdots, M;$$
$$S_i = \left[s_{ilm}\right] = C_{Ti},$$

where the elements of the $M \times M$ *fuzzy compatibility relation* C_i *are given by*
$$\left[c_{ilm}\right] = S\left(G^l(x_i), G^m(x_i)\right).$$

Step 2. *Aggregate similar fuzzy sets*

for l =1,2, ..., M

$$G_i^l = \left\{G_i^m \middle| S_{ilm} > \lambda\right\}, m \in \{1,2,\cdots,M\}$$
$$G'^l_i = Merge\left\{G_i^l\right\},$$

end

Step 3 *to* ***Step-6***. *The steps 3-6 are same as in iterative merging algorithm.*

Example 7.1

Five triangular fuzzy sets $F_1, F_2, \cdots, F_5$ shown in Figure 7.9(a) are used to partition the universe of x. Applying the ***Jaccard similarity index*** (7.3) a compatibility relation C and the corresponding similarity relation S i.e., the ***max-min*** transitive closure of C are given as follows:

$$\mathbf{C} = \begin{bmatrix} 1.0 & 0.09 & 0.06 & 0.05 & 0.0 \\ 0.09 & 1.0 & 0.73 & 0.59 & 0.06 \\ 0.06 & 0.73 & 1.0 & 0.73 & 0.06 \\ 0.05 & 0.73 & 0.59 & 1.0 & 0.09 \\ 0.0 & 0.06 & 0.06 & 0.09 & 1.0 \end{bmatrix}, \text{ and } \; S = \mathbf{C}_T = \begin{bmatrix} 1.0 & 0.09 & 0.09 & 0.09 & 0.09 \\ 0.09 & 1.0 & 0.73 & 0.73 & 0.09 \\ 0.09 & 0.73 & 1.0 & 0.73 & 0.09 \\ 0.09 & 0.73 & 0.73 & 1.0 & 0.09 \\ 0.09 & 0.09 & 0.09 & 0.09 & 1.0 \end{bmatrix}.$$

Applying a threshold $\lambda = 2/3$ to the similarity relation, we identify a set of similar fuzzy sets $F = \{F_l | S_{lm} > \lambda, l \neq m\} = \{F_2, F_3, F_4\}$. The linguistic terms (labels) represented by the three fuzzy sets $\{F_2, F_3, F_4\}$ are merged to create a generalized concept moderate represented by fuzzy set F_c. The resulting fuzzy partition is depicted in Figure 7.9(b).

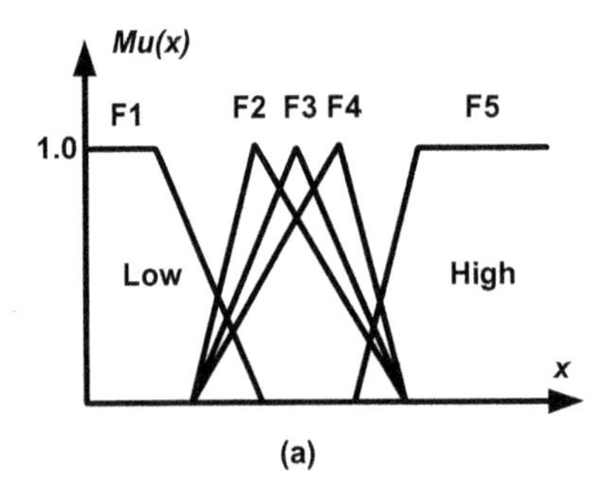

Figure 7.9(a). Fuzzy sets (initial partition)

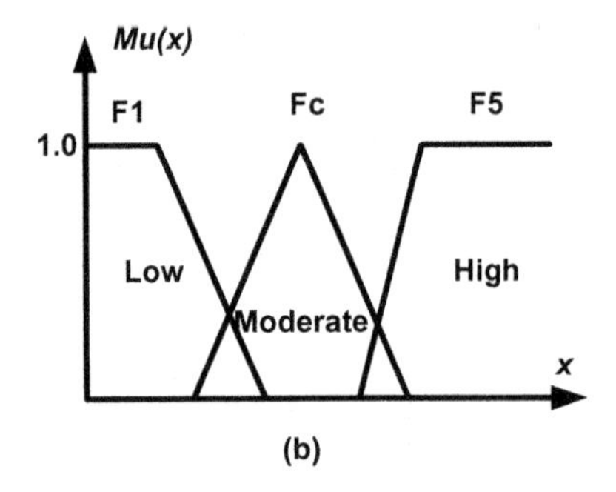

Figure 7.9(b). After merging of fuzzy sets

7.7 Model Competitive Issues: Accuracy versus Complexity

The advantage of transparent representation of the fuzzy model is paid at the cost of reduced numerical accuracy of fuzzy models compared with that of, say, a neural-networks-based model, when both models have approximately the same number of parameters. The reason is that the complexity of fuzzy models grows with the dimension of input and output spaces, which, as shown by Barron (1993), is not the case with neural networks. Therefore, for high-order and for multivariable systems a neural-network-based model might be easier to obtain and may provide a more compact representation than a fuzzy model. However, fuzzy models are less prone to overfitting, and provide better control over the interpolation and extrapolation properties of the mapping obtained.

In order to deal effectively with multivariable complex systems, hybrid approaches should be applied which can use the available prior knowledge about the system, and allow for decomposition of a large problem into a number of simpler subproblems. Furthermore, if different fuzzy models of the same type (say, of Mamdani and Takagi-Sugeno type) are only considered, then the accuracy, transparency, or complexity and compactness of the generated model may also, based on the various factors, vary. In addition, for a set of fuzzy models of the same type (Mamdani) representative of an identical process and even with identical model inputs and output(s) besides their identical domain representation for all input and output variables, the accuracy, transparency and compactness of these models, generated by the same or a different data-driven automated approach, may be totally different. This is particularly because the model accuracy, transparency, and compactness are influenced by many factors like

- Number of antecedent (or consequent) fuzzy sets assigned to each variable.
- Coverage of the antecedent (consequent) fuzzy sets.
- Number of fuzzy rules.
- Fuzzification/defuzzification or inference mechanism

The first factor suggests that the accuracy of the model may generally increase if the input universes (and also output universes for a Mamdani model) are fine partitioned using a large number of membership functions or antecedent (also consequents) fuzzy sets. In fact, it was observed in Chapter 4 that when the input and output universes of discourse are partitioned by 27 Gaussian membership functions instead of an initially chosen 17 Gaussian membership functions, the accuracy of the generated fuzzy chaotic time series forecaster model has significantly increased.

Coverage means that each domain element is assigned at least one fuzzy set with ε (non-zero) membership degree, *i.e.*

$$\forall x \in X, \exists i, \mu_{G_i}(x) > \varepsilon.$$

So, coverage actually insists on there being a certain amount of overlapping between the adjacent fuzzy set, so that entire universe of discourse is well covered by the input/antecedent (output/consequent) fuzzy sets (see Figure 7.10). Optimum selection of this coverage (small) value can result in both an accurate and a transparent model. However, large coverage may result in indistinguishable fuzzy sets, creating a model that is completely non-transparent. It is also observed that the accuracy of the model may generally increase if the number of rules are such that all possible combinations of inputs (antecedents) and output fuzzy sets are covered by at least one rule (for a Mamdani model).

Suppose that for a two-input and one-output system the first input and second input universes are partitioned respectively by antecedent fuzzy sets such as (*low*, *medium* and *high*) and (*slow*, *moderate* and *fast*). In this case at least ($3^2 = 9$) nine fuzzy rules are required to take into account all possible combinations of antecedent fuzzy sets, such as (low, small), (low, moderate), (low, fast), ..., and (high, fast) of the two input variables. However, for a multivariable system with a large number of input and output variables and with a reasonably large number of antecedent fuzzy sets this is not feasible, as it will explode the fuzzy rule base, making the model non-transparent, computationally very expensive, and non-compact.

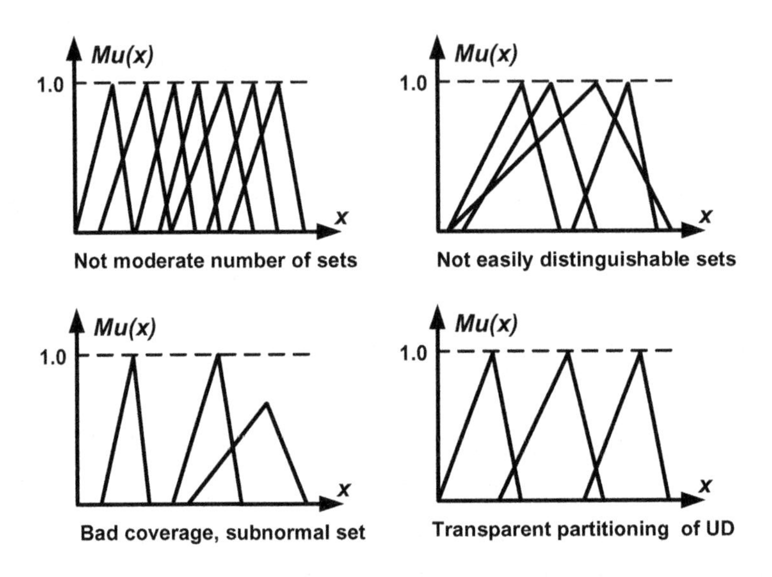

Figure 7.10. Transparent partitioning of domain by distinguishable fuzzy sets

Accuracy of the model, of course, depends on the type of fuzzification (singleton or non-singleton) and of defuzzification (mean of maxima or centre of gravity) method, as well as of the inference mechanism used. For inferencing a Mamdani-type fuzzy model one can select the product/min operator for degree of firing of rule computation with Mamdani's inferencing mechanism. Similarly, for relational matrix computation (which is used in min-max compositional rule of inference), Mamdani implication (min operator), or the alternative Larsen implication (product operator), can be used (see Chapter 4). The different choices of all those possibilities result in different accuracy of the model even though the model type (Mamdani), number of inputs and outputs, and their partitioning fuzzy sets numbers and types of membership function (Gaussian/triangular) may be the same.

However, assuming that for the identical type of model and using identical fuzzification, defuzzification, and inferencing mechanisms we obtain fuzzy model 1, which is the most accurate, model 2, which is the most transparent and model 3, which is the most compact, the question that arises now is which model is to be selected for a particular situation. There is no unique answer to this question, because each model has it's own advantage for a particular application, but is less advantageous for another one. In what follows, a few suggestions are given for selecting the fuzzy model for some applications.

An extremely complex but very accurate model (high level of similarity acceptable) can be useful for off-line simulation (function approximation) or prediction application, because in this case accuracy is more important than model transparency and compactness. On the other hand, in order to explain the operation of a particular system, *i.e.* for operator training, operator interaction, expert validation, and to understand the basic concepts of the system a transparent model with a comprehensible linguistic description (where a little similarity is accepted) is needed. In such cases, it is reasonable to trade some accuracy for extra transparency and better readability of the fuzzy model. Consequently, this actually implies the use of a lower value of similarity threshold so that more fuzzy sets can be found to meet this similarity threshold, which in turn can be merged to result in fewer fuzzy sets. A model with fewer fuzzy sets and fewer rules is also computationally less-expensive. Thereby, computationally less-expensive models are more suitable for applications like model predictive control, memory-expensive implementations, and fast, on-line model adaptation.

7.8 Application Examples

In order to illustrate the similarity-based rule simplification algorithm presented in this chapter, the second-order nonlinear plant model (Wang and Yen, 1999) that was modelled using the neuro-fuzzy approach in Chapter 6, is once again considered here.

Table 7.1. Performance comparison of fuzzy model after neuro-fuzzy network training and similar fuzzy sets merging

Training data	Evaluation data	No. of rules and no. of fuzzy sets
SSE = 0.0090 MSE = 8.972e -05	SSE = 0.0069 MSE = 6.856e -05	Rules = 5 GMFs/input = 5
MSE (after merging) = 0.0093	MSE (after merging) = 0.0147	Rules = 2 GMFs/input = 2

The neuro-fuzzy trained model generated has five Takagi-Sugeno-type fuzzy rules and the antecedent fuzzy sets generated for first input (u) and second input (y) respectively are shown in Figure 7.11(c) and Figure 7.11(d). From Figure 7.11(c) and Figure 7.11(d) it is seen that the antecedent fuzzy sets are not interpretable, as they largely overlap each other. However, the accuracy of this fuzzy model is very high, as the MSE value with the training and validation data are respectively 8.9720e -05 and 6.8560e -05 (see Table 7.1).

In order to improve the model transparency, similar fuzzy sets are merged together and the corresponding final interpretable fuzzy sets are shown in Figure

7.11(e). After merging of similar fuzzy sets, the number of fuzzy sets and rules are reduced to two. Thereafter, the Takagi-Sugeno rule's consequents are recalculated.

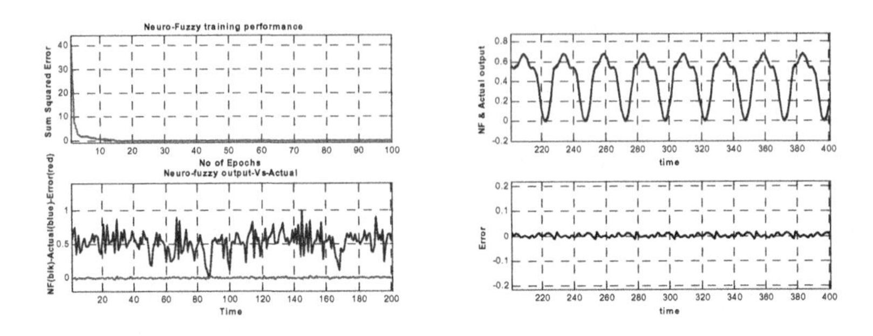

Figure 7.11(a). Neuro-fuzzy network training **Figure 7.11(b).** Neuro-fuzzy prediction

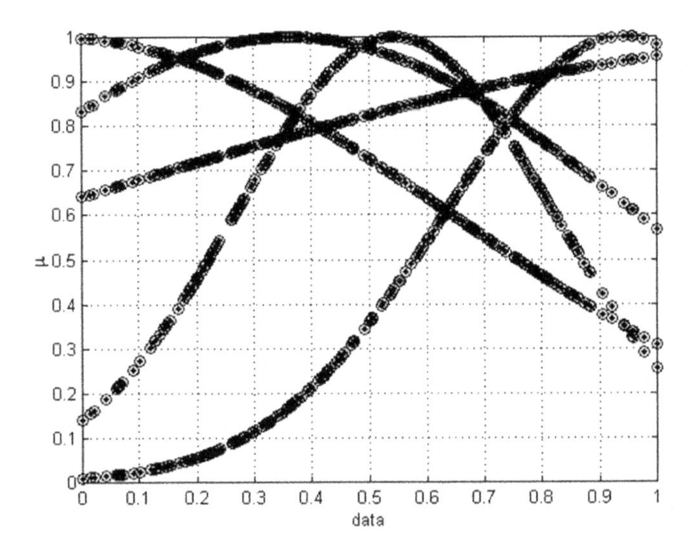

Figure 7.11(c). Fuzzy membership functions for input domain (u) partition after neuro-fuzzy training.

This resulted in the final fuzzy model, which is very much transparent to interpretation. However, the accuracy of the model is greatly hampered as the MSE values achieved with training and validation data sets are now respectively 0.0093 and 0.0147. Therefore, to improve the model accuracy while retaining its transparency the fuzzy sets have to be further tuned using genetic-algorithm-based constrained optimization, as described by Setnes and Roubos (2000), Roubos and Setnes (2001) and Panchariya *et al.* (2004b). By this way one can generate a transparent, yet accurate and compact fuzzy model.

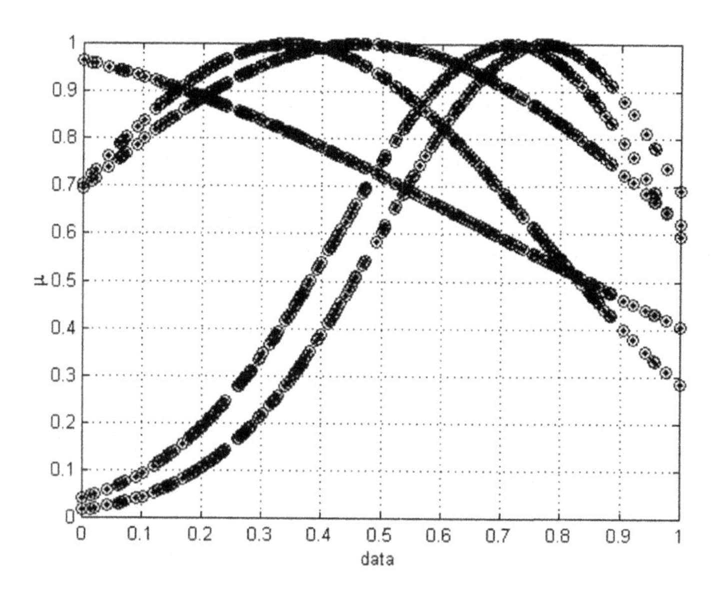

Figure 7.11(d). Fuzzy membership functions for second input (y) domain partition after neuro-fuzzy training.

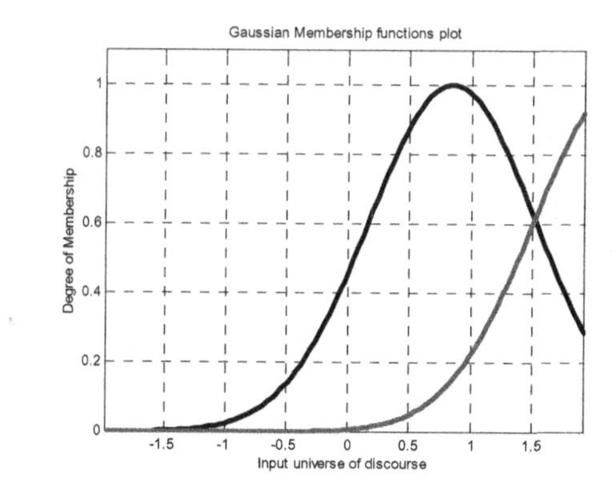

Figure 7.11(e). Fuzzy sets after merging the similar fuzzy sets for first input.

7.9 Concluding Remarks

In this chapter a similarity-driven rule base simplification method is presented. This rule base simplification method serves two practical purposes: increase in model transparency and decrease in computational cost. Furthermore, this method can be combined with any data-driven automated fuzzy modelling procedure together with genetic-algorithm-based fuzzy set tuning procedure to generate a transparent yet accurate and compact fuzzy model. However, the efficiency of the approach depends largely on three threshold parameter values which are currently set by trial and error. Genetic algorithms or evolutionary computations, in general, can possibly also be used here as a proper support tool to determine the optimum values of these three threshold parameters.

References

[1] Babuška R (1996) Fuzzy modelling and identification, Ph.D thesis, Delft University of Technology, Delft, The Netherlands.
[2] Barron AR (1993) Universal approximation bounds for superposition of a sigmoidal function, IEEE Trans. Information Theory, vol. 39: 930-945.
[3] Bezdek JC, Pal NR (1998) Some new indexes of cluster validity, IEEE Trans. on System, Man and Cybernetics, 28(3):301-315
[4] Chao CT, Chen YJ and Teng TT (1996) Simplification of fuzzy-neural systems using similarity analysis, IEEE Trans. on SMC, Part-B: Cybernetics vol. 26: 344-354.
[5] Cross V (1993) An analysis of fuzzy set aggregators and compatibility measures. Ph.D. thesis, Wright State University, Ohio, USA.
[6] Gath I, Geva AB (1989) Unsupervised optimal fuzzy clustering, IEEE Trans. Pattern Analysis and Machine Intelligence. 11(7):773-781
[7] Jain AK, Dubes RC (1988) Algorithms for clustering data, Prentice-Hall, Englewood Cliffs, New Jersey.
[8] Kaymak U, Babuška R (1995) Compatible cluster merging for fuzzy modelling, Proc. of FUZZ-IEEE/IFES, Yokohama, Japan, 897-904
[9] Klir GJ, Yuan B (1995) Fuzzy sets and fuzzy logic, Theory and Applications, Prentice-Hall Inc., Upper Saddle River, New Jersey.
[10] Palit AK and Babuška R (2001) Efficient training algorithm for Takagi-Sugeno type neuro-fuzzy network, proc. of FUZZ-IEEE, Melbourne, Australia, vol. 3: 1367-1371.
[11] Palit AK and Popović D (1999a) Fuzzy logic based automatic rule generation and forecasting of time series, Proc. of FUZZ-IEEE, Seoul, Korea, vol. 1: 360-365
[12] Palit AK and Popović D (1999b) Forecasting chaotic time series using neuro-fuzzy approach, Proc. of IEEE-IJCNN. Washington DC, USA, vol. 3:1538-1543
[13] Panchariya PC, Palit AK, Popovic D and Sharma AL, (2003a) Data driven simple fuzzy rule generation algorithm for fuzzy modelling and identification, Proc. of First Indian Internat. Conf. on AI (IICAI-03), Hyderbad, India, 1088-1097.
[14] Panchariya PC, Palit AK, Popovic D and Sharma AL, (2003b) Simple fuzzy modelling scheme for compact TS fuzzy model using real-coded Genetic algorithm, Proc. of First Indian Internat. Conf. on AI (IICAI-03), Hyderbad, India, 1098-1107.
[15] Panchariya PC, Palit AK, Popovic D and Sharma AL, (2004a) Nonlinear system identification using Takagi-Sugeno type neuro-fuzzy model, Proc. of IEEE Intern. Conf. on Intelligent Systems (IEEE-IS), Varna, Sofia, Bulgaria, vol. 1: 76-81.

[16] Panchariya PC, Palit AK, Sharma AL and Popovic D (2004b) Rule extraction, complexity reduction and evolutionary optimization, International Journal of Knowledge-Based and Intelligent Engineering Systems, vol. 8(4): 189-203.
[17] Roubos JA and Setnes M (2001) Compact and Transparent fuzzy model through iterative complexity reduction, IEEE Trans. on Fuzzy Systems, 9(4):516-524
[18] Setnes (1995) Fuzzy rule-base simplification using similarity measure, M.Sc Thesis, Delft University of Technology, Control Laboratory, Dept. of Elect. Engg.
[19] Setnes M (1999) Supervised fuzzy clustering for rule extraction, Proc. of FUZZ-IEEE, Seoul, Korea, pp. 1270-1274.
[20] Setnes M and Cross V (1997) Compatibility-based ranking of fuzzy numbers, Proc. of NAFIPS, New York, USA, pp. 305-310.
[21] Setnes M and Kaymak U (1998) Extended fuzzy *c*-means with volume prototypes and cluster merging, Proc. of EUFIT, Aachen, Germany, pp. 1360-1364.
[22] Setnes M, (2000) Supervised fuzzy clustering for rule extraction, IEEE Trans. on Fuzzy Systems, 8(5): 509-522
[23] Setnes M, (2001) Complexity reduction in Fuzzy systems, Ph.D. thesis, Delft University of Technology, Delft, The Netherlands.
[24] Setnes M, Babuška R, Kaymak U, Nauta Lemke HR (1998a) Similarity measures in fuzzy rule base simplification, IEEE Trans. on System, Man and Cybernetics, B 28(3):376-386
[25] Setnes M, Babuška R, Verbruggen HB (1998b) Transparent fuzzy modelling, International Journal of Human-Computer Studies, 49(2): 159-179
[26] Setnes M, Babuška R, Verbruggen HB (1998c) Rule based modelling: precision and transparency, IEEE trans. on System, Man and Cybernetics-part C: Applications and Reviews, 28(3): 376-386
[27] Setnes M, Roubos JA (2000) GA-fuzzy modelling and classification: complexity and performance, IEEE Trans. on Fuzzy Systems, 8(5):509-522
[28] Wang L, Yen J, (1999) Extracting fuzzy rules for system modelling using a hybrid of genetic algorithms and Kalman filters. Fuzzy Sets and Systems, 101: 353-362
[29] Wang LX, Mendel JM, (1992a) Generating fuzzy rules by learning from examples, IEEE Trans. on Systems, Man and Cybernetics, 22(6): 1414-1427
[30] Wang LX, Mendel JM, (1992b) Back-propagation fuzzy system as nonlinear dynamic system identifiers, Proc. of FUZZ-IEEE, vol. 2: 1409-1418
[31] Xie XL, Beni GA, (1991) Validity measure for fuzzy clustering. IEEE Trans. on Pattern Analysis and Machine Intelligence. 3(8):841-846
[32] Yao J, Dash M, Tan ST and Liu H (2000) Entropy-based fuzzy clustering and fuzzy modeling, Fuzzy Sets and Systems, vol. 113: 381-388.
[33] Zwick R, Carlstein E and Budescu DV (1987) Measures of similarity among fuzzy concepts: A comparative analysis. International Journal of Approximate Reasoning. vol. 1: 221-242.

8

Evolving Neural and Fuzzy Systems

8.1 Introduction

One of the main application fields of evolutionary computation, especially of genetic algorithms and evolutionary programming, has for a long time been the design or evolving of intelligent computational structures, such as neural networks, fuzzy logic systems, neuro-fuzzy systems and of their combination to implement intelligent controllers. In the following, evolving of neural networks and fuzzy logic systems using evolutionary algorithms will be presented.

8.1.1 Evolving Neural Networks

In evolving of neural networks for specific applications, the user is faced with the following two key issues:

- what network architecture (*i.e.* how many hidden layers, number of neurons in each layer, and what interconnections between them) should be selected as the most adequate
- what specific weight values should the interconnecting elements have for optimal network performance.

No standard guidelines are available for resolving the above selection problems, at best only some recommendations and some hints could be found in some publications. In this chapter we will take a closer look at these selection difficulties and we will describe some approaches that have been used successfully in evolving of optimally design neural networks.

In the past, most very frequently a trial-and-error approach has been used in developing the neural network structures, which have afterwards been optimized by simulation or by some optimization methodologies. For the process of network development, two basic approaches have been used.

- ***Constructive approach***, an approach that starts with a minimal network architecture and continues by its stepwise growth through adding new neurons and new interconnection links between the neurons, under permanent evaluation of network performance, until the optimal network structure has been achieved.
- ***Destructive approach***, which starts with a "large enough" architecture and continues by its stepwise reduction of size through removal of some individual neurons and the related links between them, under continuous evaluation of network performance, until the optimal network structure has been achieved.

Both approaches, however consider, through incremental changes of network structure, only a limited (neighbouring) topological space, instead of considering the entire search space of possible network structures. This deficiency definitely restricts the overall possible optimal network structure that could be developed.

In the last decade or so, a way out of network development, by arbitrarily adding and deleting of neurons and connecting weights, has been found in using some more systematic evolutionary approaches. During this period of time, researchers have succeeded in elaborating evolutionary methods capable of covering most of the basic requirements in developing, training, and application of neural networks. Using the new methods the following network evolving issues have been supported:

- evolving optimal interconnection weights
- evolving global network architecture
- evolving pure network architecture
- evolving activation function
- evolutionary network training.

This is the main subject of the paragraphs that follow.

8.1.1.1 Evolving Connection Weights

Traditionally, optimal values of interconnection weights have from the very beginning been determined through network training, usually by using a gradient-based parameter-tuning algorithm, like the backpropagation algorithm. Yet, the substantial risk of all gradient-based algorithms to be trapped in a local minimum was a good enough reason to avoid their use in optimization problems and to look for gradient-free search algorithms.

Jurick (1988) suggested that the network training process to be understood - within the frame of the given network architecture and the objectives of learning task - as an evolutionary process through which the optimal values of connection weights can be determined. Montana and Davis (1989) decided to take the genetic algorithms, instead of backpropagation algorithm, in searching the optimal weights values. Using the new search strategy, they were able to find the global optimal values of connection weights, without gradient implications. The results achieved have been confirmed by Kitano (1990), who also accelerated the network training convergence using an improved version of the genetic approach.

In 1990s, the awareness was spread out among the experts that the evolutionary algorithms could, in the future, become the most efficient tools for neural networks training, so that since that time the evolutionary approaches have been very successfully used in training of backpropagation neural networks (Johnson and Frenzel, 1992; Porto et al., 1995; Schwefel, 1995), and later in training of recurrent neural networks (Angeline *et al.*, 1994; McDonnell and Waagen, 1994).

Surprisingly, although the evolutionary approaches, while based on extensive computations are generally slower than the gradient methods, it was reported by some investigators that in network training the evolutionary algorithms have been considerably faster than the gradient methods (Prados, 1992; Porto *et al.*, 1995; Sexton *et al.*, 1998).

McInerney and Dhawan (1993) pursued an alternative way of network training by combining two different search algorithms for network training, namely the backpropagation and the genetic algorithms. They in this way created two alternative ***hybrid training algorithm***s:

- algorithms that use genetic programming to bring the search process close to the global optimum and then the backpropagation algorithm has to locate it more exactly
- an algorithm that first finds (based on backpropagation search) "all" local minima and then leaves the task for the genetic algorithm to find the smallest one as the global minimum.

In both algorithms the backpropagation algorithm is used because it is relatively fast, but it suffers with the inherent troubles associated with gradient methods being prematurely trapped in local minima. Genetic algorithms, although being relatively slow, are used because they are robust in finding the global optimum. Their combination, as expected, profits from the advantage of one algorithm and from the possibility of counterbalancing the disadvantages of the other. In addition, in the evolutionary algorithms, unlike in the gradient-based training algorithms, the error function, *i.e.* the fitness function, does not require any differentiation and even need not be continuous.

The joint application of genetic algorithms and gradient methods has been the subject of extensive research in the 1990s (Kinnebrock, 1994; Zhang *et al.*, 1995; Yang *et al.*, 1996; Yan *et al.*, 1997).

Nevertheless, in practical applications of genetic algorithms the encoding of weight values in chromosomes has proven to be the most crucial problem (Balakrishnan and Honavar, 1995; Curran and O'Riordan, 2003). However, further research in this area has borne a great number of possible solutions that can be classified into two categories:

- ***direct encoding approaches***, in which all parameters that define the neural network (*i.e.* weight values, number of nodes, connectivities, *etc.*) or some of them are encoded in gene code
- ***indirect encoding approaches***, which represent a neural network in terms of assembly instructions or of recipes.

Direct encoding approaches facilitate the reverse operation of decoding that consists of back-transformation of ***genotypes*** into ***phenotypes.*** The best illustration for direct encoding represents the ***connection matrix*** that exactly specifies the architecture of the network to be evolved. For direct encoding the following approaches have been recommended:

- *connectionist encoding*
- *node-based encoding*
- *layer-based encoding*
- *S-expressions based encoding.*

Indirect encoding needs much more work in styling the phenotypes adequately, because here, in encoding of phenotypes, the ***rewrite rules*** and ***construction rules*** are also applied recursively. For indirect encoding the following approaches are recommended:

- *matrix re-writing*
- *edge encoding*
- *cellular encoding*
- *growth encoding*

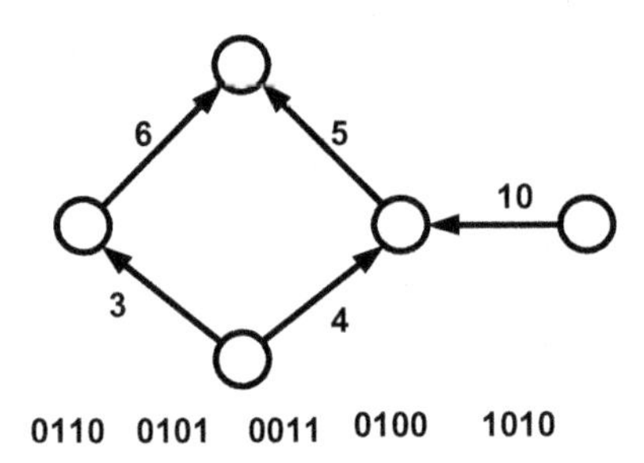

Figure 8.1. Binary representation of parameter values

Figure 8.1 shows the transparency and the simplicity of binary representation of a neural network, whose architecture is given and whose connection weights are represented as a 4-bits binary chain. Binary representation enables a direct acting of crossover and mutation operators on the coding structure. But still, the serious drawback of binary coding is that the total length of the ***concatenated strings*** grows steadily with the number of interconnections to be considered. This, increasingly slows down the computational speed of the genetic algorithm. The total length of concatenated strings grows even more if the higher computational accuracy is required, because in this case more bits need to be represented in binary. This can be mastered by using the real numbers for connection weight representation, so that each individual in the evolving population becomes a real vector. However, new circumstances are faced here, since it is difficult to use directly the binary-encoded crossover and mutation operators. A better way to evolve the real vectors is to use the evolutionary strategies or evolutionary programming (Fogel *et al.*, 1990; Yao, 1993) rather than genetic algorithms.

8.1.1.2 Evolving the Network Architecture

The first action in evolving the network architecture is to lay down the network topological structure, *i.e.* the proper number of nodes, the interconnection pattern of the nodes, the activation function to be assigned to each node, *etc*. This activity, if properly carried out, is very promising in leading, through the process of evolution, to a final network architecture, optimally shaped for the given problem to be solved.

The adequacy of the selected topology generally depends on the network task. For example, if the network to be evolved is to be used for identification of nonlinear interdependencies between the collected data of a time series and to process them, then it must be a ***multilayer network*** because a single-layer network is not capable of doing this. Similarly, if the network has to be able to discover and to handle the temporal dependencies in the environment, then it must be a ***recurrent network*** because the feed-forward networks are not capable of doing this.

A further important decision to be made when evolving neural networks is to select the appropriate initial network topology size. For example, if the selected network topology size is too small, then the evolved network might fail to learn the desired input-output mapping. In contrast, if it is too large, then the generalization capability of the network will be very poor (Sietsma and Dow, 1991).

All this indicates that, for adequate selection of initial network topology, much expert knowledge and practical experience is needed, because here also we are short of a well-paved way for systematic topology selection. Therefore, for the less experienced network developer, the only way left is to select different initial network topologies and, using a trial-and-error strategy, to find the most appropriate one.

The next critical issue of an evolving neural network architecture is the decision to be made about the encoding strategy to be used. Encoding strategies help in transforming the network structure into specific representations, called ***genotypes***, on which the evolutionary operators (mainly mutation and recombination) act during the process of network evolution. Both the selected genotypes and the evolutionary operators to be used belong to the crucial issues to be resolved before the evolving process is initiated. This is needed because the application success of a neural network in solving the problem for which it is evolved depends predominantly on the selected ***genotypic representation*** and on the evolutionary operations.

For genotypic representation, two alternative encoding strategies are available:

- ***direct encoding strategies***, in which all architectural aspects of the network are encoded by direct transformation of genotypes to phenotypes, for instance through building a ***connection matrix***
- ***indirect encoding strategies***, in which grammatical or morphological encoding is used, based on a compressed description of the network to be evolved.

For evolving the network architecture, there are also two alternative approaches available:

- evolving the ***pure network architecture***, without interconnecting weights, which presumes that weight values are to be determined through network training
- simultaneous evolution of both architecture and weights.

8.1.1.3 Evolving the Pure Network Architecture

Evolving a genuine network architecture requires a decision about the degree to what extent the genotypes (*i.e.* the chromosomes) should bear the detailed information related to the targeted network architecture. This depends on the representation scheme to be used. Should it be a direct representation that includes all the details of every node, or should it be an indirect representation in which only some dominant nodes are represented by some details like the number of hidden layers and the number of neurons in the layers?

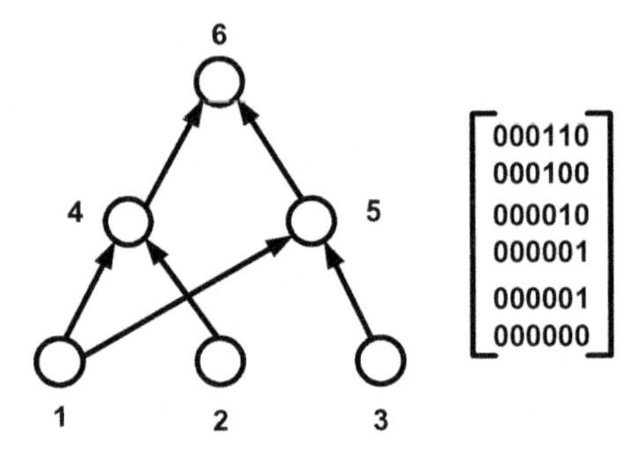

Figure 8.2. Example of low level encoding

If the genotypes should not contain any statements about the connecting weights, then a random set of initial weight values can be taken. In this case the risk exists that the weight values finally determined could be noise spoiled. This is because the fitness values of genotypes will be represented by fitness values of phenotypes, which could be due to the randomization of initial values of training runs (Angeline *et al.*, 1994). To reduce this noise the use of one-to-one mapping between the genotypes and phenotypes is recommended (McDonnel *et al.*, 1994).

Otherwise, when using the direct encoding scheme in evolving the pure network architecture, each network connection is represented by a binary string of a specified length. Once accepted for representation, the strings should be concatenated to build corresponding ***chromosomes***. The set of chromosomes belonging to the same network can then form the ***connectivity matrix*** that itself depicts the network architecture in terms of network interconnection pattern. This is shown in Figure 8.2. The matrix, again, could also be interpreted in the inverse direction in the sense that, given the desired structure of the network to be evolved, the corresponding connectivity matrix can be generated and translated into the corresponding binary string.

A direct encoding strategy, although transparent and easy to implement, still suffers from the ***scalability problem***, which hampers its application in evolving complex network configurations, because in this case a large connectivity matrix and much computing time for network evolution are required. In addition, the potential difficulty of direct encoding strategy is the ***permutation problem*** that disturbs the evolution of proper network architecture.

Indirect encoding strategies are especially popular because they help in reducing the length of genotypic representation of architectures; this is achieved, however, at the cost of a reduced feasible search space. In this kind of strategy, only some characteristics of the architecture (*i.e.* those to be evolved) are binary encoded, which enables a more compact, modular overall network description.

Because being based on a restricted initial information, the indirect decoding strategies obviously pursue the principle of a growing network, termed ***grammatical encoding***. Their major advantage is that they favour the modular design of network structure. Much of pioneer work in this area was done by Kitano (1990), particularly in defining the ***matrix rewriting*** encoding strategy. This strategy, however, was soon abandoned for the reason that it failed to deliver better results than the direct encoding strategies.

8.1.1.4 Evolving Complete Network

We now come to the most challenging design task in which the network topology along with the interconnection weights are simultaneously evolved. The advantages of such a design approach are, however, accompanied by the difficulties in finding an adequate representation of genotypes. In the past, apart from the direct binary encoding that is also applicable here, two additional encoding strategies have been in use:

- ***parameterized encoding***, in which (instead of a connectivity matrix) the compact network description is stored in terms of number of layers, number of neurons within the layers, number of connections between the layers, *etc*.
- ***grammar encoding*** (Vonk *et al.*, 1995), particularly ***matrix grammar encoding*** (Kitano, 1990).

In a parameterized encoding network the parameters can be freely encoded. Some recommendations on this issue have been elaborated by Harp *et al.* (1990).

Grammar encoding roots in the research achievements of Lindenmayer (1968) in the area of encoding strategies. Using the biological principle of information exchange between the cells, Lindenmayer has introduced the so-called ***L-systems***. To implement this, he defined a special grammar with the parallel representation of production rules that Boers and Kuiper (1992) later used to evolve neural networks. The benefits of grammatical encoding are the identification possibility of network building blocks and the general reusability of development rules. Kitano (1990) used the productions as the ***grammar rewriting rules*** to develop his matrix rewriting encoding strategy.

Gruau (1994) represented neural networks as grammar trees, called ***cellular encoding***, which is similar to the ***edge encoding*** strategy.

In the ***graph grammar encoding***, the network is understood as a lattice made up of functions and terminals. Each node of the lattice, which is seen as a function (neurons) or a terminal (input variables), is provided with the information concerning the connections to other nodes, the weights of the connections, bias, *etc.*

An entirely different indirect encoding strategy was proposed for encoding the ***developmental rules*** that are to be optimized instead of direct optimization of the network architecture. The development rules are similar to the IF-THEN rules used in production systems, written in recursive form.

Some interesting findings in evolving the ***learning rules*** have been reported. Chalmers (1990) was the first to report on automatic evolving of the ***delta learning rule***, and Harp and Samad (1991) reported on evolving the rules that can learn and adapt the network training parameters, such as training speed and network training accuracy.

However, the inherent problem of encoding neural networks in gene code is still the well-known ***permutation problem***, created by the fact that different genotypes can produce equivalent networks, because the fitness and the network function could produce the permutation of hidden nodes. This is evident from Figure 8.3, which represents the differently encoded network shown in Figure 8.1. Both networks are topologically equivalent (Tettamanzi and Tomassini, 2001).

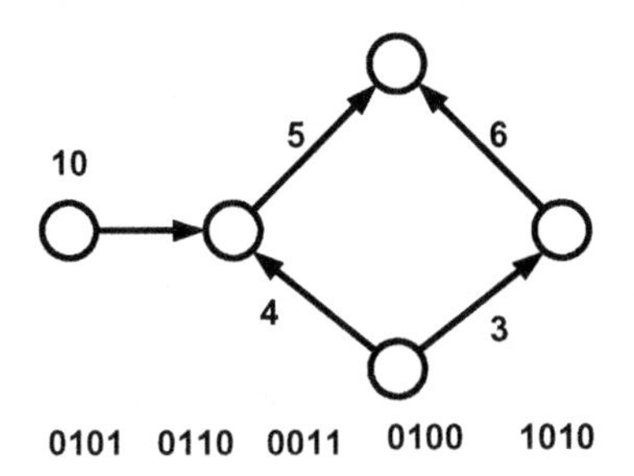

Figure 8.3. Differently encoded network presented in Figure 8.1

The permutation problem considerably decreases the suitability of the genetic algorithm as a training tool for feedforward networks.

8.1.1.5 Evolving the Activation Function

So far, we have ignored the evolving issue of the neuron activation function, assuming silently that it is given in advance by the network expert, preferably as a sigmoidal function. This is indeed not always the case, but it is assumed for simplicity that the activation functions of all neurons in a layer or in the entire network are equally shaped. The first trial to evolve both the activation functions, placed in nodes as a node transfer function, and the network architecture was done by Stork *et al.* (1990). A series of improvements followed, the most interesting of them being the achievement of Hwang *et al.* (1997) in evolving the network topology, connection weights, and the node transfer function simultaneously.

8.1.1.6 Application Examples

Evolved neural networks have found a wide application in time series forecasting, because the network evolutionary process has contributed the network structure and the network parameters (connecting weights, activation functions, hidden nodes, *etc.*) that are optimal. The only difficulty that accompanies the application of such networks is the selection of the optimal initial population that will guarantee the shortest search time. To reduce it's influence on the problem at hand, Prudencio and Ludermir (2001) have advocated using the ***case initialized genetic algorithm*** (Louis and Johnson, 1999), based on experience in optimizing the solution of some similar problems. The solution concept was applied to the problem of river flow prediction, where the time series models NARX (Nonlinear Auto-Regressive model with eXogenous variables and the NARMAX (Nonlinear Auto-Regressive Moving Average model with eXogenous variables) have been employed. In the models, the following parameters have been optimized: length of time window, length of context layer, and the number of hidden layers. The network was trained with the Levenberg-Marquardt method (Marquardt, 1963). The objective of the case study was to forecast the monthly river flow of a hydrographic reservoir, based on 144 available flow values acquired within a period of 12 years. During the experiments, about 20 neural network architectures were developed in order to find the best one. The software system developed, although tailored for forecasting purposes, is suitable for application in other problem classes.

8.1.2 Evolving Fuzzy Logic Systems

In evolving fuzzy logic systems, two principal decisions should be made:

- selection of the fuzzy rule base that could be considered as the most promising one to solve optimally the given problem and the selection of strategy for their genetic encoding
- definition of membership function parameters.

Optimal definition of membership functions to be used in the process of systems evolution is also a crucial problem here that, to be well-solved, needs much skill and computational efforts. This is because the performance of the system to be developed is very sensitive to the shapes of the membership functions. The early proposals on how to manage these problems (Shao, 1988) did not bring a significant success in performance improvement, until it was recognized that for solving this problem the optimal parameter tuning of membership function shape should be used, for instance by being carried out using an evolutionary algorithm. Tettamanzi (1995) has proven that the integration of evolutionary algorithms and fuzzy logic could cover the following application fields:

- optimum search capabilities of evolutionary algorithms can help design and optimally tune the parameters of fuzzy logic systems
- during the evolutionary processes the rule base of the fuzzy system could be used to automatically tune the algorithm parameters in order to avoid its premature convergence and other undesired behaviour of the search process
- the fuzziness can be embedded into the algorithms for internal calculations of fitness function, *etc*.

The work on the design of fuzzy logic systems using evolutionary computation was effectively initiated in early 1990s and was made public by the reports of Thrift (1991) and Karr (1991) on the use of genetic algorithms in synthesis of fuzzy logic controllers. This was later extended to the synthesis of a model-reference adaptive controller (Hwang and Thompson, 1994). In the early considerations of the evolving procedures, triangular membership functions were preferred because their encoding within the chromosome as finite-length bit strings was relatively simple. This kind of membership functions is parameterized by the left and the right base and by the distance from the previous centre point. For evolving purposes, the same triangular form for all membership functions and the same number of membership functions for each variable were taken.

Some researchers (Hwang and Thompson, 1994) encoded all the rules and the fixed membership functions into the chromosome. Under this condition, the evolving process, however, did not evolve an optimal fuzzy system, because the shape of the membership functions is strongly related to the character of the rules. As a consequence, both the rules and the membership functions have to be evolved simultaneously. Homaifar and McCormick (1995) solved the problem of simultaneous tuning of the membership functions and evolving the rule set by encoding all the rules and the base length of each triangular membership function into chromosomes.

Thrift (1991) pleaded for building the fuzzy rule base in tabular form by assigning to each input variable a number of partition domains, say n, that are to be specified in detail. This, however, was not applicable, because in this way a huge number of detailed data are generated that cannot be stored in a transparent form. The idea of Thrift, of representing the generated data in matrix form, was acceptable only for small fuzzy systems, say for systems with two input variables for which an $n \times n$ matrix is to be built. However, for a system with a higher number of input variables an $n \times n \times ... \times n$ dimensional matrix has to be built.

To avoid the super-dimensionality problem, Lee and Takagi (1993a) recommended numerating the rules instead of tabulating them. They also encoded the membership functions and the rule set into the chromosomes, but they took another route to encoding the triangular membership function by restricting the adjacent membership functions from fully overlapping and by some additional restrictions. This considerably reduced the total number of membership functions required. Further reduction is still possible by grouping the given rules into relevant (needed) and non-relevant ones, and by encoding only the relevant rules. This enables fuzzy systems of higher dimensionality to be evolved.

While considering the Takagi-Sugeno model, in which the consequent part is made up of a linear combination of the input values, Lee and Takagi (1993b) found it more advantageous to encode both the membership and the fitness functions in chromosomes. To each rule with N input variables and n membership functions in the genotype they assigned a gene to encode the N+1 weights in a linear combination of the input variables for the rule concerned. The drawback of the encoding approach is that n^N combinations have to be encoded.

Tettamanzi (1995) implemented his fuzzy control evolving system on a WARP fuzzy processor capable of supporting up to 256 rules with up to four antecedent clauses and one consequent clause, as well as antecedent membership functions of arbitrary shapes. To define the appropriate fitness function he used the ***concept of competition***, defined later (Tettamanzi, 1994). The concept registers the number of competitions c undergone by an individual, the number of its wins w, and the number of successes s. Using this statistical data the membership function of fitness for a given individual is defined as

$$\mu_{\mathrm{f}}(x) = N(a,b)x^{n}(1-x)^{b}$$

with

$$N(a,b) = \frac{(a+b)^{(a+b)}}{a^{a}b^{b}}$$

as a normalization factor, in which $a = w + s$ and $b = c - s$.

Recently, a new evolutionary road to fuzzy systems design was paved by Shi *et al.* (1999), who, along with the membership function shapes and the fuzzy rule set, also encoded the membership function type and the number of rules inside the set. Two types of membership function have been considered: linear and nonlinear (Gaussian, triangle and their combination). Each membership function was completely defined by its start point, its end point, and the function type.

In order to make the evolving process easier, the fitness function, which measures the performance of the system, was carefully defined. Depending on the application, the fitness functions taken are

$$E_{\mathrm{f}} = \frac{1}{N}\sum_{i=1}^{N}(o_i - t_i)^2$$

and

$$E_{\mathrm{f}} = \frac{1}{N}\sum_{i=1}^{N}\left(\frac{o_i - t_i}{t_i}\right)^2$$

o_i and t_i being the ith obtained and target outputs respectively.

For control of crossover and mutation as the most critical parameters, an adaptive tuning approach, made up of eight fuzzy rules, was integrated into the genetic algorithm. The completed evolutionary fuzzy system was written in C++ code and was compiled with the Borland C++4.5 compiler. The benefits of the developed system have been demonstrated on examples of iris data classification, but the system could be successful for a large range of similar problems.

Several researchers have focused their attention on integration of fuzzy logic and evolutionary approaches in optimal tuning of the parameters of a fuzzy logic controller by adapting the fuzzy membership functions by learning the IF-THEN rules (Varsek *et al.* 1993; Mohammadian and Stonier, 1994; Herrera *et al.* 1995). For instance, Zeng and He (1994) evolved a fuzzy controller with a self-learning feature for approaching the optimality conditions of the given control task. Thereafter, the integrated genetic algorithm took over the initiative to tune the controller parameters optimally. The modified fuzzy controller was successfully applied to control an unstable nonlinear system that demonstrated high accuracy and robustness of the evolved fuzzy controller.

Wong and Chen (2000) elaborated a genetic-algorithm-based approach to fuzzy systems construction directly from collected input-output data. The basic idea of the approach is that each individual in the population determines the number of fuzzy rules and that the consequent part of the evolved fuzzy system is determined by a recursive least-squares method. The effectiveness of the approach was demonstrated on construction of some nonlinear systems.

In the recent past, some reports have been published on evolving and/or tuning a fuzzy controller implemented using neural networks. Kim *et al.* (1995) introduced a genetic-algorithm-based computationally aided design methodology for ***rapid prototyping*** of control systems. As an example, they designed a ***fuzzy net controller*** (FNC), with the intention to use genetic algorithms for optimizing the fuzzy membership functions capable of meeting various operational specifications. Seng *et al.* (1999) described a genetic-algorithm-based strategy for simultaneously tuning the parameters of a fuzzy logic controller implemented on an RBF network, named NFLC (neuro-fuzzy logic controller). Belarbi and Titel (2000) presented an alternative approach to designing all parameters of fuzzy logic controllers (*i.e.* the parameters of the membership functions of both the input and the output variables, and the rule base) using genetic algorithms. The fuzzy logic controller designed was implemented in neuro-technology. The application of binary-coded genetic algorithm was reported by Palit and Popovic (2000) in order to train a fixed structure neuro-fuzzy network that used the singleton type of rules consequent. Thereafter, the genetic-algorithm-trained neuro-fuzzy network was applied to forecast the future values of a chaotic time series. In addition to the above applications Setnes and Roubos (2000), Roubos and Setnes (2001) also applied the a genetic-algorithm-based fuzzy logic system for identification and modeling of a nonlinear plant. In their method, a real-coded genetic algorithm was mainly used to fine tune the fuzzy antecedent memberships (triangular) that were obtained by similarity-based fuzzy set merging. It was reported that the genetic-algorithm-tuned fuzzy model was transparent, accurate but compact. Similar applications of real-coded genetic algorithms were reported by Panchariya *et al.* (2003, 2004) for improving the fuzzy model transparency. In the last case, using a distance (entropy)-based fuzzy clustering algorithm, an initial Takagi-Sugeno fuzzy model with high accuracy was obtained. However, the initial fuzzy model was not compact and transparent. In order to improve the model transparency, and also the accuracy, antecedent (Gaussian) fuzzy sets were consequently merged (as described in Chapter 7) and thereafter real-coded genetic algorithms were applied. Finally, the evolved fuzzy model of Panchariya *et al.* (2003, 2004) was also, in this case, applied for nonlinear plant modeling and reported to have much better accuracy than that reported by contemporary literature on the same benchmark problem.

References

[1] Angeline PJ, Saunders GM, and Pollack JB (1994) An Evolutionary Algorithm That Constructs Recurrent Neural Networks. IEEE Trans. Neural Networks. 4: 54-65.

[2] Balakrishnan K and Honavar V (1995) Evolutionary Design of Neural Architectures. Technical Report CS TR 95-01, Dep. of Computer Science, Iowa State University, Ames, Iowa, US.

[3] Balarbi K and Titel F (2000) Genetic Algorithm for the Design of a Class of Fuzzy Controllers: An Alternative Approach. IEEE Trans. on Fuzzy Systems 8(4): 398-405.

[4] Boers EJW and Kuiper H (1992) Biological Metaphors and the Design Artificial Neural Networks. Master's Thesis, Niels Bohrweg 1, 2333 CA, Leiden, The Netherlands.

[5] Chalmers DJ (1990) The evolution of learning: An experiment in genetic connectionism. In D.S. Touretzky, JL Elman, and GE Hinton, eds. Connectionist models: Proc. of the 1990 Summer School, Morgan Kaufmann, San Mateo, CA: 81-90

[6] Curran D and O'Riordan C (2003) Applying Evolutionary Computation to Design Neural Network: A Study of the State of the Art. The 7th Int. Conf. on Knowledge-Based Intelligent Information and Engineering Systems, Sept. 3-5, Oxford, UK.

[7] Fogel DB, Fogel LJ, and Porto VW (1990) Evolving neural networks. Biological Cybern. 63(6): 487-493.

[8] Gruau F (1994) Neural network synthesis using cellular encoding and the genetic algorithm, Ph.D Thesis, Ecole Normale Superieure de Lyon.

[9] Harp SA and Samad T (1991) Genetic synthesis of neural network architecture. In L. Davis, ed. Handbook of Genetic Algorithms, Van Nostrand : 202-221.

[10] Harp SA, Samad T, and Guha A (1990) Designing application-specific neural Networks using the genetic algorithm. In: Advances Neural Information Processing Systems 2, Touretzky DS, Ed. Morgan Kaufmann, San Mateo, CA.

[11] Herrera F, Lozano M, and Veregay L (1995) Tuning Fuzzy Logic Control by Genetic Algorithms. Intl. J. of Approximate Reasoning 12(3/4): 299-315.

[12] Homaifar A. and McCormick E (1995) Simultaneous design of membership functions and rule sets for fuzzy controllers using genetic algorithms. IEEE Trans. Fuzzy Syst, 3: 129-139.

[13] Hwang MW, Choi JY and Park J (1997) Evolutionary projection neural networks, Proc. IEEE-CEC '97, pp. 667-671.

[14] Hwang WR and Thompson (1994) Design of intelligent fuzzy logic controllers Using genetic algorithms. Proc. IEEE Intl. Conf. Fuzzy Syst.: 1283-1388.

[15] Johnson DJ and Frenzel JF (1992) Application of genetic algorithms to the training of higher order neural networks. J. Syst. Eng., vol.2: 272-276

[16] Jurick M. (1988) Back error propagation: A critique. Proc. IEEE COMPCON 88, San Francisco, CA: 387-392.

[17] Karr CL (1991) Genetic Algorithms for fuzzy logic controllers. AI Expert 6(2): 26 - 33.
[18] Kim J, Moon Y, and Zeigler BP (1995) Designing Fuzzy Net Controllers Using Genetic Algorithms. IEEE Control Systems Magazine15(3): 66-72.
[19] Kinnebrock W (1994) Accelerating the standard backpropagation method using a genetic approach. Neurocomputing 6(5-6): 583-588.
[20] Kitano H (1990) Designing neural networks using genetic algorithms with graph generation system. Complex Systems 4(4): 461-476.
[21] Lee M and Takagi H (1993a) Dynamic control of dynamic algorithms using fuzzy logic techniques. In S. Forrest, editor, Proc. of the Fifth Intl. Conf. on Genetic Algorithms: 76-83, Morgan Kaufmann, San Mateo, CA.
[22] Lee M and Takagi H (1993b) Embedding a priori knowledge into an integrated Fuzzy system design method based on genetic algorithms. Proc. of the 5th IFSA World Congress IFSA'93, II: 1293-1296.
[23] Lindenmayer (1968) Mathematical models for cellular interaction in development, Pt. I and II, Journal of the Theoretical Biology, 18: 280-315
[24] Louis S and Johnson J (1999) Robustness of Case Initiated Genetic Algorithm. http://citeseer.nj.nec.com/92649.html.
[25] Marquardt D (1963) An Algorithm for least-squares estimation of nonlinear parameters SIAM J. Applied Mathematics. 11: 431-441.
[26] McDonnell JR and Waagen D (1994) Evolving Recurrent Perceptrons for Time - Series Modeling. IEEE Trans. Neural Networks 5: 24-38.
[27] McInerney M and Dhawan AP (1993) Use of Genetic Algorithms with Back Propagation in Training of Feed-Forward Neural Networks. Proc. 1993 IEEE Int. Conf. Neural Networks, ICNN '93, San Francisco, CA: 203-208.
[28] Mohammadian M and Stonier RJ (1994) Generating fuzzy rules by genetic algorithms. Proc. Of 3rd IEEE Intl. Workshop on Robot and Human Communication, Nagoya, pp. 362-367.
[29] Mohammadian M and Stonier RJ (1994) Generating fuzzy rules by genetic algorithms, 3rd IEEE Internat. Workshop on Robot and Human Communication, Nagoya: 362-367.
[30] Montana DJ and Davis L (1989)Training feedforward neural networks using genetic algorithms. Proc. 11th Int. Joint Conf, Artif. Intelligence 1:789-795.
[31] Palit AK, Popovic D (2000), Intelligent processing of time series using neuro-fuzzy adaptive Genetic approach, Proc. of IEEE-ICIT Conference, Goa, India, ISBN: 0-7803-3932-0, 1:141-146.
[32] Panchariya PC, Palit AK, Popovic D, Sharma AL, (2003) Simple fuzzy modeling scheme for compact TS fuzzy model using real coded Genetic algorithm, Proc. of First Indian internat. Conf. on AI (IICAI-03), Hyderbad, India, 1098-1107.
[33] Panchariya PC, Palit AK, Sharma AL, Popovic D (2004) Rule extraction, complexity reduction and evolutionary optimization, International Journal of Knowledge-Based and Intelligent Engineering Systems, 8(4): 189-203.
[34] Porto VW, Fogel DB, and Fogel LJ (1995) Alternative Neural Network Training Methods. IEEE Expert, June: 16-22.
[35] Prados DL (1992) New learning algorithm for training multilayered neural networks that uses genetic algorithm techniques. Electron. Lett., 28: 1560 -1561.
[36] Prudencio RBC and Ludermir TB (2001) Evolutionary Design of Neural Networks: Application to River Floe Prediction. Proc. of the IASTED Int. Conf. on Artificial Intelligence and Applications, AIA 2001, Marbella, Spain.
[37] Roubos JA, Setnes M (2001) Compact and Transparent fuzzy model through iterative complexity reduction, IEEE Trans. on Fuzzy Systems, 9(4): 516-524
[38] Schwefel H-P (1995) Evolution and Optimum Seeking. Springer-Verlag, Berlin

[39] Seng TL, Khalid MB, Yusof R (1999) Tuning of a Neuro-Fuzzy Controller by Genetic Algorithm. IEEE Trans. on SMC, 29(2): 226-236.

[40] Setnes M, Roubos JA, (2000) GA-fuzzy modeling and classification: complexity and performance, IEEE trans. on fuzzy systems, 8(5):509-522

[41] Sexton RS, Dorsey RE, and Johnson JD (1998) Toward global optimization of neural networks: A comparison of the genetic algorithm and backpropagation. Decision Support Syst. 22(2): 171-185.

[42] Shao S (1988) Fuzzy self-organizing controller and its application for dynamic processes. Fuzzy Sets and Systems, 26: 151-164.

[43] Shi Y, Eberhart R, and Chen Y (1999) Implementation of Evolutionary Fuzzy Systems. IEEE Trans. On Fuzzy Systems, 7(2):109-119.

[44] Sietsma J and Dow RJF (1991) Creating neural networks that generalise. Neural Networks 4(1): 67-79.

[45] Stork DG, Walker S, Burns M, and Jackson B (1990) Pre-adaptation in network circuits. Proc. Int. Joint Conf. Neural Networks, Washington DC 1: 202-205.

[46] Tettamanzi A (1994) A distributed model for selection in evolutionary algorithms Technical Report 110-94, Dipartimento Di Scienze dell'Informazione – Università degli Studi di Milano, Milano, Italy.

[47] Tettamanzi A (1995) Evolutionary algorithms and fuzzy logic: A to-way integration. Proc. of the 2nd Annual Joint Conf. on Information Sciences.: 464-467, Duke University, Durham, NC.

[48] Tettamanzi A and Tomassini M (2001) Soft computing: Integration of evolutionary, neural and fuzzy systems, Springer-Verlag, Berlin.

[49] Thrift P (1991) Fuzzy logic synthesis with genetic algorithms. In R.K. Belew and L.B. Booker, editors, Proceedings of the Fourth Intl. Conf. On Genetic Algorithms: 509-523, Morgan Kaufmann, San Mateo, CA.

[50] Varsek A, Urbancic T, and Filipic B (1993) Genetic Algorithms in Controller Design and Tuning. IEEE Trans. on Systems, Man, and Cybernetics 23(5) : 1330-1339.

[51] Vonk E, Jain LC, and Johnson R (1995) Using genetic algorithms with grammar encoding to generate neural networks. Proc. 1995 IEEE Int. Conf. Neural Networks, 4: 1928-1931.

[52] Wong C-C and Chen C-C (2000) A GA-Based Method for Constructing Fuzzy Systems Directly from Numerical Data. IEEE Trans. on Systems, Man, and Cybernetics, Part B 30(6): 904-911.

[53] Yan W, Zhu Z, and Hu R (1997) Hybrid/genetic/BP algorithm and its application for radar target classification. Proc. 1997 IEEE Natnl. Aerospace and Electronics Conf., NAECON, 2: 981-984.

[54] Yang J-M, Cao C-Y, and Horng J-T (1996) Evolving neural induction regular language using combined evolutionary algorithms. Proc. 1996 1st Joint Conf. Intelligent Systems ISAI-IFIS: 162-169.

[55] Yao X (1993) An empirical study of genetic operators in genetic algorithms. Microprocessing and Microprogramming 38(1-5): 707-714.

[56] Zeng S and He Y(1994) Learning and Tuning Logic Controllers Through Genetic Algorithm. Proc. IEEE World Congress on Computational Intelligence: IEEE Intl. Conf. on Neural Networks (WCCI/ICNN '94): 1632-1637, Orlando, Florida.

[57] Zhang P, Sankai Y, and Ohta M (1995) Hybrid adaptive learning control of nonlinear system. Proc. 1995 American Control Conf. 4: 2744-2748.

9

Adaptive Genetic Algorithms

9.1 Introduction

The genetic algorithms, or in general the various evolutionary computations, have been introduced to the reader in Chapter 5 along with their important implementation aspects. Genetic algorithms (GAs) are often described as a gradient-free, robust search and optimization technique, where the search direction, unlike a gradient-based optimization method, is not biased towards a local optimum, but, at the same time, GAs can also be applied to an ill defined complex problem for optimization. However, the above advantages of GAs may be totally jeopardized because of the extremely long run time required for a complex optimization problem. Furthermore, even at the end of an extremely large number of generations the solution obtained from the GA run may be completely unacceptable. This being the main motivation why GA researchers are constantly trying to improve GAs in order to obtain an acceptable solution within a reasonable number of generations of a GA run. With the above objectives in mind, the present chapter furnishes a few important possibilities, collected from various publications, for the improvement of a standard GA run.

The most typical features of genetic algorithms (GAs) are:

- genetic representation or encoding of data to be optimized
- initial population of encoded data
- control parameters of the algorithm
- fitness function.

In practical applications, the adequate selection of GA features substantially influences its performance; and, *vice versa*, the non-adequate selection of GA features might lead to nonacceptable problem solutions. To prevent the latter situation, this was a challenging task of researchers in the 1980s, who tried to implement various practical concepts to facilitate the feature selection process. In association with the concepts of adaptive control of dynamic systems, the idea was widely accepted to work on adaptive GAs by incorporating the parameter adaptation mechanism in conventional GAs. Different researchers have concentrated their efforts on implementing genetic algorithms with different parameters tuned. Herrera and Lozano (1996) later classified the proposed adaptive GA systems as systems with

- *adaptive parameter setting*
- *adaptive genetic operators*
- *adaptive operator selection*
- *adaptive representation*
- *adaptive fitness function.*

In practical realizations of adaptive GA approaches it should first be decided at what algorithm level the adaptation should work (Smith and Fogarty, 1997), *i.e.* should it be at the

- ***population level***, at which the global GA parameters of all individuals of the population are on-line adapted
- ***individual level***, at which the strategy parameters, usually mutation and crossover, are adapted only in some elected population individuals in order to effectuate only elected individuals
- ***component level***, at which the strategic parameters of some components or of some genes of population individuals are locally varied?

9.2 Genetic Algorithms Parameters to Be Adapted

Adaptive versions of genetic algorithms are particularly needed because, in the process of the evolutionary search, the algorithm should converge to the global optimum with a high speed of convergence, so that the global optimum value is found in the minimum number of steps, *i.e.* it should be finished after a minimum number of generations treated. This is usually achievable by on-line adapting of the ***control parameters*** of the algorithm, such as the probability of crossover, mutation, or of reproduction. Several empirical and theoretical studies devoted to identifying the optimal mode of parameter settings for genetic algorithms (DeJong, 1985; Grefenstette, 1986; Hesser and Manner, 1990) have resulted in the following general assessments:

- ***Crossover***. This parameter controls the rate at which the solutions are subjected to crossover effects. When its value is increased, new solutions are more rapidly introduced into the population. Through this, the search process can become so fast that it can be disrupted.
- ***Mutation.*** This parameter restores the genetic material and transforms the GA – when its value is increased too much - into a purely random search algorithm, whereas, a small value of the mutation parameter is required to prevent the premature convergence of the GA to a suboptimal solution.

- ***Reproduction.*** This parameter determines the rate at which the old solution will be copied into the new population. When its value is increased the chance of survival of a solution in the subsequent generation will also be increased. This subsequently increases the number of "super-fit" individuals in the next generation, which is not always desirable.

Apart from the above genetic parameters and their probabilities, two additional parameters can be used for GA adaptation:

- ***Population size.*** This GA parameter can be adapted to the problem to be solved. During the search process, the proper population size is the most critical factor that strongly influences the convergence speed of the search process, in the sense that too small a population size speeds up the search convergence and leads eventually to a premature solution. On the contrary, a very large population size could stretch the search process *ad infinitum* (Baker, 1985).
- ***Fitness function.*** As a performance index, this helps in carrying out the selection process optimally and has to be defined adequately with respect to the problem to be solved.

9.3 Probabilistic Control of Genetic Algorithms Parameters

In the early 1980s it was a general view that the on-line adjustment of ***crossover probability***, or ***crossover rate***, can be favourable for optimal progress in the search process, because it can help in avoiding the premature end of the search process through the higher loss of the alleles. Using the entropy measure over the entire population, Wilson (1986) was able to quantify the benefit of crossover adjustment. To compensate for this, the value of ***mutation probability*** should be increased. This indicates that, when the GA parameters are adaptively tuned, the following two tendencies have to be balanced out:

- convergence to the solution optimum, after the region that contains the solution optimum or nearly the optimum has been traced
- searching for new regions of the solution space in order to find a real global optimum.

This illustrates that the genetic algorithm operates by a permanent balancing between the best result that can be achieved and searching for the possibility to achieve some better results. For monitoring the status of the balance the ***exploitation-to-exploration relation*** (EER) has been introduced to serve as a ***diversity measure*** of the search process. In the above case, the balance between the values of the crossover probability $p(c)$ and the mutation probability $p(m)$ should be kept at an optimal level. In practice, moderate values of crossover probability ($0.5 < p(c) < 1.0$) and small balancing values of the mutation probability ($0.001 < p(m) < 0.05$) are commonly used.

Li *et al.* (1992) proposed an EER-based ***dynamic GA***, capable of balancing ideally the GA behaviour by adjusting the crossover and mutation probabilities, by which the EER is defined. They used two diversity functions for adjustment purposes.

The dynamic GA presented by Li *et al.* operates in the following three stages:

- in the ***initial stage***, in which the diversity measures follow the initial conditions and the initial parameter values of GA
- in the ***search stage***, in which the dynamic GA varies its parameters to enable a broad search and improved exploitation
- in the ***refinement stage***, in which the balance is adapted to manage the search process more efficiently, while the best chromosome is already close to the optimum problem solution in the search space.

Srinivas and Patnaik (1994) also used the manipulation of crossover and mutation probabilities to retain the population diversity and still to support the convergence capability of the algorithm. In order to vary the crossover, mutation and reproduction probabilities, *i.e.* to vary $p(c)$, $p(m)$ and $p(r)$ adaptively with the objective of preventing the premature convergence of the GA to a local optimum, they first tried to identify whether the GA is converging to a local or to a global optimum at all. For this, they recommended the observation of the relation between the average fitness value f_{avg} across the population and the maximum fitness value $f_{\max}$ within the population (*i.e.* the fitness of the best chromosome in the population). The value of the difference $(f_{\max} - f_{\text{avg}})$ is likely to be less for a population that has converged to an optimum solution than that of the population scattered in the solution space. The same property has been observed in experiments with the GA. This is obvious, because convergence of the GA means that the majority of the population has a similar high fitness value. This alternatively implies that the average fitness of the population is high and is most possibly close to the maximum fitness of the population. Therefore, this justifies the difference $\left(f_{\max} - f_{\text{avg}}\right)$ being used here as the measure of convergence of the GA.

Usually in the adaptive GA experiments, the probability values $p(c)$, $p(m)$, and $p(r)$ are varied, depending on the difference value $\left(f_{\max} - f_{\text{avg}}\right)$, *i.e.* on search results (Palit and Popovic, 2000). Since the probability values of $p(c)$ and $p(m)$ have to be increased (in order to bring more genetic diversity into the population) when the GA converges to the local optimum, *i.e.* when the difference $\left(f_{\max} - f_{\text{avg}}\right)$ decreases, both $p(c)$ and $p(m)$ have to be varied inversely with $(f_{\max} - f_{\text{avg}})$. Therefore, the same expression can be written mathematically as follows:

$$p(r) = k_{\text{rep}} \Big/ \left(f_{\max} - f_{\text{avg}}\right) \tag{9.1}$$

$$p(m) = k_{\text{mu}} \Big/ \left(f_{\max} - f_{\text{avg}}\right) \tag{9.2}$$

$$p(c) = k_{\text{cross}} \Big/ \left(f_{\max} - f_{\text{avg}}\right) \tag{9.3}$$

where the numerators k_{rep}, k_{mu} and k_{cross} in the right-hand side expressions are some constants of the (respective) variations.

From the above expressions it is evident that all three probabilities do not depend on the fitness of any particular solution and have the same value for all solutions of the populations. Consequently, solutions both with high and with low fitness values are subjected to the same level of reproduction, mutation, and crossover. Also, when the population converges to an optimal global or local solution, the increase of $p(c)$ and $p(m)$ may eventually cause disruption of the near-optimal solutions. Therefore, the population will neither converge to a local optimum nor converge to the global optimum. Therefore, though we may prevent the GA from getting stuck at a local optimum solution, the performance of the GA – in terms of generations required for convergence will be very large - will certainly deteriorate.

To overcome this problem, we need to preserve the "good" solutions of the population by using some higher value of $p(r)$ and lower values of $p(c)$ and $p(m)$ for higher fitness solutions and some higher values of $p(c)$ and $p(m)$ for lower fitness solutions. This is because high fitness values support the convergence speed of the GA, whereas low fitness solutions prevent the GA from getting stuck at local optima.

Thereby, the value of $p(m)$ should not only depend on $(f_{max} - f_{avg})$ but also on the fitness value of the solution. Similarly, the $p(c)$ value should not only depend on the difference $(f_{max} - f_{avg})$ but also on the fitness of the two parent solutions. Furthermore, if the value of the difference $(f_{max} - f_{avg})$ can identify whether the GA is converging or not, then the difference $(f_{avg} - f_{min})$ will possibly also identify the convergence of the GA because, in our experiment (Palit and Popovic, 2000), all the populations for subsequent generations are selected from the mating pool that consists of the best 50% populations of the current generation. Therefore, by using both of them as a measure of GA convergence, the adaptive values of the control parameter of the GAs are set as follows. For

$$f_{selrep} \geq f_{avg}$$

it is

$$p(r) = k_{1rep}\left(\frac{f_{selrep} - f_{avg}}{f_{max} - f_{avg}}\right) + k_{1repbias} \tag{9.4}$$

and for

$$f_{selrep} < f_{avg}$$

it is

$$p(r) = k_{2\text{rep}} \tag{9.5}$$

where f_{selrep} represents the fitness value of the chromosome (individual) selected by the roulette wheel selection mechanism for further genetic (reproduction) operation. The values of other constant terms in the right-hand side expression, *e.g.* $k_{1\text{rep}} = 0.9$, $k_{2\text{rep}} = 0$, and $k_{1\text{repbias}} = 0.1$, have been selected in the adaptive GA experiment (Palit and Popovic, 2000). Note that, in (9.4), when the best individual with highest fitness $(f_{\max})$ is selected by the roulette wheel, *i.e.* $f_{\text{selrep}} = f_{\max}$, the probability of reproduction *p*(*r*) = 1 and that for an average individual (f_{avg}) is only 0.1 (set by the bias term in (9.4)). On the other hand, the sub-average individual cannot be reproduced at all, as per (9.5), since *p*(*r*) = 0.

Similarly, for the adaptive probability of mutation, *i.e.* for *p*(*m*), the following mathematical expressions were used. If

$$f_{\text{selmu}} \geq f_{\text{avg}}$$

then

$$p(m) = k_{1\text{mu}} \left(\frac{f_{\max} - f_{\text{selmu}}}{f_{\max} - f_{\text{avg}}} \right) + k_{1\text{mubias}} \,. \tag{9.6}$$

Otherwise, when

$$f_{\text{selmu}} < f_{\text{avg}}$$

then,

$$p(m) = k_{2\text{mu}} \left(\frac{f_{\text{avg}} - f_{\text{selmu}}}{f_{\text{avg}} - f_{\min}} \right) + k_{2\text{mubias}} \,, \tag{9.7}$$

where the following values for the constant terms have been selected: $k_{1\text{mu}} = 0.01$, $k_{2\text{mu}} = 0.09$, $k_{1\text{mubias}} = 0.005$, and $k_{2\text{mubias}} = 0.005$. Here, as per (9.6), the best individual will undergo the lowest mutation (since *p*(*m*) = 0.005), whereas the average individual will undergo a moderate level of mutation (since *p*(*m*) = 0.015). On the other hand, the worst chromosome, as per (9.7), will have the highest possibility of mutation (since *p*(*m*) = 0.095).

Furthermore, for the crossover probability *p*(*c*) the following relations hold. If

$$f_{\text{selcross2}} \geq f_{\text{avg}}$$

then,

$$p(c) = k_{1\text{cross}} \left(\frac{f_{\text{selcross2}} - f_{\text{avg}}}{f_{\max} - f_{\text{avg}}} \right) + k_{1\text{crossbias}} \tag{9.8}$$

elseif,

$$f_{\text{selcross2}} < f_{\text{avg}}$$

It is

$$p(c) = k_{2\text{cross}} \left(\frac{f_{\text{selcross}} - f_{\min}}{f_{\text{avg}} - f_{\min}} \right) + k_{\text{crossbias}} \tag{9.9}$$

with $k_{1\text{cross}} = 0.5$, $k_{2\text{cross}} = 0.2$, $k_{1\text{crossbias}} = 0.5$, and $k_{2\text{crossbias}} = 0.3$. Here, $f_{\text{selcross2}}$ corresponds to the fitness of parent 2 and the same is only selected through roulette wheel, since parent 1 is always the best individual selected of all generations for the crossover operation. This is because, when the crossover is performed between the best individual of all generations and worst individual of current generation, the possibility of generating better individuals is generally low, hence, as per (9.9), the probability of crossover for such a case is low (set by $k_{2\text{crossbias}} = 0.3$).

9.4 Adaptation of Population Size

Genetic algorithm starts with an initial population that is randomly generated so that it - as far as possible - uniformly represents the entire search space. This assumes that the knowledge about the search space and the problem to be solved is *a priori* available. This also helps – using an efficient heuristics – drive the initial population in the direction of the most promising problem solution.

The initial population size potentially defines the size of the search space to be considered and it directly influences the convergence speed and the achievable solution accuracy. This is closely related to the problem of premature convergence of the search process and to the problem of search crashes. Therefore, it is advisable to adapt the population size steadily while executing the search process. Baker (1985) was the first to show how this could be done. He noticed that the chromosomes that produce a large number of offspring during the process of crossover and mutation contribute considerably to the acceleration of convergence speed. Owing to the limited population size, this forces the rest of the population to produce a reduced number of offspring, even it prevents some chromosomes from contributing any offspring at all. This causes a rapid decrease in the population diversity, which leads to premature convergence of the search process. In order to monitor this phenomenon, Baker introduced the ***percent involvement*** as a measure that indicates the percentage of a generation contributing the offspring for the next generation. Based on this measure he could control a dynamic population size by adding or deleting additional population chromosomes in order to balance out the contribution percentages over the entire current population.

Entirely different approaches to resolving the premature convergence problem have been proposed by Arabas *et al.* (1994) and by Kubota and Fukuda (1997), based on the concepts of ***age of chromosome*** and of ***age structure of population*** respectively. In the age of chromosome concept, the number of generations that a chromosome has survived is taken as an indicator that replaces the plain selection mechanism. The concept assigns to every created chromosome its ***lifetime***, which determines the age at which the chromosome will die. The lifetime length is calculated by taking into account the minimum, average, and the maximum fitness values within the current population and the minimum and maximum fitness values in the past generations. The chromosomes with the outstanding fitness values get a longer lifetime assigned.

The concept of ***age structure of population*** maintains the genetic diversity of the population by deleting the aged individuals. This mimics nature by removing individuals from the population by reaching the lethal age. Defining the natural ***life cycle*** as the time interval between the birth of parents and the birth of offspring, there are two conceptual possibilities to be used

- the parents and the children may not simultaneously live as long as the parents live (AGA algorithm)
- both the parents and the children may coexists for a period of time (ASGA algorithm), which is the most natural case.

In the aged genetic algorithm concept, each individual is characterized by its ***age*** and its ***lethal age*** as parameters. As soon as an individual is born it is assigned a lethal age and the zero value of its age parameter. Thereafter, its parents die immediately. The remaining individuals increase their age parameter value by one in every generation. Starting with an initial generation in which all individuals have a zero-value age parameter, an ***age operator*** manages the aging and dying process.

The effect of the proposed genetic algorithm with age structure was tested on a simulated knapsack problem. The simulation results have shown that the new concept can prevent individuals with a large fitness value from overrunning the population and maintain a considerable genetic diversity in the population. The introduced age concept also helps in solving optimization problems with a relatively small population size. There are, however, some unpleasant effects that accompany the age concept application (Knappmeier, 2003):

- there is an increased possibility of weak individuals surviving as long as their lethal time is not expired
- there is an enlarged possibility for strong individuals to die formally earlier, *i.e.* before they become bad, when their lethal time has expired.

9.5 Fuzzy Logic Controlled Genetic Algoithms

A number of scientists, after experimenting with probabilistic approaches for improving GA performance, were not satisfied because, by pursuing this research track, much vague and ill-structured knowledge and some highly exhaustive computational procedures have to be used. They, therefore, started searching for more comfortable and more efficient alternatives for solving this problem. To escape from the probabilistic concepts and to by-pass the long-lasting calculations they selected fuzzy logic as a possible tool for on-line adaptation of GA parameters and for GA resources management. Lee and Takagi (1993) took this route in their study and worked out a dynamically controlled genetic algorithm using a fuzzy logic technique. Soon thereafter, Arnone *et al.* (1994) reported on fuzzy government of a genetic population, and Bergmann *et al.* (1994) published their experience with GA parameter adjustment using fuzzy control rules.

Dynamically controlled genetic algorithm is an algorithm that uses a fuzzy knowledge-based system to control the GA parameters dynamically, mostly the crossover, mutation rate, and the population size. In fact, it is a typical rule-based expert system the inputs of which can be a combination made up of a genetic algorithm and performance measures, such as the ratio of average to best fitness, current population size or the mutation rate. The rules stored in the system reason about the state of the measure values and recommend adequate actions. The authors give a rule example: an increase in the present population causes the sensitivity to mutation rate to decrease, along with the best mutation rate to use. This can be programmed as follows:

IF the ratio of average fitness-to-best fitness is HIGH
THEN population size should INCREASE
IF the ratio of worst fitness-to-average fitness is LOW
THEN population size should DECREASE
IF mutation is SMALL and population is SMALL
THEN population size should INCREASE

The system developed was validated through a simulation example of an inverted pendulum control, where it has shown much better behavioural results in pendulum control than a GA with fixed parameters.

Government of the genetic population is a concept coined by Arnone *et al.* (1994) for describing the process of on-line tuning GA parameters using a fuzzy knowledge base. The concept is based on a ***fuzzy government module*** whose inputs are statistical data periodically collected from the genetic algorithm and whose outputs are the control parameters of the GA. In the concept, a facility is embedded for monitoring the evolutionary process in order to avoid its possible undesired behaviour.

Herrera and Lozano (1996) summarized the steps in building adaptive GAs using fuzzy logic controllers as follows:

- define some firm measures related to the GA behaviour, its setting parameters, and operators, *e.g.* example the diversity indices, maximum, average, and minimum fitness values, as system inputs
- define as system outputs the values of control parameters or of their changes
- define the database as a collection of membership functions and the boundary values of input and output variables
- build the rule base in which the fuzzy rules describe the relations between the input and output variables.

The statistical data generated by the genetic algorithm concern the genotypes of individuals of a population as well as the phenotypes related to the fitness and other properties of individual performance for the problem to be solved. Two typical examples for the above statistics are the

- ***genotypic diversity measure***, representing the variations of similarity within the genetic material (like chromosomes, alleles, *etc.*)
- ***phenotypic diversity measure***, which mainly concerns the fitness of chromosomes.

9.6 Concluding Remarks

In this chapter, three possibilities of adaptive versions of genetic algorithms are presented, the first of which dynamically controls the basic tuning parameters, such as probability of crossover, mutation and reproduction *etc.*, based on the on-line measurement of GA convergence. Other methods control mainly the population size, based either on the concept of the age of the chromosome, the age structure of the population, or by application of the average-fitness -to- best-fitness ratio, worst-fitness -to- best-fitness ratio, besides the mutation- and crossover-rates-based fuzzy *IF-THEN* rules. The efficiencies of the various methods are demonstrated on application examples that can be found in the corresponding publications list.

References

[1] Arabas J, Michalewicz J, and Mulawka (1994) GAVaPS – a Genetic Algorithm with varying population size. Proc. of the 1st IEEE Conf. on Evolutionary Computation: 73-78.

[2] Arnone S, Dell'Orto M, and Tettamanzi A (1994) Toward a fuzzy government of genetic populations. Proc. of the 6th IEEE Conf. on Tools with the Artificial Intelligence TAI'94, IEEE Computer Press, Los Alamitos, CA.

[3] Baker J (1985) Adaptive selection methods for genetic algorithms. In: Proc.1st Intl. Conf. on Genetic Algorithms (J.J. Grefenstette, ed.): 101-111. Lawrence Erlbaum Associates, Hillsdale , NJ.

[4] BergmannA, Burgard W, and Hemker A (1994) Adjusting parameters of genetic algorithms by fuzzy control rules. In K.-H. Becks and D. Perret-Gallix, editors, New Computing Techniques in Physics Research III. World Scientific Press, Singapore.

[5] DeJong KA (1985) Genetic Algorithms: A 10 year perspective. Proc. of Intl Conf. on GAs and Applications: 169-177

[6] Grefenstette JJ (1986) Optimization of Control Parameters for GAs. IEEE Trans. On Systems, Man, and Cybernetics 16(1): 122-128.

[7] Herrera F and Lozano M (1996) Adaptation of Genetic Algorithm Parameters Based on Fuzzy Logic Controllers. In: F. Herrera and J.L. Verdegay Genetic Algorithms and Soft Computing, Physica-Verlag: 95-125.

[8] Hesser J and Manner R (1990) Towards an optimal mutation probability for system learning of a Boole an GAs. Proc. of the 1st Workshop, PPSN-I: 23-32.

[9] Knappmeier N (1993) Genetic algorithms with age structure and hybrid populations, Final report for the research project 416, Univ. of Darmstadt.

[10] Kubota N and Fukuda T (1997) Genetic algorithms with age structure. Soft Computing 1: 155-161.

[11] Lee M and Takagi H (1993) Dynamic control of genetic algorithms using fuzzy logic techniques. In S. Forrest, editor, Proceedings of the 5th Intl. Conf. on Genetic Algorithms, Morgan Kaufmann, San Mateo, CA.

[12] Li T-H, Lukasius CB and Kateman G (1992) Optimization of calibration data with the dynamic genetic algorithm, Analytica Chimica Acta, 2768: 123-134.

[13] Palit AK and Popovic D (2000), Intelligent processing of Time series using neuro-fuzzy adaptive Genetic approach, in Proceedings of IEEE-ICIT conference, Goa, India, ISBN: 0-7803-3932-0, v. 1: 141-146.

[14] Smith JE and Fogarty TC (1997) Operator and Parameter Adaptation in Genetic algorithms

[15] Srinivas and Patnaik (1994) Adaptive probabilities of crossover and mutation in genetic algorithms. IEEE Trans. on Systems, Man and Cybernetics 24(4): 656-667.

[16] Wilson SW (1986) Classifier System learning of a Boolean function. Research Memo RIS-27r, Rowland Institute for Science, Cambridge, MA.

Part IV

Recent Developments

10

State of the Art and Development Trend

10.1 Introduction

In the previous chapters we have presented the main issues of computational intelligence that, in our opinion, are of outstanding interest to practising engineers in the industry. The issues presented, to our knowledge, also make up the main part of syllabus of postgraduate courses on fuzzy logic, neural networks, computational intelligence, and soft computing in computer science and engineering.

In the last decade, some additional issues have become of growing interest, such as ***support vector machines***, ***wavelet neural networks*** and ***fractally configured neural networks***, which will be presented below. However, there are some advanced issues like ***stochastic machines***, ***neurodynamics***, and ***neurodynamic programming*** that have been left out of scope of the book because they are, in our belief, primarily of interest to informatics scientists and mathematicians.

The new development trends in computational intelligence are multiple. In the following, some development trends that, in our opinion, are the most promising for applications in engineering will be presented.

In the area of neuro-technology, the progress and the expectations in the area of ***bioinformatics*** and ***neuroinformatics*** are tremendous (Chen *et al.*, 2003). The advances have prevalently been possible due to the availability of sophisticated computer facilities for collection and management of tremendous amounts of complex experimental data, required for analyzing of brain infrastructure (see the *Proceedings of the IEEE*, special issues on bioinformatics, November 2002 and December 2002). The foremost research goal here is to understand the synaptic communication pathway of neurons and of supporting cellular elements. The pivotal achievements thus far are the understandings of how the computational processes in the living cells are performed by interaction of molecules and how the ***stochastic biochemical networks*** are built. Although the research work for the time being is predominantly carried out by biologists and bioneurologists, there is still expectation that, at a certain point in development, it will attract the attention of engineers. Here, it should be recalled that the learning rule of Hebb, himself a neurophysiologist, was formulated after his study of the learning principle of neurons.

A modest step in this direction is the special issue of *Control Systems Magazine*, August 1994, devoted to biological networks and cell regulation.

In the area of ***fuzzy logic technology***, the intensive research trend towards ***knowledge extraction from data*** or ***data understanding*** using rule-based systems (Duch *et al.*, 2004) is remarkable. The beginning of this research has roots in the achievements in ***image interpretation*** using the methods of artificial intelligence. The aim behind this was to explain the meaning of the images from the collected data, mainly using

- ***perceptual knowledge***, which supports interpretation in terms of lines, patterns, areas, *etc.*
- ***semantic knowledge***, which enables the use of some abstract concepts like the object shapes, the relationships between the objects, *etc.*
- ***functional knowledge***, *i.e.* the problem-oriented knowledge that finds out the best image interpretation by conducting intelligently the inference process.

Thus far, the rule-based data understanding approach has chiefly been used for data-based medical diagnostics, like for diagnosis of cancer and diabetes diseases (Setiono, 2000; Mertz and Murphy, 1993). However, it is highly possible that the approach can also find application in the production industry for material analysis, product quality inspection, and for production performance identification. What is interesting from our point of view is that rule-based data understanding can help to elucidate more inherent knowledge from the time series to be analyzed in this way than using the approaches described in previous chapters.

In the area of ***genetic algorithms***, the trend toward development of new, more advanced search algorithms is noteworthy. The most prominent example represents the development trend in ***particle swarm optimization***, invented by Kennedy and Eberhart (1995), *i.e.* by a social psychologist and an electrical engineer. It is a population-based search approach placed somewhere between genetic algorithms and evolutionary programming because it works in the following way:

- Each particle (representing a potential problem solution) keeps track of its coordinates in the problem space related with the best solution, called ***pbest***, based on the fitness obtained thus far. By evaluation of pbest values across the particle's population the global best temporary solution, called ***gbest***, is found and the parameter's adjustments are performed. This, in principle, corresponds to the crossover operation of GAs.

- Like evolutionary programming, the particle swarm concept also relies on the stochastic processes within the population.

The advantage of the particle swarm algorithm, compared with both of its precursors, *i.e.* with genetic algorithms and evolutionary programming, lies in the programming simplicity, which is due to the simplicity of its underlying concept.

Usually, a few computer program lines are needed to define the algorithm and the search objectives.

Initialization of the particle swarm algorithm starts with the random generation of particles that at this stage represent the potential problem solutions. Through the search for a final optimal problem solution the initial solutions will be improved by updating the values of each particle generation, and this will be performed without using the evolutionary operators such as crossover and mutation. During the search process, the particles fly through the solution space towards the current pbest, changing their velocity after each evaluation step.

It is interesting to add that the concept of particle swarm optimization was worked out by its inventors through the observation of bird flocking and fish schooling behaviour, and in the attempt to simulate birds seeking food through social cooperation of neighbouring birds.

Presently, the main application of particle swarm optimization is in solving the constrained optimization problems, such as optimization of nonlinear functions (Hu and Eberhart 2002a), multiobjective optimization (Hu and Eberhart, 2002b), dynamic tracking, *etc.* He *et al.* (1998) have even shown a way how to extract the rules from fuzzy-neural networks using the particle swarm optimization approach. In the meantime, the term ***swarm engineering*** was also coined (Kazadi, 2000), dealing with the ***multi-agent systems***.

Finally, some useful information about the development trends in this area of research can be found in the special issue on particle swarm optimization, *IEEE Transactions on Evolutionary Computation* (June, 2004).

10.2 Support Vector Machines

Over the last decade or so, increased attention has been paid to support vector machines, based on the computational approach termed the ***principle of structural risk minimization***, formulated by Vapnik (1992). This principle is of fundamental relevance to ***statistical learning theory*** and represents an innovative methodology for development of neural networks (Vapnik, 1998 and 1995) for applications in function approximation, regression estimation, and signal processing (Vapnik *et al.*, 1996). The applications are also extended to include pattern recognition (Burges, 1998), and time series forecasting (Cao, 2003) and prediction (Muller *et al.*, 1997).

Originally, support vector machines were designed for solving pattern recognition problems by determining a hyperplane that separates positive and negative examples, by optimization of the separation margin between them. This is generally based on the method of structural risk minimization and the theory of statistical learning, where the error rate of learning of test data is limited by the training error rate and by the ***Vapnik-Chervonenkis dimension*** (Vapnik and Chervonenkis, 1968).

The fundamental concept of a support vector machine relies on ***Cover's theorem*** (Cover, 1965), which states that the mapping of an input vector $\boldsymbol{x}$ into a sufficiently high-dimensional space, called a ***feature space***, using a nonlinear mapping function $\varphi(x)$ could more probably be linearly separable than in the low-dimensional input space of the vector (see Figure 10.1).

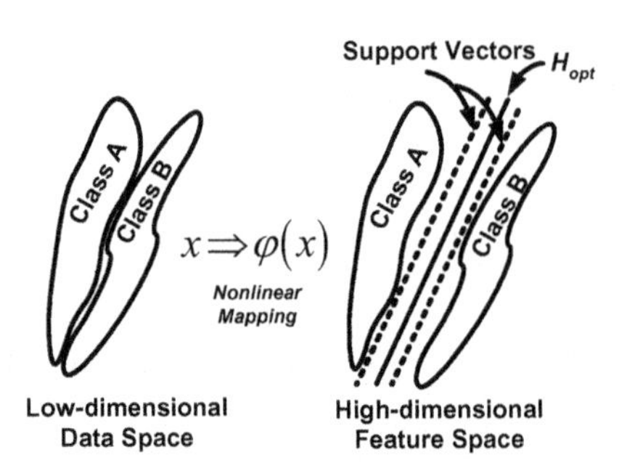

Figure 10.1. Nonlinear mapping from input space into feature space

In the high-dimensional feature space, if the data are nonlinearly separable in the low-dimensional data space, then linear separability of features could be achieved by constructing a hyperplane as a linear discriminant. In this way, a data classifier can be built, as illustrated by the following example.

Let a set of labelled training patterns (x_i, y_i) be available, with i = 1, 2, ..., N, where x_i is an n-dimensional pattern vector and y_i is the desired output corresponding to the input pattern vector x_i, the values of which belong to the ***linearly separable*** classes A with y_i = -1 and B with y_i = +1. The postulated separability condition implies that there exists an n-dimensional weight vector w and a scalar b such that

$$w^T x_i + b < 0 \quad \text{for} \quad y_i = -1, \tag{10.1}$$

$$w^T x_i + b \geq 0 \quad \text{for} \quad y_i = +1, \tag{10.2}$$

whereby the parametric equation

$$w^T x_i + b = 0 \tag{10.3}$$

defines the n-dimensional separating hyperplane. The data point nearest to the hyperplane is called the ***margin of separation***. The objective is to determine a specific hyperplane that maximizes this margin between the two classes, called the ***optimal hyperplane*** H_{opt}, defined by the parametric equation for H_{opt} in the feature space

$$w_0{}^T x_i + b_0 = 0 \tag{10.4}$$

as shown in Figure 10.1. From all the possible hyperplanes separating the two classes, H_{opt} is at equal distance between the data points that are nearest to the boundary between the two classes. Such data points, satisfying one of the following conditions

$$w_0^T x_i + b_0 = +1, \qquad \text{for } y_i = +1$$
$$w_0^T x_i + b_0 = -1, \qquad \text{for } y_i = -1,$$

are called ***support vectors*** (see Figure 10.1).

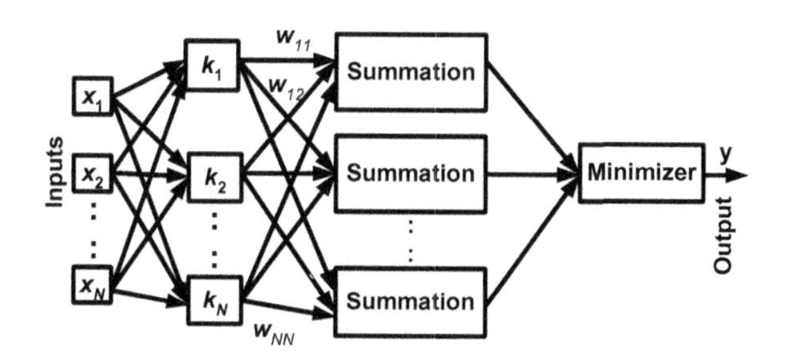

Figure 10.2. Structure of a kernel-based machine

Nonlinear mapping from the input space $x = [x_1, x_2, ..., x_N]$ to the higher dimensional feature space is carried out using the ***kernel function family***

$$k(x) = [k_1(x), k_2(x), ..., k_N(x)]$$

which helps in defining the linear discriminant function

$$\sum_{i=1}^{N} w_i k_i(x) + b = 0 \tag{10.5}$$

in the feature space, where w_i, i = 1, 2, ..., N, are parameters of the discrimination function. Assuming that $b = w_0 k_0(x) = 1$, then Equation (10.5) can be rewritten in the more compact form

$$\sum_{i=0}^{N} w_i k_i(x_i) = 0 \tag{10.6}$$

which is equivalent to the vector form

$$w^T k(x) = 0 . \tag{10.7}$$

Equation (10.6), as shown in Figure 10.2, can be used directly for implementation of a ***kernel-based machine*** (Principe *et al.*, 1999).

In the following, we will seek for the optimal separating hyperplane (Haykin, 1999) using the set of training data samples (x_i, y_i) and the constraint $y_i(w^T x_i + b) \geq 1$, $i = 1, 2, \ldots, N$. This is achieved by optimal selection of the value of b and by determination of the optimal value of w by minimizing the cost function

$$J(w) = \frac{1}{2} w^T w . \tag{10.8}$$

Using for this purpose the method of Lagrange multipliers, we have to minimize the Lagrangian function

$$J(w, b, \lambda) = \frac{1}{2} w^T w - \sum_{i=1}^{N} \lambda_i [y_i (w^T x_i + b) - 1] \tag{10.9}$$

with respect to w and b and to maximize with respect to λ by solving the equations

$$\frac{\partial J(w, b, \lambda)}{\partial b} = 0 \tag{10.10}$$

and

$$\frac{\partial J(w, b, \lambda)}{\partial w} = 0 . \tag{10.11}$$

As a result, the values of the weight vector w are found as

$$w = \sum_{i=1}^{N} \lambda_i y_i x_i \tag{10.12}$$

under the condition that

$$\sum_{i=1}^{N} \lambda_i y_i = 0 , \tag{10.13}$$

holds. Taking into consideration the nonlinearly transformed value of x_i, *i.e.* $k(x_i)$, the optimal value of w found above becomes

$$w = \sum_{i=1}^{N} \lambda_i y_i k(x_i), \tag{10.14}$$

where the transformed value $k(x_i)$ represents the feature vector corresponding to the input vector x_i.

After replacing the last Equation by equation (10.7), the separating surface in the feature space is found as

$$\sum_{i=1}^{N} \lambda_i y_i k^T(x_i) k(x) = 0, \tag{10.15}$$

or as

$$\sum_{i=1}^{N} \lambda_i y_i K(x, x_i) = 0, \tag{10.16}$$

where

$$K(x, x_i) = k^T(x_i) k(x) = k^T(x) k(x_i) \tag{10.17}$$

is the ***inner product kernel***, which is a symmetric function, *i.e.*

$$K(x, x_i) = K(x_i, x). \tag{10.18}$$

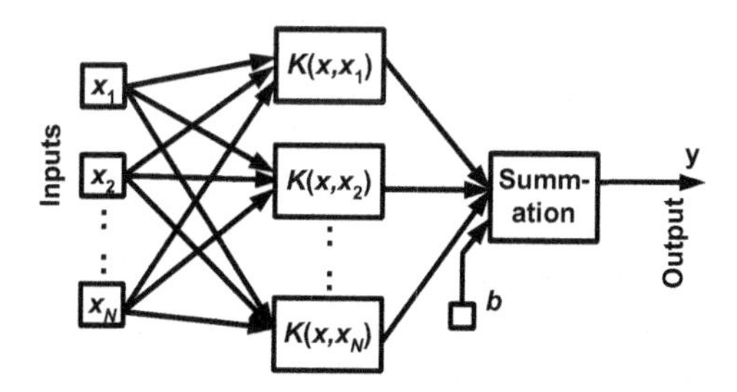

Figure 10.3. Basic architecture of a support vector machine

In practice, kernels of various shapes have been used:

polynomial kernels $\quad K(x, x_i) = [(x^T, x_i) + 1]^n$

RBF kernels $K(x, x_i) = e^{-\frac{\|x-x_i\|^2}{2\sigma^2}}$

sigmoid kernels $K(x, x_i) = \tanh[c(x^T, x_i) + \theta]$.

The condition for a selected kernel to be acceptable as an ***inner product kernel*** and to be useful for building a support vector machine is defined by ***Mercer's theorem***, which states that the proposed kernel function must be a symmetric function, as defined by Equation (10.17). Furthermore, an inner product kernel to be used in building the basic architecture of the ***support vector machine*** shown in Figure 10.3 must be expandable in the series

$$K(x, x_i) = \sum_{i=1}^{\infty} \lambda_i k_i(x) k_i(x_i), \tag{10.19}$$

where λ_i are eigenvalues and $k_i(x)$ are the eigenfunctions of the expansion.

10.2.1 Data-dependent Representation

Auflauf and Biehl (1989), using a ***data-dependent representation***, have worked out a simple and fast convergent sequential algorithm for finding the optimal parameters of a discriminant function with the largest margin. The algorithm that they called ***adatron*** considers the discriminant function in terms of

$$f(x) = \operatorname{sgn}(\sum_{i=0}^{N} \alpha_i x^T x_i + b), \tag{10.20}$$

where N is the number of samples and α_i the multipliers of individual samples that should be selected so that the quadratic form

$$J(\alpha) = \sum_{i=1}^{N} \alpha_i - \frac{1}{2} \sum_{i=1}^{N} \sum_{j=1}^{N} \alpha_i \alpha_j d_i d_j \langle x_i, x_j \rangle, \tag{10.21}$$

is optimized subject to the constraint

$$\sum_{i=1}^{N} \alpha_i d_i = 0, \tag{10.22}$$

for $\alpha_i \geq 0$, $i = 1, 2, \ldots, N$, where $\langle .,. \rangle$ represents the inner product of x_i and x_j.

In order to understand the building of a machine using the adatron algorithm, the discriminant function, relying on N data samples x_i and the corresponding weight multipliers w_i, should be written as

$$f(x) = x^T w + b = \sum_{i=1}^{N} \alpha_i x^T x_i + b\,, \tag{10.23}$$

and the machine output function as

$$y(x) = \mathrm{sgn}[f(x)]\,. \tag{10.24}$$

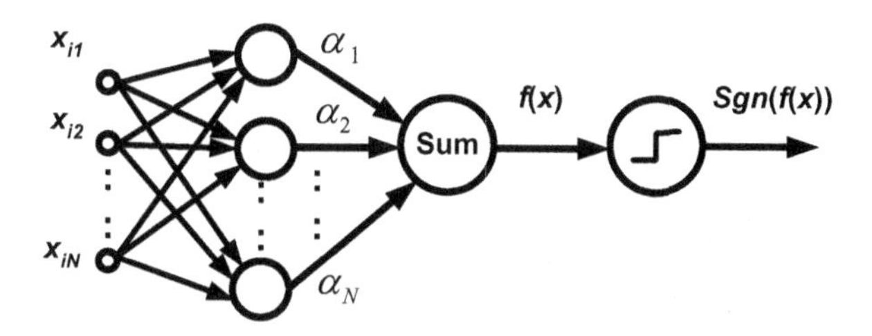

Figure 10.4. Adatron-algorithm-based perceptron

Equations (10.23) and (10.24) define the structure of a data-dependent machine, shown in Figure 10.4, in accordance with a perceptron with $b = +1$ as its bias input.

The idea of adatron was born during the search for a perceptron with ***optimal stability.*** Among the best iterative computational proposals for the design of such a perceptron, the adatron algorithm has proven to be the best one, since it theoretically promises – if the problem solution exists – to deliver an optimal solution with an exponential speed of convergence. The adatron algorithm is a kernel-based on-line algorithm for a learning perceptron under the premise that it operates in a feature space in which it is supposed that a maximal margin hyperplane exists.

10.2.2 Machine Implementation

After presenting the support vector machines concept and the aspects of its implementation, we would now like to summarise some essential issues and give a typical example of a support vector machine based on the RBF function as its kernel function (Figure 10.5). In doing this, we would first like to remind that the decision methodology of a support vector machine is based on implementation of the following two successive steps:

- mapping the training points by a nonlinear function φ to a sufficiently high-dimensional feature space in which the training points are linearly separable

- determination of the optimal separation hyperplane that maximizes the margin, *i.e.* the distance to the points.

We also recall that the adatron algorithm is capable of maximizing the margin. This can be used to implement a kernel-based machine. In the specific case of an RBF kernel of Gaussian style, the discriminant function $f(x)$, represented by Equation (10.23), takes the form

$$f(x) = \sum_{i=0}^{N} \alpha_i e^{-\frac{\|x - x_i\|^2}{2\sigma^2}} + b, \qquad (10.25)$$

which can be implemented as shown in Figure 10.5.

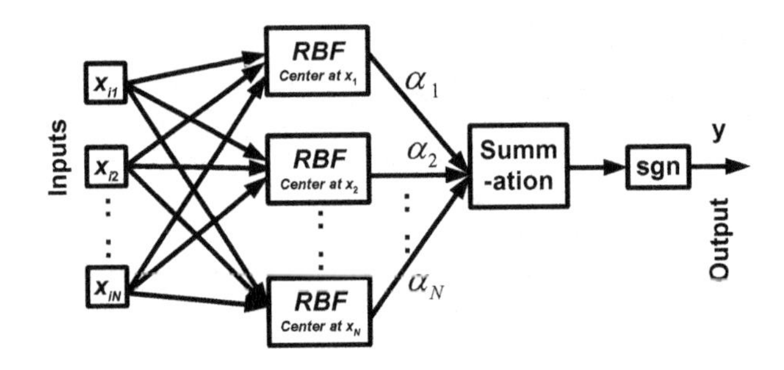

Figure 10.5. Architecture of an RBF-based support vector machine

This implementation effectively represents the structure of an RBF-based support vector machine in which the Gaussian activation functions are centred at sampled values, and the multipliers α_i play the role of interconnecting weights.

10.2.3 Applications

In engineering, support vector machines have found useful applications in nonlinear regression estimation and in time series forecasting and prediction.

Nonlinear regression estimation addresses the problem of estimating a function given by a set of data (x_i, y_{di}), i= 1, 2, …, N, generated by an unknown function to be estimated, where x_i are the sampled values of data set and y_{di} are the desired values to be estimated using the approximating function

$$f(x, a) = \sum_{i=0}^{N} a_i \phi_i(x) + b.$$

In the above function, the functions $\phi_i(x)$ are called features and a_i are coefficients to be estimated from given data by minimizing the functional

$$J(w) = \frac{1}{N}\sum_{i=1}^{N}\left|y_i - f(x_i, a)\right|_a + \lambda\left\|a\right\|^2 ,$$

where λ is a constant.

Mukherjee *et al.* (1997) experimentally investigated the performances of a support vector machine in nonlinear regression estimation of the database of a chaotic time series, and compared the results with those achieved with other techniques, such as with polynomial and rational approximations, radial basis functions, and with neural networks. They reported that the support vector machine performs better than any of the techniques taken for comparison. Cao and Tay (2001) concentrated their research on application of support vector machines in financial time series forecasting using the S&P daily index as the data set. They showed that, compared with neural networks, support vector machines performed better because of their better generalization capabilities.

10.3 Wavelet Networks

10.3.1 Wavelet Theory

The origin of the wavelet concept lies at the begin of the last century, as an extension of the Fourier transform. The real application of the new concept, however, began many decades later, sometime in the 1980s. It was soon realized that the wavelet concept, as a unified framework of various methodologies, could provide an efficient tool for signal processing, speech and image compression, *etc.* Moreover, wavelets became very popular in statistical time series analysis (Nasin and Sachs, 1999). Of more advanced use is the *wavelet transform* (WT) in analysis of non-stationary processes.

Presently, various types of wavelet transform are in use, such as the continuous, discrete, and discrete-time wavelet transform, which are appropriate for various applications. For instance, the continuous wavelet transformation

$$T_x(\tau, a) = \frac{1}{\sqrt{|a|}}\int x(t)h\left(\frac{t-\tau}{a}\right)dt$$

is seen as an alternative to the short-time Fourier transform

$$T(\tau, f) = \int x(t)h(t-\tau)e^{-j2\pi f_0 t}\,dt$$

and to the *Gabor transform*. Evidently, the wavelet transform is a kind of *signal decomposition* in a family of *basis functions* called *wavelets*, whereby wavelets of a family are obtained from a *prototype wavelet* or *mother wavelet* ψ as

$$\psi_{\alpha,\beta}(t) = \frac{1}{\sqrt{\alpha}}\psi\left(\frac{t-\beta}{\alpha}\right),$$

whereby ψ is a fixed time-frequency function meeting the restrictions

$$|\psi(t)| \leq \gamma\left(1+|t|\right)^{-1-\varepsilon}$$

and

$$|\psi(\omega)| \leq \gamma\left(1+|\omega|\right)^{-1-\varepsilon},$$

where $\psi(\omega)$ represents the Fourier transform of $\psi(t)$ and ε meets the condition $\varepsilon > 0$.

10.3.2 Wavelet Neural Networks

The wavelet decomposition approach, formulated at the end of 1980s, became a powerful tool for function approximation, and it was also applicable to time series analysis. Based on this decomposition, some structural representations of wavelet neural networks have been developed. In the first half of the 1990s, a number of publications reported on the synthesis and applications of wavelet neural networks (Zang and Benveniste, 1992; Pati and Krishnaprasad, 1993; Zhang *et al.*, 1995). The initial idea of Zang and Benveniste (1992) was to depict the wavelet neural network as an approximator of continuous functions using the universal approximation capability of wavelet decomposition.

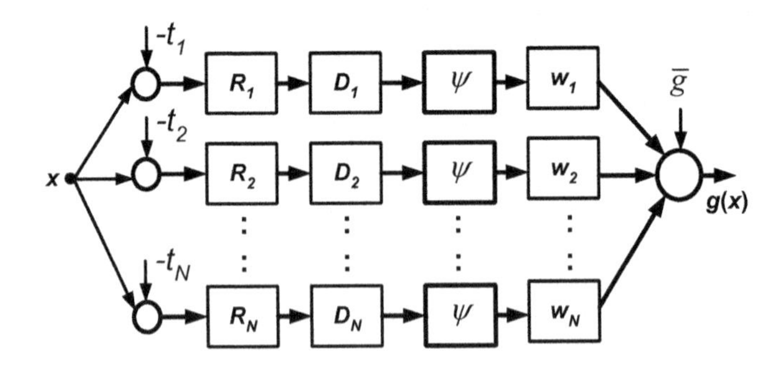

Figure 10.6. Wavelet network as a function approximator

Hence, they proposed a neural network structure described by the decomposition algorithm

$$g(x) = \sum_{i=1}^{N} w_i \psi[D_i R_i (x - t_i)] + g^*,$$

where the D_i values are diagonal matrices built from dilation vectors and R_i, $i = 1, 2, \ldots, N$, are some rotation matrices. The redundant parameters g^* are introduced to deal with non-zero-mean functions, because the wavelet $\psi(x)$ is a zero-mean function. The network's equivalent structure is shown in Figure 10.6. Rao and Kumthekar (1994) worked out the structure of recurrent wavelet networks using the equivalence between the

- statement of Cybenko (1989) that, if $\sigma(.)$ is a continuous discriminating function, then finite sums of the form

  $$f(x) = \sum_{i=1}^{N} w_i \sigma(a_i^T x + b_i)$$

  are dense in the space of continuous functions, so that any continuous function $f(.)$ may be approximated by a weighted sum of $\psi(.)$ functions

- analogous results of wavelet theory, which state that arbitrary functions can be written as a weighted sum of dilated and translated wavelets

$$f(x) = \sum_{i=1}^{N} w_i \det\left(D_i^{1/2}\right) \psi(D_i x - i_i).$$

A more transparent wavelet network representation was proposed by Chen *et al.* (1999). In this network, the wavelets are used as activation functions in the network's hidden layer, replacing the sigmoid functions, whereby the wavelet shape and the wavelet parameters are adaptively determined to deliver the optimal value of an energy function. In analogy with the input-output mapping of a one hidden-layer perceptron, generally written as (see Chapter 3)

$$y = f_o\left(\sum w_h f_h \left[\sum_{i=1}^{N} f_i(w_i^T x)\right]\right),$$

Chen *et al.* (1999) proposed a similar wavelet neural network structure

$$y_i(t) = \sigma\left\{\sum\nolimits_{j=0}^{n} w_{ij} \varphi_{ab}\left[\sum\nolimits_{k=0}^{m} w_{jk} x_k(t)\right]\right\}$$

for $i = 1, 2, \ldots, N$, where x_k and y_i are the input and the output vectors respectively, and w_{jk} are the connecting weights between the output unit i and the hidden unit j. In the last equation, the factors a_j and b_j represent the ***dilation*** and the ***translation*** coefficients of the wavelet in the hidden layer respectively. Similarly, w_{jk} represents the connecting weight between the hidden unit j and the input unit k. Relying on the above representation of neural networks, Cybenko (1989) and Hornik *et al.* (1989) proved – using the Stone-Weierstrass theorem – that any arbitrary function can be approximated with a given accuracy, thus designating the single hidden-layer neural network as a ***universal approximator***.

The proposed wavelet neural network is trained using the backpropagation algorithm with the cost function

$$E = \frac{1}{2}\sum_{p=1}^{P}\sum_{i=1}^{N}\left(d_i^p - y_i^p\right),$$

where d is the desired network output of pth input pattern. Furthermore, P represents the sum of input sample and m, n, and N the sum of input, hidden, and output nodes respectively.

Pati and Krishnaprasad (1993) developed an alternative structure of feedforward network, based on the ***discrete affine wavelet transform***. This is possible because the sigmoid activation function can be viewed as being composed of ***affine wavelet decompositions*** of mappings.

Zhang *et al.* (1995) described a wavelet neural network structure similar to that of a radial basis function network in which the radial basis functions are replaced by orthonormal scaling functions that are not necessarily radially symmetric. The wavelets used for network implementation are functions whose translations and dilations build an orthonormal basis of $L^2(R)$, which encompasses all square integrable functions of R, with the ***mother wavelet*** of the form

$$\psi_{m,n}(t) = 2^{m/2}\psi(2^m t - n).$$

The objective of the proposed network is that, given a training data set

$$T_N = \{t_i, f(t_i)\},$$

where $i = 1, 2, \ldots, N$, the optimal estimate of $f(t)$ could be found using

$$f(t) = \sum_k \langle f, \varphi_{M,k}\rangle \varphi_{M,k}(t).$$

For a given set of M and k, the wavelet network implements the mapping

$$g(t) = \sum_{k=-K}^{K} c_k \varphi_{M,k}(t)$$

which can be used to approximate $f(t)$ when the weights c_k are properly chosen.

Mukherjee and Nayer (1996) proposed a methodology for automatic generation of RBF networks based on the ***integral wavelet transform***. In fact, they concentrate on automated construction of a ***generalized radial basis function network***. To solve the problem considered, there is a general question to be answered: Can a multivariate function $f(x)$ be represented by the sums and products of univariate functions? The answer is to be found in approximation theory, which for this purpose recommends minimizing the cost functional

$$H\left[F(W,x)\right] = \int_{-\infty}^{+\infty} \left[f(x) - F(W,x)\right]^2 dx$$

with respect to W. In order to make the approximation problem ***well posed***, ***regularization techniques*** have to be used by introducing smoothness constraints into the approximation problem, so that the extended cost functional becomes

$$H\left[F(W,x)\right] = \sum_{i=1}^{N} [f(x_i) - F(W,x_i)]^2 + \lambda \left\| PF(W,x) \right\|^2 .$$

Solving this problem (for details see Chapter 3), the approximation function for the generalized radial basis network is defined by

$$F(W,x) = \sum_{j=1}^{n} c_j G(x; z_j),$$

where z_j, $j = 1, 2, \ldots, n$, are the centres of the new basis functions, which can be computed - along with the coefficients in the last equation – by minimizing the cost functional

$$H\left[F(W,x)\right] = \sum_{i=1}^{N} [F(W,x_i) - f(x_i)]^2 .$$

Based on the results of Zang (1997) in the use of wavelet network in non-parametric estimation, Li and Chen (2002) proposed a ***robust wavelet network***, based on the theory of ***robust regression***.

10.3.3 Applications

As mentioned earlier, wavelets have been widely used in various application fields of engineering. Some remarkable achievements have been reported in the *Proceedings of the IEEE*, special issue on wavelets, in April 1996. A state-of-the art report on wavelet applications in signal processing was compiled by Rioul and Vetterly (1991). Also, Li *et al.* (2000) have presented a real-life application of the wavelet transform in manufacturing for tool wear condition monitoring and tool breakage, based on measurements of spindle and feed motor currents. For decomposition of power inputs to the spindle and to the feed motor servos, both continuous and discrete wavelet transforms were used, and for detection of tool wear state a fuzzy classification method was developed relying on mathematical models of relationships between the current signals and the cutting parameters in the various tool wear states.

Recently, the results of wavelet application in time series forecasting and prediction have been published. Zhang *et al.* (2001) used wavelet decomposition for multi-resolution forecasting of financial time series. For this purpose, the time series was decomposed into an invariant scale-related representation and the individual wavelet series modelled by a separate multilayer perceptron. In order to build the overall time series forecast, the individual forecasts are recombined by a linear reconstruction property of the inverse transform with the chosen autocorrelation shell representation. Also, for time series preprocessing, a combined Bayesian and wavelet-based approach was used. Wavelet decomposition was also used by Soltani (2002) for nonlinear time series prediction. To produce improved prediction values, he used a combination of wavelet decomposition (as a filtering step) and neural networks. The most difficult problems to be solved here are the selection of an appropriate model order and the determination of optimal estimator complexity. Chen *et al.* (1999), again, used the ***multiresolution learning*** capability of a feedforward wavelet neural network described above for single- and multi-step predictions of chaotic time series and for systems modelling. Finally, in his Ph.D. thesis, Lotric (2000) used wavelet-based smoothing in time series prediction with neural networks and applied it to process quality control.

10.4 Fractally Configured Neural Networks

Engineering, information science, and mathematics have learnt much from biology and physiology. Examples are the creation of genetic and evolutionary searches, the discovery of Hebbian learning, reinforcement learning, associative memories, *etc*. From the complexity points of view, all arts of learning are categorized as ***elementary learning processes*** used for recognition and classification of patterns from given data. With the progress of time, the attention was shifted towards ***higher level learning processes*** or ***cognitive functions***, which are based on a set of elementary learning processes. As a tool for solving problems involved in higher level processes that, for instance, conventional neural networks cannot solve, ***fractally configured neural networks*** (or simply ***fractal networks***) have been proposed. The primary reason for this was because the higher cognitive functions, such as ***consciousness***, are basically hierarchically organized complex systems that cannot be modelled by a simple neural network, but rather they need several sub-networks (Takeshi and Akifumi, 1999).

In ***general systems theory***, various concepts have been elaborated for modelling of ***hierarchically organized modular systems***, among them the concept of ***partially bounded open systems***, in which the system itself and it's modules interact with their environment through their inputs and outputs. In the same way, the modules interact with each other at each hierarchical level as well as with the modules at a higher hierarchical level, so that they can be seen as partial open systems themselves. This interaction creates new, more complex open systems having a "higher order" intelligent behaviour, which is analogous with the capabilities of biological modules building higher level systems (multi-cellular organisms) out of lower level modules (cells) that, within the higher level system, behave as partially bounded open systems with mutual interaction.

The core issue, however, is: How should the modules interact mutually? This is the issue that was irrelevant for general systems theory. Furthermore, the question also arises as to what internal models should be embedded in individual modules. At least now, contemporary intelligent technology, particularly neuro-technology, is called for help. For instance, in analogy with the modules of biological systems, modules made up of neural networks should be structured as kinds of ***nested networks*** made up of networks that themselves build the individual modules capable of mutual communication. This indicates that the overall hierarchically organized modular system should have some fractal structure.

The operational principle of fractally configured neural networks is as follows. The modules at the lowest hierarchical level primarily have a sensing function. While interacting with the environment, the basic function is to collect the input data and to learn their characteristic features. The modules thereafter interact with modules of the next higher hierarchical level by sending the results of learning to them. The higher level modules receive from more than one lower level module the information learnt and perform a "higher level abstraction" that is forwarded to higher level modules, *etc*. This procedure is repeated until the central module of the system receives the combined information needed for final recognition and interpretation of the environment situation.

Following this operational principle, the entire neural network to be built becomes fractally configured. The problem now is what types of neural network should be used for system implementation. Because the modules should transfer the learning results towards to higher level modules, the feed-forward networks could be appropriate for this function. These types of network, however, do not have the storage capacity that, for example, the recurrent networks have. They can also perform self-organized learning, but, again, cannot be easily organized hierarchically. For this purpose, Morita (1993) proposed using what he called ***non-monotone neural networks***, capable of "abstracting" the input signals and of building the ***associative memory***.

Finally, the structure of the hierarchically organized modular neural network was worked out as shown in Figure 10.7, in which the ***sensory level***, ***recognition level***, ***abstraction (generalization) level***, and the ***final interpretation and decision level*** are chained hierarchically. This depicts the ***cerebral cortex hierarchy*** made up of ***sensory cortices***, ***association cortices***, ***frontal association cortices***, and the ***central motor cortex*** on the top of the hierarchy. From the figure it is evident that the fractal neural networks are tree-structured neural networks made up of hierarchically distributed sub-network clusters.

All the modules presented in Figure 10.7 are made up of non-monotone neural networks, the simplified structure of which is shown in Figure 10.8. In fact, the internal neural networks of modules consist of non-monotone networks, represented as circles. The non-monotone networks themselves consist of a number of pairs of ***inhibitory neurons*** and ***excitatory neurons***, so that they are structured as multi-input, multi-output neurons.

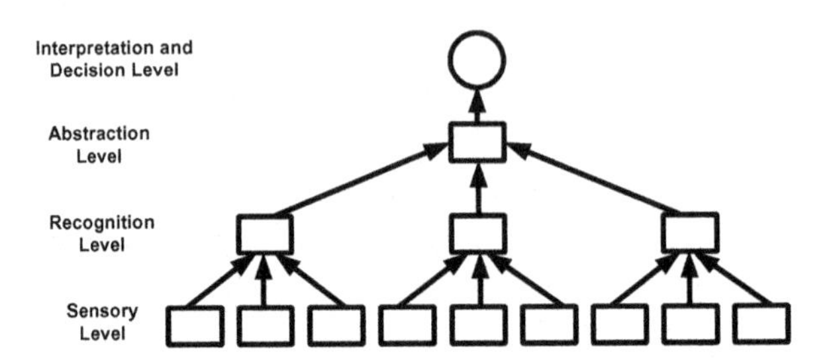

Figure 10.7. Hierarchically organized modular neural network

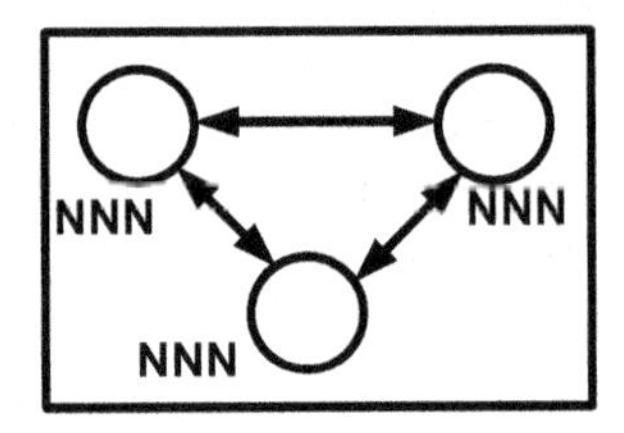

Figure 10.8. Simplified structure of a non-monotone neural network

It should finally be mentioned that although the discovery of fractally based neural networks was introduced in the late 1980s, the subsequent work on their implementation and application was rather dilatory.

10.5 Fuzzy Clustering

In Chapter 4 we have already described various fuzzy clustering algorithms, such as the ***fuzzy c-means algorithm*** that relies on fixed distance norm and the ***Gustafsson-Kessel algorithm*** that takes into account the adaptive version of distance norms for various geometrical shapes of clusters. Here, two other fuzzy clustering algorithms will be described, one that relies on the neural self-organizing network of Kohonen and the other is an entropy-based method.

Once the data clustering algorithm is applied in the product space of X and y, where a regression matrix $X^T = [x_1, x_2, ..., x_N]$ and the corresponding output vector $y^T = [y_1, y_2, ..., y_N]$ are constructed from a given set of time series data, the identification of a nonlinear time series model is simply a two-step procedure.

From the generated fuzzy partition matrix $U = \left[\mu_{g,s}\right]_{c \times N}$ that contains the membership degrees of the data object $z_s \in Z$, s = 1, 2, ..., N, and $Z^T = \left[X, y\right]$, in the cluster group g = 1, 2, ..., c, the one-dimensional antecedent fuzzy sets are constructed from the point-wise projection of the rows of matrix U. Thereafter, the Takagi-Sugeno (TS) rule's consequents are estimated from the training data, using the antecedent fuzzy sets, by the least squares error method. After validating the time series model with the validation data, the future values of the time series can be predicted easily by applying the generated Takagi-Sugeno rules.

10.5.1 Fuzzy Clustering Using Kohonen Networks

A Kohonen network is a self-organizing neural network, usually trained in unsupervised competitive mode. It is very well suited for data clustering. The network is closely related to the c-means clustering algorithm (Huntsburger and Ajjimarangsee, 1989). This was demonstrated by Bezdek *et al.* (1992) in their proposal of a data clustering algorithm that was based upon the Kohonen networks.

The ideas from the fuzzy c-means (FCM) algorithm are basically integrated into the learning rate and weight-updating strategies of the Kohonen-type networks, while implementing the *fuzzy Kohonen clustering network* (*FKCN*). The new algorithm can be viewed as a Kohonen-type fuzzy c-means (*FCM*) algorithm.

A Kohonen network (Kohonen, 1982) basically performs on some specific heuristic procedures, the termination of which does not represent the optimization of any model. In this kind of network, the final weight vectors depend on the input sequence. As a consequence, different initial conditions usually lead to different results.

Bezdek *et al.* (1992) introduced a new class of networks called FKCNs. In FKCNs, fuzzy membership values of output categories are incorporated into learning rates. In addition, FKCNs are self-organizing networks, since the size of the update neighbourhood is automatically adjusted during the learning process. Also, FKCNs usually terminate in such a way that the FCM objective function is approximately minimized. An FKCN is non-sequential and, therefore, it is independent of the sequence of feed of the input data.

The learning algorithm of an FKCN can be described as follows.

A data set that consists of observations of n measured variables (*e.g.* pressure, temperature, flow, *etc.* of a process) grouped into n-dimensional column vectors $z_s = \left[z_{1s}, z_{2s}, \cdots, z_{ns}\right]^T$, $z_s \in \mathbb{R}^n$, and a set of N such observations (*e.g.* at time instants 1, 2, ..., N *etc.*) can be denoted as $Z = \{z_s \mid s = 1, 2, ..., N\}$ and represented by the $n \times N$ matrix $Z = \left[z_{rs}\right]_{n \times N}$, where the rows and columns are indicated respectively by r = 1, 2, ..., n and s = 1, 2, ..., N. The rows and columns of this Z matrix are called features (attributes) and patterns (objects) respectively. For a given data set Z, c fuzzy clusters (groups) $\{\mu_g : Z \rightarrow [0,1]\}$ are fuzzy partitions of data Z in the $c \times N$ values of $\mu_{gs} = \mu_g(z_s)$, with $1 \le g \le c$ and $1 \le s \le N$, that satisfy the three conditions (see Section 4.7.1.3). Here, $\mu_{gs} = \mu_g(z_s)$ represents the degree of membership of data object z_s in the cluster group g. Based on the above representation of data and membership degree, the following steps implement the FKCN algorithm.

Step 1:

- Initialize the constants c, m and ε, where c represents the number of clusters sought in the data, m is the fuzziness exponent and ε is the termination tolerance, such that

$$1 \le c \le N$$
$$1 \le m \le \infty$$
$$\varepsilon > 0$$

- Initialize the cluster centre vectors

$$V_0 = \left[v_{1,0}, v_{2,0}, \cdots, v_{c,0}\right], \quad v_{g,0} \in \mathbb{R}^n$$

  where $v_{g,0}$ represents the prototype vector for cluster group g.
- Select the fuzziness exponent $m > 1$, and m is usually set to 2. Select also $T_{\max}$, the number of maximum allowed iterations.

Repeat for iteration $t = 1, 2, 3, \ldots, T_{\max}$;

Step 2:

- Compute all learning rates using

$$\alpha_{gs} = \left(\mu_{gs}\right)^m, \text{ where } \mu_{gs} = \left(\sum_{h=1}^{c}\left(\frac{\|z_s - v_g\|}{\|z_s - v_h\|}\right)^{2/(m-1)}\right)^{-1}, 1 \le g \le c,\ 1 \le h \le c,$$

  where α is the learning rate, μ are the membership values and c is the number of clusters.

Step 3:

- Update the weight vectors with

$$v_g(t) = v_g(t-1) + \sum_{s=1}^{N} \alpha_{gs}\left(z_s - v_g(t-1)\right) \Big/ \sum_{s=1}^{N} \alpha_{gs}$$

where t is the iteration number.

Step 4:

- Test for the terminating condition, *i.e.* calculate

$$E_t = \left\| v_g(t) - v_g(t-1) \right\|^2,$$

if $E_t \leq \varepsilon$ or, $t \geq T_{\max}$
then stop
else go to step 2.

It is to be noted that very often a termination tolerance ε = 0.001 is selected, even though ε = 0.01 works well in most cases. In the above algorithm, the weight vector v_g of the winning unit is closest to the input vector z_S. During the learning, the weight vector corresponding to the winning unit is adjusted so as to move further closer to the input vector. Most importantly, for a fixed value of m, the FKCN updates the weight vectors, using the conditions that are necessary for FCM and, in fact, with a fixed value of fuzziness exponent m, Bezdek *et al.* (1992) showed that the FKCN is equivalent to the fuzzy c-means clustering algorithm. However, particularly for m = 1, the FKCN behaves as a hard c-means clustering. As an illustration, they used an FKCN for clustering of iris data.

10.5.2 Entropy-based Fuzzy Clustering

The fuzzy c-means clustering methods, proposed by Bezdek (1974), and it's variant, the Gustafson-Kessel clustering algorithms (Babuška, 2002), based on an adaptive distance metric, although being very popular and powerful, both had to undergo some modifications (Yuan *et al.*, 1995; Medasani *et al.*, 1995; Babuška *et al.*, 2002), particularly the improvement of their performance and the reduction of their computational complexities.

One of the most important issues here is the determination of the number and initial location of cluster centres. In the original versions of both the above approaches the initial locations are selected randomly. Setnes and Kaymak (1998) in their extended version of both approaches have advocated selecting a large number of clusters initially and by compatible cluster merging reducing their number. Babuška (1996) and Setnes (2000) have suggested using a cluster validity measure, such as ***Xie and Benie's index***, to select the optimum number of clusters. Yager and Filev (1994) and Chiu (1994) proposed methods that automatically determine the number of clusters and locations of cluster centres. Chiu's method is a modification of Yager and Filev's mountain method, in which the potential of each data point is determined based on it's distance from other data. A data point is considered to have a high potential if it has many data points nearby and the data point having the highest potential is selected as the first cluster centre. Thereafter, the potentials of all other data points are recalculated according to their distance from the selected cluster centre. This procedure is repeated until no data point has it's potential above a threshold. This method requires values of three parameters:

- the radius beyond which data points have little influence on the calculation of the potential
- the amount of potential to be subtracted from each data point as a revision after a cluster centre is determined
- the threshold that potential uses to stop selecting cluster centres.

Although these methods are simple and effective, they are computationally heavy because, after determining each cluster centre, the potential values of all other data points have to be revised. The problem of recalculating the potential values is aggravated with an increase in the number of cluster centres, because the values of all three of the above parameters vary considerably from one data set to another.

In order to overcome the above difficulties, Yao *et al.* (2000) proposed using the ***entropy measure*** instead of the ***potential measure***, and in this way one avoids any revision after finding a cluster centre. The entropy at each data point is calculated based on a similarity measure. Note that the similarity measure here indicates the similarity between the data points and *not* between the fuzzy sets as described in Chapter 7. Data points in the middle of the clusters will have lower entropy than other data points. In other words, they have a better chance of being selected as cluster centres. The data point having the lowest entropy is chosen as the first cluster centre. Data points having similarity with this cluster centre less than a threshold are removed from being considered as cluster centres in the rest of the iterations. The rationale here is that the data points having high similarity with the chosen cluster centre should belong to the same cluster with a high probability, and are not likely to be centres of any other clusters. This is repeated until there are no data points left. An advantage of this method compared with other methods is its lower computational complexity. This is because, in this method, the calculation of entropy values is done only once. Also, the method requires a fewer number of parameters and the parameters assume values within a narrow range. In the following, an entropy measure for fuzzy clustering is introduced and a fuzzy clustering algorithm, based on entropy measure, is presented.

10.5.2.1 Entropy Measure for Cluster Estimation

Consider a set of N data points in an M-dimensional hyperspace, where each data point z_s, $i = 1, 2, \ldots, N$, is represented by a vector of M components $(z_{s1}, z_{s2}, \ldots, z_{sM})$. The values of each dimension are normalized in the range [0.0, 1.0]. Let us now assume that there are several clusters (groups) in the data. Now, for a data point to be a cluster centre, the ideal situation is when it is close to the data points in the same cluster centre and away from the data points in other clusters. This situation restricts the data points in the border of the cluster from becoming cluster centres.

10.5.2.1.1 The Entropy Measure

Yao *et al.* (2000) postulate that the data set has an ***orderly configuration*** if it has distinct clusters, and a ***disorderly configuration*** or ***chaotic configuration*** otherwise. From entropy theory (Fast, 1962) it is known that entropy (or probability) is lower for orderly configurations and higher for disorderly configurations. Therefore, if we try to visualize the complete data set from an individual data point, then an orderly configuration means that for most of the individual data points there are some data points close to it (*i.e.* they probably belong to the same cluster) and others away from it. In a similar reasoning, a disorderly configuration means that most of the data points are scattered randomly. So, if the entropy is evaluated at each data point then the data point with minimum entropy is a good candidate for the cluster centre. This may not be valid if the data have outliers, in which case they should be removed first before determining the cluster centres. The next section addresses this issue more.

The entropy measure between two data points can assume any value within the range [0, 1]. It shows very low values (close to zero) for very close data points, and very high values (close to unity) for those data points separated by the distance close to the mean distance of all pairs of data points. The similarity measure S is based on distance, and assumes a very small value (close to zero) for very close pairs of data points that probably fall on the same cluster, and a very large value (close to unity) for very distant pairs of data points that probably fall into different clusters. Entropy at one data point with respect to another data point is defined as

$$E = -S\log_2 S - (1-S)\log_2(1-S). \tag{10.26}$$

From the above expression it can be seen that entropy assumes the maximum value of 1.0 when the similarity $S = 0.5$ and the minimum value of 0.0 when $S = 0.0$ or 1.0 (Klir and Folger, 1988). The total entropy value at a data point z_i with respect to all other data points is defined as

$$E = -\sum_{j\in Z}^{j\neq i}\left\{S_{ij}\log_2 S_{ij} - (1-S_{ij})\log_2(1-S_{ij})\right\}, \tag{10.27}$$

where S_{ij} is the similarity between the data points z_i and z_j, normalized to [0.0, 1.0]. It is defined as

$$S_{ij} = e^{-\alpha D_{ij}}, \tag{10.28}$$

where D_{ij} is the distance between the data points z_i and z_j. If we represent the similarity against the distance graphically, then the representative curve will have a greater curvature for a larger value of α. The experiments with various values of α suggest that it should be robust for all kinds of data sets. Yao *et al.* (2000) proposed calculating the α value automatically by assigning a similarity of 0.5 in Equation (10.28) when the distance between two data points is mean distance of all pairs of data points. This produced a good result, as confirmed in various experiments (Yao *et al.*, 2000). Mathematically, this can be expressed as

$$\alpha = \frac{-\log_e(0.5)}{\bar{D}}, \tag{10.29}$$

where $\bar{D}$ is the mean distance among the pairs of data points in a hyperspace. Hence, α is determined by the data and can be calculated automatically.

10.5.2.2 Fuzzy Clustering Based on Entropy Measure

In order to determine the first cluster centre, the entropy at each data point is evaluated. The data point that has the lowest entropy value is selected as a potential cluster centre. Thereafter, this first cluster centre and all the data points that have similarity with it greater than a threshold value of β are removed, so that they are ignored as possible subsequent cluster centres in the next iterations. The procedure is continued with the search for the next cluster, which is selected as the point with the minimal entropy value among the remaining data points and, again, this cluster centre and the associated data points having similarity greater than β are similarly removed. This process is repeated until no data points are left.

The parameter β can be viewed as a threshold of ***similarity value*** or as ***association value*** among the data points in the same clusters. It assumes a value within the range (0.0, 1.0), whereby the value of $\beta = 0.7$ is quite robust, as shown experimentally in Yao *et al.* (2000). In the algorithm described below, T is the input data with N data points, each of which has M dimensions.

Algorithm 10.1. Entropy-based fuzzy clustering: EFC(T)

- *Step 1: calculate the entropy for each* z_i *in* T *for* $i = 1, 2, \ldots, N$.
- *Step 2: choose* z_{iMin} *that has lowest entropy*
- *Step 3: remove* z_{iMin} *and all the data points that have similarity greater than* β *with the cluster centre* z_{iMin} *from the data set* T.
- *Step 4: continue step 2 to 3 till* T *is not empty.*

If the data set has outliers that are very distant from the rest of the data, then the EFC algorithm described may select these data points for the cluster centres because the entropy value for these data points will also be very low. To overcome this problem, a new parameter γ is introduced in Yao *et al.* (2000) that acts as a threshold between potential clusters and the outliers. Before selecting a data point as cluster centre the number of data points are counted that have similarity greater than β with that cluster centre. If the number of counts is less than the value of γ, then that data point is unfit to be a cluster centre and should be rejected from the data set, so that it is not considered further for the next iteration. In the work of Yao *et al.* (2000) $\gamma = 0.05N$ is selected as the threshold for outliers detection. The selection of γ and, therefore, the corresponding removal of outliers also prevent the data overfitting.

10.5.2.3 Fuzzy Model Identification Using Entropy-based Fuzzy Clustering

In this section, ***entropy-based fuzzy clustering*** (EFC) will be presented to construct a fuzzy model for predicting values of output variables. The fuzzy modelling approach presented here is proposed by Yao *et al.* (2000) and differs slightly from the other modelling approach described in Chapter 4 and elsewhere in the book.

In the EFC-modelling approach Takagi-Sugeno-type rules with singleton consequents are considered. A fuzzy rule is based on a fuzzy partition of the input space. In each fuzzy subspace one input-output relation is formed. For a data point with an unknown value of output variable the values of input variables of the data point are applied to all rules and each rule gives a value by fuzzy reasoning. The predicted output value is then obtained by aggregation of all the values given by the rules.

Consider now a set of c cluster centres (v_1*, v_2*, ..., v_c*) in M-dimensional hyperspace that is generated by the EFC algorithm. Now, suppose that the last L dimensions of a kth cluster centre (v_k*) are output dimensions, whereas the first $(M-L)$ dimensions are input dimensions. Then, each cluster centre v_k* can be decomposed into two vectors: x_k* in $(M-L)$-dimensional input space and y_k* in L-dimensional output space. Then, a fuzzy model is a collection of c rules of the form

Rule k: IF X is close to x_k* *THEN Y is close to* y_k*,

where X is the input vector consisting of $(M-L)$ input variables $[x_{s1}, x_{s2}, ..., x_{s(M-L)}]$ and Y is the output vector consisting of L output variables $[y_{s1}, y_{s2}, ..., y_{sL}]$ of a data point z_s, with $s = 1, 2, ..., N$, training (input-output) samples. The membership function, representing the degree to which rule k is satisfied, is given as

$$\mu_k = \exp\left(-\sigma_k \left\|x - x_k *\right\|^2\right),$$

where x is the input vector, $X = x$, and σ_k is automatically calculated from the data. In the above, the symbol ||.|| denotes the Euclidean distance. The output vector, $Y = y$, is calculated as

$$y = \frac{\sum_{k=1}^{c} \mu_k y_k *}{\sum_{k=1}^{c} \mu_k}.$$

We can now write a fuzzy rule in a more specific form as

IF x_1 *is* A_{k1} *and* x_2 *is* A_{k2} *and ... and* $x_{(M-L)}$ *is* $A_{k(M-L)}$ *THEN Y is y,*
for $k = 1, 2, ..., c$.

where x_j is the jth input variable and A_{kj} is given by

$$A_{kj} = \exp\left\{-\sigma_k \left(x_j - x_{kj}{*}\right)^2\right\},$$

where $x_{kj}*$ is the jth element of kth cluster centre v_k* and the "and" operator is implemented by multiplication.

The parameter σ_k is crucial for the fuzzy model to perform well. It's initial value can be estimated from

$$\sigma_k = \frac{-\log_e(0.5)}{0.5 \cdot D_{\min}}$$

For each cluster centre we find its closest cluster centre and calculate the distance D_{min} between these two cluster centres. This formula implies that, in the fuzzy set around a cluster centre, if there is a data point midway between the cluster and its closest neighbouring cluster centre then the membership value of this data point belonging to the fuzzy set should be 0.5. This estimation is further verified and confirmed with the experimental evidence by Yao *et al.* (2000).

References

[1] Auflauf J and Biehl (1989) The Adatron: An adaptive perceptron algorithm Europhysics Letters 10(7):687–692.

[2] Babuška (1996) Fuzzy modelling and identification, Ph.D Thesis, Control Laboratory, Delft University of Technology, the Netherlands.

[3] Babuška R, Van der Veen PJ, and Kaymak, U (2002) Improved Covariance Estimation for Gustafson–Kessel Clustering, FUZZ-IEEE 2002, vol. 2: 1081-1085.

[4] Bezdek JC (1974) Cluster validity with fuzzy sets. J. Cybernet.: 58–71.

[5] Bezdek JC, Tsao EC, and Pal N R (1992) Fuzzy Kohonen Clustering Networks. Proc. of the IEEE Conf. on Fuzzy System: 1035–1043.

[6] Burges CJC (1998) A Tutorial on Support Vector Machines for Pattern Recognition. Data Mining and Knowledge Discovery. Preprint: Kluwer Academic Publishers, Boston.

[7] Cao L (2003) Support vector machines experts for time series forecasting, Neurocomputing, 51:321–339.

[8] Cao L and Tay EHF (2001) Financial Forecasting Using Support Vector Machines. Neural Computation & Application. 10:184–192.

[9] Chen J, Li Huai, Sun K, and Kim B (2003) How will Bioinformatics impact signal processing research. IEEE Signal Processing Maga. 20(6): 16–26.

[10] Chen Z, Feng TJ, and Meng QC (1999) The Application of wavelet neural network in time series prediction and system modelling based on multi-resolution learning.

[11] Chiu SL (1994) Fuzzy model identification based on cluster estimation: J. Intell. Fuzzy Systems: 2: 267–278.

[12] Cover TM (1965) Geometrical and statistical properties of systems of linear inequalities with applications in pattern recognition, IEEE Trans. on Electronic Computers, vol. 14: 326–324.

[13] Cybenko G (1989) Approximation by superposition of a sigmoidal function. Mathematics of Control, signals and systems 2:303–314.

[14] Duch W, Setiono R, and Zurada JM (2004) Computational Intelligence Methods, for Rule-Based Data Understanding. Proc. of the IEEE 92(5): 771–805.
[15] Fast JD (1962) Entropy: the significance of the concept of entropy and it's applications in science and technology. In: The statistical significance of the entropy concept, Philips Technical Library, Eindhoven.
[16] Haykin S (1999) Neural Networks: A Comprehensive Foundation.2nd Edition. Prentice Hall International, Inc., Hamilton, Ontario, Canada.
[17] He Z, Wei C, Yang L, Gao X, Yao S, Eberhart R, and Shi Y (1998) Extracting rules from fuzzy neural networks by particle swarm optimization. IEEE Intl. Conf. on Evolutionary Computation, Anchorage, Alaska, USA.
[18] Hornik K, Stinchcombe M, and White H (1990) Multilayer feedforward networks are universal approximators. Neural Networks 2: 359–366.
[19] Hu X and Eberhart RC (2002a) Solving constrained nonlinear optimization problems with particle swarm optimization. 6th World Multiconference on Systems, Cybernetics and Informatics (SCI 2002), Orlando, USA.
[20] Hu X and Eberhart RC (2002b) Multiobjective optimization using dynamic neighbourhood particle swarm optimisation. Proc. of the 2002 Congress on Evolutionary Computation, Honolulu, Hawai, May 12–17, 2002.
[21] Huntsberger T and Ajjimaransee P (1989) Parallel self organizing feature maps for unsupervised pattern recognition, Intl. Journal of General systems, 16: 357–372.
[22] IEEE : Special Issues on Wavelets: Trans. on Information Theory (March 1992), Trans. on Signal Processing (December 1993), and Proceedings of the IEEE (April 1996).
[23] IEEE: IEEE Control Systems Magazine, August 2004
[24] Kazadi S (2000) Swarm Engineering. Ph.D. Thesis, California Institute of Technology.
[25] Kennedy J and Eberhart RC (1995) Particle Swarm Optimization. Proc. IEEE Conf. on Neural Networks, vol. 4: 1942–1948.
[26] Klir GJ and Folger TA (1988). Fuzzy sets, Uncertainty, and Information. Prentice Hall International Editions.
[27] Kohonen T (1982) Self-organizing function in neural computing. Applied optics, 26: 4910–4918.
[28] Li ST and Chen SC (2002) Functional Approximation using Robust Networks. IEEE Intl. Conf. on Tools with Artificial Intelligence (ICTAI'02)
[29] Li X, Tso SK, and Wang J (2000) Real-Time Tool Condition Monitoring Using Wavelet Transforms in Fuzzy techniques. IEEE Trans. on Systems, Man, and Cybernetics, Pt. C 30(3): 352–357.
[30] Lotric U (2000) Using Wavelet Analysis and Neural Networks for Time Series Prediction, PhD thesis, University of Ljubljana, Slovenia.
[31] Medasni S, Kim J, and Krishnapuram R (1995) Estimation of membership functions for pattern recognition and computer vision. In: Fuzzy Logic and it's application to engineering. Information Sciences and Intelligent systems, Kluwer Academic publishers, Dodrecht: 45–54.
[32] Mertz J and Murphy PM (2003) UCI respiratory of machine databases. (visit: http://www.ics.uci.edu/pub/machine-learning-data -bases)
[33] Morita M (1993) Associative memory with nonmonotone dynamics. Neural Networks 6: 115–126.
[34] Mukherjee S and Nayar S (1996) Automatic Generation RBF Networks Using Wavelets. Patter Recognition 29(8):13691383.
[35] Mukherjee S, Osuna E, and Girosi F (1997) Nonlinear Prediction of Chaotic Time Series Using Support Vector Machines. Proc. of IEEE NNSP ′97, Amelia Island, Fl, 24–26 Sept.

[36] Muller KR, Smola JA, Ratsch G, Scholkopf B, Kohlmorgen J, and Vapnik VN (1997) Predicting time series with support vector machines. Proc. of the 7th Intl. Conf. on Artificial Neural Networks (ICANN'97):999–1004. Lausanne, Switzerland.
[37] Nason GP and Sachs R (1999) Wavelets in time series analysis. Phil. Trans. R. Soc. London. A: 1–16.
[38] Pati YC and Krishnaprasad PS (1993) Analysis and Synthesis of Feedforward Neural Networks Using Discrete Affine Wavelet Transformations. IEE Trans. on Neural Networks 4(1): 73–85.
[39] Principe JC, Euliano NR, and Lefebvre WC (1999) Neural and Adaptive Systems: Fundamentals through Simulations. Wiley, NY, USA.
[40] Rao SS and Kumthekar B. (1994) Recurrent Neural Networks. 0-7803-1901-x/94 IEEE 1994:3143–3147.
[41] Rioul O and Vetterli M (1991) Wavelets in Signal Processing. IEEE Signal Processing Magazine: 14–38.
[42] Setiono R (2000) Generating concise and accurate classification rules for breast cancer diagnosis. Artificial Intell. Med. vol. 18: 205–219.
[43] Setnes M (2000) Supervised fuzzy clustering for rule extraction, IEEE Trans. on Fuzzy Systems, 8(4): 416–424.
[44] Setnes M and Kaymak U (1998) Extended fuzzy *c*-means with volume prototypes and cluster merging, Proc. of EUFIT, Aachen, Germany, pp. 1360–1364.
[45] Soltani S (2002) On the use of the wavelet decomposition for time series prediction. Neurocomputing 48: 267–277.
[46] Takeshi I and Akifumi T (1999) Modularity and Hierarchy in Cerebral Cortex; A Proposal of Fractal Neural Networks. Proc. of 4th Intl. Workshop on Neural Networks in Applications NN'99: 23–29.
[47] Vapnik VN (1992) Principles of risk minimisation for learning theory, Advances in Neural Information Processing Systems, 4: 831–838. Morgan Kaufmann, San Mateo, CA.
[48] Vapnik VN (1995) The Nature of Statistical Learning Theory. Springer, New York.
[49] Vapnik VN (1998) Statistical Learning Theory. Wiley, New York.
[50] Vapnik VN and Chervonenkis AY (1968) On the uniform convergence of relative frequencies of events to their probabilities, Doklady Akademi Nauk USSR (in Russian)
[51] Vapnik VN, Golowich SE, and Smola AJ (1996) Support Vector Method for Function Approximation, Regression Estimation, and Signal Processing. Advances in Neural Information Processing Systems, 9:281–287. Morgan Kaufmann, San Mateo, CA.
[52] Yager RR, Filev, DP (1994) Generation of fuzzy rules by mountain clustering. J. Intell. Fuzzy Systems 2: 209–219.
[53] Yao J, Dash M, Tan ST, and Liu H (2000) Entropy based fuzzy clustering and fuzzy modelling, Fuzzy Sets and Systems, 113: 381–388.
[54] Yuan B, Klir GJ, and Swan-Stone J.F (1995) Evolutionary fuzzy *c*-means clustering algorithm. FUZZ-IEEE 1995: 2221–2226.
[55] Zang Q (1997) Using wavelet network in nonparametric estimation, IEEE Trans. on Neural Networks, vol. 8(2): 227–236.
[56] Zang Q and Benveniste A (1992) Wavelet Networks, IEEE Trans. on Neural Networks 3(6): 889–898.
[57] Zhang BL, Richard C, Jabri MA, Derssch D, and Flower B (2001) Multi-resolution Forecasting for Futures Trading Using wavelet Decompositions. IEEE Trans. on Neural Networks. 12(4): 765–774.
[58] Zhang J, Walter GG, Miao Y, and Lee WNW (1995) Wavelet Neural Networks for Functional Learning. IEEE Trans. on Signal Process. 43(6):1485–1496.

Index